Innovation has become a major field of study in economics, management, sociology, science and technology, and history. Case studies, empirical models, appreciative analyses and formal theories abound. However, after several decades of study on innovation, and so many different types of contribution, there are still many phenomena we know very little about. The debate on innovation still has much to deliver; important questions remain unanswered and many problems require solution. Bringing together many leading figures in the field, this collection aims to address these concerns by offering detailed analyses of topics that are crucial for understanding innovation. In addition, it offers discussions of topics that researchers are just beginning to explore and of topics that continue to defy our efforts to understand and systematize. This important and wide-ranging collection will be essential reading for academic researchers and graduate students who wish to gain a broad overview of frontier-research in innovation.

Franco Malerba is Professor of Industrial Economics and Director of CESPRI at Bocconi University, Milan.

Stefano Brusoni is Associate Professor of Applied Economics and Deputy Director of CESPRI at Bocconi University, Milan.

Perspectives on innovation

Editors

Franco Malerba and Stefano Brusoni

CAMBRIDGE UNIVERSITY PRESS
Cambridge, New York, Melbourne, Madrid, Cape Town, Singapore, São Paulo

Cambridge University Press
The Edinburgh Building, Cambridge CB2 8RU, UK

Published in the United States of America by Cambridge University Press,
New York

www.cambridge.org
Information on this title: www.cambridge.org/9780521685610

First published 2007

Printed in the United Kingdom at the University Press, Cambridge

A catalogue record for this publication is available from the British Library

Library of Congress Cataloging in Publication Data

Perspectives on innovation / editors, Franco Malerba and Stefano Brusoni.
p. cm.
Includes bibliographical references and index.
ISBN 978-0-521-68561-0 (hardback: alk.paper)
ISBN 978-0-521-86664-4 (pbk.: alk.paper)

1. Technological innovations Economic aspects. 2. Technological innovations –
Social aspects. I. Malerba, Franco, 1950- II. Brusoni, Stefano.
HC79.T4P48 2006
338'.064 – dc22   2006034602

Contents

List of figures

List of tables

List of contributors

MASAHIKO AOKI is Professor Emeritus of Japanese Studies in the Economics Department, Senior Fellow of Stanford Institute of Economic Policy Research (SIEPR) and Freeman Spogli Institute for International Studies (FSI) at Stanford University.

ASHISH ARORA is Associate Professor in the Heinz School of Public Policy and Management at the University of Carnegie Mellon.

JANET BERCOVITZ is Associate Professor of Management at the Fuqua School of Business, Duke University.

TIMOTHY BRESNAHAN is Professor of Economics at Stanford University.

STEFANO BRUSONI is Associate Professor of Applied Economics at CESPRI, Bocconi University.

PARASKEVAS CARACOSTAS is Advisor in the "Science, Economy and Society" Directorate, Directorate General (DG) Research, European Commission, in Brussels.

WESLEY M. COHEN is Professor of Economics and Management at the Fuqua School of Business, Duke University.

ANDREA CUOMO is Executive Vice President and Chief Strategy and Technology Officer at ST Microelectronics of Genève.

GUSTAVO CRESPI is Research Fellow at SPRU, University of Sussex.

PAUL A. DAVID is Professor of Economics and Senior Fellow of the Institute for Economic Policy Research at Stanford University.

GIOVANNI DOSI is Professor of Economics at Sant' Anna School of Advanced Studies

YVES L. DOZ is Timken Chaired Professor of Technology and Innovation, Professor of Business Policy at INSEAD.

JAN FAGERBERG is Professor of economics at the University of Oslo.

MARYANN FELDMAN is Professor of Business Economics at the Joseph L Rotman School of Management, University of Toronto.

PAUL A. GEROSKI has been Chairman of the Competition Commission and Professor of Economics at the London Business School from 1991 to 2005.

ALDO GEUNA is Director of the Master Programme in Industry and Innovation Analysis and Senior Lecturer at SPRU–University of Sussex.

STEVEN KLEPPER is Professor of Economics and Social Science at the University of Carnegie Mellon.

DANIEL A. LEVINTHAL is Professor of Management at the Wharton School of University of Pennsylvania.

BENGT-ÅKE LUNDVALL is Professor of Economics at the University of Aalborg.

FRANCO MALERBA is Professor of Industrial Economics and Director of CESPRI at Bocconi University.

MAUREEN MCKELVEY is Professor of Technology Management at Chalmers University of Technology.

STAN METCALFE is Professor of Political Economy at the University of Manchester.

PIERA MORLACCHI is Lecturer at SPRU–University of Sussex.

RICHARD R. NELSON is George Blumenthal Professor of International and Public Affairs, Business and Law at the Columbia University.

LUIGI ORSENIGO is Full Professor of Applied Economics at the University of Brescia.

HENRY SAUERMANN is Ph.D. student in Management at The Fuqua School of Business, Duke University.

W. EDWARD STEINMUELLER is Professor at SPRU–University of Sussex.

PETER THOMPSON is Associate Professor, Department of Economics at the Florida International University, Miami.

BART VERSPAGEN is Professor of 'Economics and Technology' at the Eindhoven University of Technology.

SIDNEY G. WINTER is Professor of Management at the Wharton School of University of Pennsylvania.

ULRICH WITT is Professor of Economics and Director of the Evolutionary Economics Group at the University of Jena.

JULIE WRAZEL is Strategic Alliance Manager at the Hewlett–Packard Company.

CRISTIAN ZELLNER is a Post-doc at the Ecole Polytechnique Fédérale de Lausanne.

Prefatory note

In the course of the publication of this volume, Paul Geroski died. Paul was an outstanding scholar in the field of innovation and industrial economics, with major contributions in industrial dynamics, entry, firms growth, market structure and innovation, technology diffusion, determinants of corporate performance and many, many other topics. He was a man of deep intellectual curiosity, high integrity and great enthusiasm. We will miss him dearly as a scholar, colleague and friend.

Introduction

Franco Malerba and Stefano Brusoni

1   Why a book on "perspectives on innovation"?

Innovation is nowadays a pervasive issue in both the academic literature and policy debates. It plays a central role in firms' strategies. It is
a fundamental element in public policies for growth and competitiveness. It is core to many university programs. Innovation has
become a major field of study in economics, management, sociology,
science and technology studies, and history. Cases, empirical models,
appreciative analyses, and formal theories abound. In economics,
various ways of examining innovation have been developed, ranging
from the neoclassical paradigm to evolutionary theory, to more institutionalist approaches, to innovation-system views.

In a way, the breadth, length, scope, and sheer visibility of the
"innovation problem" have become an issue. The need to articulate
what we know about innovation has led to the publication of a variety of
textbooks. Also, handbooks of innovation, each with a different twist
or focus, are now abundant and well reflect the variety of approaches
that scholars in different disciplines have followed to make sense of
innovation.

So, why a book about "perspectives on innovation"? Because after
several decades of studies on innovation, and so many different types of
contribution, it is useful to take stock and ask ourselves again which the
main topics of research are. Which are the main unanswered questions?
What are the main challenges? It is our belief that the debate about
innovation has still much to deliver. Quite simply, there are still many
phenomena we know very little about, many questions without an
answer, many problems without a solution. And this is precisely the sort
of issue on which the work presented in this book focuses. Thus, our aim
in this book is not summarizing the state of our knowledge: this is not a
handbook. And our aim is not to focus on a single aspect of innovation,
either. Rather, the aim is to provide, within a single book, a broad
overview and a comprehensive map of frontier-research in innovation.

This means several things: to highlight those areas that we consider key to understanding innovation, and those topics that researchers are just beginning to explore, those subjects that still defy our efforts to understand and systematize.

In a way, the title of this book aims at drawing an explicit connection to Nathan Rosenberg's seminal book, *Perspectives on Technology* (1976). That book provided a fresh perspective on the study of technology, moving away from the traditional economist's view of it as a black box (a title of another book by Rosenberg) and entering in depth into the analysis of the main features, dynamics, and consequences of technological evolution. Moreover, one of the key ideas of that book was that the direction of technological change is influenced by a number of "focusing devices," which provide a target for scientists' and engineers' problem solving efforts. Similarly, in this book on innovation, we intend to identify those topics that are attracting increasing attention from scholars, researchers, and policy-makers. Many of the contributors to this book stress how rich, diverse and multiperspective the study of innovation has become, as it inexorably entails the joint analysis of institutions, organizations, business strategies, selection processes, and technologies.

This is the reason for the title "Perspectives on Innovation." However, our "perspectives" will not be oriented toward developments in technologies, operations, or engineering practices. They will refer instead mainly to the economic, organizational, strategic, and institutional aspects of innovation. As a consequence the book is organized around themes and issues of research that cut across these different approaches. The themes refer to (i) innovation and economic growth, (ii) the microdynamics of the innovation process, (iii) innovation and industrial dynamics, (iv) innovation and institutions, (v) innovation, firms' organization and business strategies, (vi) innovation and entrepreneurship, (vii) innovation and the evolution of the university system, and (viii) innovation and public policy. Some of these topics have been present for a long time, such as technology and economic growth, but have gone through a period of recent rediscovery. Others are quite new and at the center of the research and policy debate, such as the role of the university system. Having chosen these topics in a subjective way within such a burgeoning area of research, we have decided not to include – for limit of space – other topics that are also important, such as the role of finance, and labor markets and innovation.

When one looks at the themes we selected, it is evident that there is a common reference to Schumpeter's (1934, 1939, and 1950) work and to broad Schumpeterian themes. In his work, Schumpeter was interested in a full understanding of the main drivers of change and growth: he addressed the relationship between innovation and competition, the role of entrepreneurship and new companies in innovation, the rise of R&D in large corporations, and the structural transformation of the economy as a result of the emergence of some new sectors and technologies and of the decline of others. And the Schumpeterian legacy in the last fifty years has developed these themes along several lines.

In sum, in this book we do not aim at providing a guide to the existing literature. Rather, we offer a collection of contributions by scholars engaged in asking new questions, or answering old questions in a new way, on the nature, role, and impact of innovation in firms, sectors, and countries. In this book we adopt an eclectic and multidisciplinary view, trying to put in touch approaches that come from different schools of thought or disciplines. One thing we are convinced of is that no one discipline has the magic wand for solving all issues when dealing with innovation. Accordingly, this book is organized in "pairs": each topic is approached by two, or more, scholars who look at the same issue from different angles. Moreover, each pair of chapters is commented upon by another contributor, whose task is to highlight similarities, differences, and complementarities.

2   The themes of this book

2.1   Innovation and economic growth

This is one of the oldest areas of research in economics. For a long time it has been recognized that technology is a major source of economic growth. However, after the path-breaking contributions of Marx and Schumpeter, a long time passed before this topic attracted again the attention of the economic profession. Only with the work by Christopher Freeman, Nathan Rosenberg, Richard Nelson, Moses Abramovitz, and others, followed later on by the contributions of new growth theory, has a major spurt of intensive research on economic growth taken place. However, as of today we have to recognize that we still know very little. Questions to be answered include: In which ways and through which processes innovation affects economic growth? This is an old yet always crucial question, at the base of our understanding of why some countries are ahead in terms of income levels and productivity, and others lag behind. And a related question is: Which kinds of analytical tools are needed for a full understanding of economic growth? What kinds of theory are necessary for enriching our understanding? Is modern new growth theory the appropriate answer? Is it correct that for economic growth we focus at the macro level without understanding better both the micro and meso levels? Or is the opposite true? Which are the new challenges for research on this topic? Which of the various approaches is better suited for advancing our understanding? These are the questions that Richard Nelson, Bart Verspagen, and Jan Fagerberg address in this volume.

2.2   *The micro dynamics of the innovation process*

Also, the analysis of the innovation process has always been at the center of scholarly attention, from the old debate on demand pull and technology push to the more recent contrast between the linear model and the chain-linked model. And over the years the analysis of the innovation process has taken into account both the incentives to innovate as well as the actual process of introducing new products and processes and the impact on the organization of firms and industries. Nowadays, however, the research questions have to become more subtle and deeper. Undoubtedly, one question regards incentives, the common factor affecting the innovation process that economists refer to. Is it correct to narrow all incentives to pecuniary ones only? Or do other dimensions (such as the psychological, social, and organizational ones) play an important role in the analysis of a broad phenomenon such as innovation? And should the innovation process be examined in isolation, or should it be linked to specific ways of organizing the innovative activity? In other words, under which circumstances is the power of creative destruction enhanced, and under which circumstances is it blocked? And which are the relevant dimensions to examine in this respect? These topics are examined in this volume by Wesley Cohen and Timothy Bresnahan. Their chapters are commented on by Ashish Arora.

2.3   *Innovation and industrial dynamics*

An area of analysis that has grown enormously in the last thirty years is innovation and industrial dynamics. Innovation has been identified as a major engine in the emergence and growth of industry, and as a key factor affecting the dynamics of industry population in terms of entry, growth, and exit. Over the years progress has involved case studies of innovation and industry evolution (following the so-called "SPRU tradition" initiated by Christopher Freeman and Keith Pavitt), with the recognition that industries have specific dynamics and life cycles. With the availability of advanced computer technology and new, firm level data, econometric analyses have moved from cross-sectional work during the 1960s and 1970s to panel data and longitudinal analyses since the early 1990s, shedding light on some aspects of industrial demography, entry and innovation, firms' growth, stability of firms' size distribution, and persistence of firms' asymmetric performance. Also at the modeling level, the evolutionary tradition and the neoclassical approach have developed different types of models of innovation and industrial dynamics. However, there is still major progress to be made. In this book some questions are tackled. Is it possible to identify some robust regularities in industrial dynamics? Could we say something more on the relationship between innovation, profitability, firms' growth, and selection at the industry level? And what do we know about the entry of new innovators? Is it still appropriate to identify entrants as firms coming from nowhere and bearing no relationship with the existing industrial structure? Or is there a strong link between innovative entrants, established firms, and existing organizational structures? In a sense, does the existing industrial structure generate the new innovators endogenously (from within)? In this book, these topics are examined by Giovanni Dosi, Steven Klepper, and Peter Thompson. Their contributions are discussed by Luigi Orsenigo.

2.4   *Institutions for innovations and innovation in institutions*

This topic it an extremely important one, but it has been disregarded for quite some time. Nowadays, however, everybody agrees that innovation does not happen in the vacuum and that the institutional environment plays a key role. Much has been said by the national system of innovation literature, with case studies and appreciative discussion. However, the specific feedback loops between different institutions, sets of rules, governance structures, and innovation are still little understood. We do know that these connections matter, but how, and to what extent, is yet hard to tell. The key issue here is understanding the extent to which firms have their behavior determined by the institutional environment in which they are embedded, as opposed to the extent to which they are free to navigate, and influence, the dynamics of such an environment. A fundamental dimension of the institutional framework is the way intellectual property rights are managed. Another (and related) one is how the university system behaves in the modern knowledge economy. More generally, a major topic is examining not just how institutions affect innovation, but also how innovation affects institutions and what the basic mechanisms of institutional change are. In this book these issues are examined by Masahiko Aoki and

Paul David and their chapters are commented on by Bengt-Ake Lundvall.

2.5   *Innovation, firms' organization, and business strategies*

The organization and competition aspects related to innovation have gained increasing importance in recent years. Scholars of technical change have stressed that the analysis of learning processes in uncertain environments requires conceptualizing industries as made up of persistently heterogeneous organizations. At one level, firm-level heterogeneity is a requirement for guaranteeing that the economic system runs enough experiments to explore alternative technological trajectories to be selected in the market place. At the other level, firm-level heterogeneity is the outcome of a slow, incremental process of learning and competence-building in organizations. This is why we expect firms, in the same industry, to be persistently different. According to this perspective, firms are the main agents of variety generation through their idiosyncratic competences. Such an emphasis on variety generation through competence-building has, however, often overshadowed the fact that selection does not occur only in the marketplace but also within firms. Moreover firms differ not only with respect to their technological capabilities but also in terms of the selection criteria which they use. How does selection work within innovative firms? And besides, is it not too simplistic to claim that there is only one selection process at work within innovating organizations? Or are different levels – and criteria – of selection interacting at all times? Moreover, nowadays cooperation among firms has become a central focus of the analysis of innovation. But how do cooperating organizations develop joint criteria for selecting joint strategies and evaluating the outcome of joint development projects? In this volume, these questions are raised in the chapters by Dan Levinthal and by Yves Doz, Andrea Cuomo, and Julie Wrazel, and these chapters are commented on by Sid Winter.

2.6   *Innovation and entrepreneurship*

A central figure in the innovative process is the Schumpeterian entrepreneur, who plays a key role in introducing new combinations. Entrepreneurship is analyzed in a large set of disciplines, such as economics, management, and sociology. But how to analyze fully and understand entrepreneurship? One could claim that too much emphasis has been put on entrepreneurship as such, and that entrepreneurship has not been seen as an action inserted in technological, sectoral, local, or national contexts. More specifically, much has been said about either the antecedents to entrepreneurial behavior (e.g. what kinds of person or organization are more likely to become entrepreneurs?) or its outcomes (e.g. what are the effects of entrepreneurial behavior on industry structure?). We still know relatively little about the entrepreneurial process. How does it happen? What are the key characteristics of entrepreneurial action? What is the role of human agency, of rational reasoning, and of intuition and creativity? What kinds of career paths lead certain individuals to act in an entrepreneurial way and why? It is necessary to grasp and understand these processual dimensions of entrepreneurship in order to make it amenable to policy and managerial interventions. In this book, these questions are addressed in the chapters by Piera Morlacchi and by Ulrich Witt and Christian Zellner, both commented on by Maureen McKelvey.

2.7   *Innovation and the evolution of the university system*

The critical role played by universities in the innovation process has emerged in recent years as a major stream of research. Universities have always been considered major sources of scientific advances and human-capital formation. However, a broad discussion is taking place on what different (additional?) roles universities should play in the knowledge-based society. In particular, universities are expected to be active in the creation of new "knowledge," and also in its diffusion and application to commercial uses. It is claimed that universities should act more entre-preneurially (either directly or through spin-offs). Therefore, what could we say about university-based entrepreneurship? Is it different from other types of entrepreneurship? What distinguishes academic entrepreneurs from nonacademic entrepreneurs? How can universities nurture and retain scholars interested in disclosing their findings for commercial purposes? A related issue touches upon the question of how we evaluate university performance. What do we mean by performance in the first place (e.g. scientific publications, patents, spin-offs, trained scientists, or technologists)? What kind of data, indicators, and methodologies can we deploy to measure the output of university research? These topics are the focus of the chapters by Janet Bercovitz and Maryann Feldman, and by Gustavo Crespi and Aldo Geuna, commented on by Ed Steinmueller.

2.8   *Innovation and public policy*

Last but not least, public policy. How can one transform the broad discussion developed in this book into policies? And more importantly, how to develop policies that take into account the dynamic and uncertain environment in which innovation takes place? These are key and still quite open issues. How does the emphasis on knowledge and the knowledge-based economy reshape our conception and the foundations of public policy? How do traditional policies – such as competition policies – interact with innovation policies and intellectual property rights (IPRs) protection? And what is really the role played by the policy-maker who acts at the interface between the domain of science and the domain of politics? What are the skills and the competences necessary to play such a role? With all the emphasis on learning and competences on the part of firms, is it necessary to use similar categories – cognition, competence, and behavior – for the policy-maker? In this book these key questions are addressed by Stan Metcalfe, Paul Geroski, and Paraskevas Caracostas.

All these issues and topics were examined at the Schumpeter 2004 international conference in Milan, organized by the Centre of Research on Innovation and Internationalization (CESPRI) at Bocconi University in June 2004. The drafts of the chapters in this volume emerged from that conference. However, the final chapters are the result of additional research and extensive redrafting and rewriting, in order to address the main themes of this volume.

3   The contributions to the volume

3.1   *Innovation and economic growth*

In this book, innovation and economic growth have been discussed in a broad way by two contributions that assess the progress and validity of the modern theories of economic growth, and point to the challenges of further research in these areas.

Richard Nelson in "Understanding economic growth as the central task of economic analysis". claims that understanding economic growth should be the central focus of economics and that innovation is a major engine of growth. He focuses on the two main approaches to examining economic growth – the modern neoclassical economics and evolutionary theory – and compares their methodology and results. He concludes that evolutionary economics – by being directly focused on economic dynamics and by having long-run economic growth at the center of its research agenda – is at its core a theory of economic growth. However, Nelson notes that much of recent research within this tradition has stayed too close to certain features of the early work related to the Nelson and Winter's book *An Evolutionary Theory of Economic Change*, and may run into diminishing returns. He stresses that further progress in evolutionary growth theory will be significant if research is pursued in three different ways. The first one concerns the roles of cognition and conscious problem-solving in the evolution of practice, of science in the advancement of technologies, and of knowledge (with its uneven growth across different fields) in human activities. The second one is regarding the role of business practice, organizational forms, and institutions – called "social technologies" by Nelson – and their relations with technological change. Finally, the third one is related to the multisector nature of economic activity and the recognition of industry differences in the patterns of growth, industrial dynamics, and long waves.

Bart Verspagen in "Innovation and economic growth theory: a Schumpeterian legacy and agenda" discusses the Schumpeterian legacy in the modern field of economic growth, and examines first the work on economic growth during the 1950s and 1960s, in which the Schumpeterian legacy was lost, and then the work during the 1980s and 1990s, in which two competing paradigms – the neoclassical one related to endogenous growth theory with a relatively homogeneous set of inter-related models and the evolutionary one with a more loosely connected set of contributions – aimed at explaining the relationship between technology and growth. In way of conclusions, Verpagen claims that endogenous growth theory has recently shifted toward more realistic models that accommodate a range of phenomena previously introduced or examined by the evolutionary approach (such as the notion of technology as driver of economic growth, the importance of business R&D, and the stochastic nature of technological advance) so that some convergence between the two paradigms has taken place. However, the two approaches remain quite distant in terms of microeconomic behavioral patterns (optimizing agents vs. bounded rationality) and the nature of economic process (ahistorical and Newtonian vs. historical and transformational). Verspagen concludes by saying that each of the two approaches contains a range of important and interesting lines of research: for endogenous growth the direction is the development of empirically relevant models instead of new explorations motivated by technical problems encountered by existing models, while the evolutionary tradition confronts the challenge of developing models in close interaction with non formal work, stylized facts, and historical research.

These two contributions, as *Jan Fagerberg* claims in his *Comments*, in addition to providing a detailed and critical guide on the modern theory of economic growth, provide a very broad concept of it, and call for different methodological approaches and levels of analysis for a full understanding of economic growth. In this respect, Fagerberg stresses that qualitative theory is also to be widely used, because the applications of mathematical methods to growth theories has tended to change the questions that are addressed in less-interesting and relevant directions. He finally emphasizes that growth theory has another dimension that needs to be addressed – the political dimension.

3.2   The micro dynamics of the innovation process

As far as the innovation process is concerned, the book concentrates on two cornerstones of the innovation process: incentives and creative destruction. The first one represents the traditional way followed by economists when they examine the forces that affect innovation: in this book incentives are seen in an original and sophisticated, pointing to their different dimensions and roles. Creative destruction represents the basic message of Schumpeter's view of innovation: in this book creative destruction is interpreted as a process that cannot be separated by the way industries are organized and by the extent of the division of labor.

Wesley Cohen and *Henry Sauermann* in "Schumpeter's prophecy and individual incentives as a driver of innovation" move from a view of innovation as driven by the returns earned by firms and by pecuniary benefits, to individual-level incentives to innovate and to nonpecuniary benefits. They claim that richer notions of incentives are needed in the analysis of technical advance, and that research in social psychology and organizational behavior is useful in this respect. Cohen and Sauermann propose three types of individual-level incentives: extrinsic, intrinsic, and social. Extrinsic incentives are usually considered by economists, and include pecuniary benefits. Intrinsic incentives originate within the individual or the task and are often a function of the interaction between the characteristics of the task and of the individual. Social incentives originate outside the tasks from the individual's perceived social relations. Cohen and Sauermann show that individual incentives matter for innovation, and provide a variety of examples in this respect, ranging from the innovative performance of different regions such as Silicon Valley or Route 128 to open source software or to Digital's Alpha Chip development. However, they claim that there are still few empirical studies of the impact of individual-level incentives on innovation: here the focus has to be on a finer-grained analysis of the types of incentives; the recognition of the differences of incentives across engineering and scientific fields and R&D types; the analysis of the alignment of individual incentives and firms' objectives; and the examination of the recruitment, integration, and management of different people with different incentives, skills, and talent. The implications of explicitly considering individual incentives for management and policy are relevant. In particular, the management of R&D may prove quite difficult and require the deployment of various mechanisms for the provision of different types of individual incentive. Policies such as patent policy or technology transfer should consider individual incentives explicitly.

Timothy Bresnahan in "Creative destruction in the PC industry" examines waves of innovation in a highly dynamic industry – PC – over twenty years (1980s and 1990s). Bresnahan suggests that after a founding stage based on entrepreneurial innovation in the second half of the 1970s, a wave of creative destruction took place in the early 1980s with two innovations – the spreadsheet and the word-processing program – the introduction of IBM PC, and the focus on a new market: the white collar one. Waves of complement innovations and positive-feedback systems were possible because openness, modularity, and vertical disintegration characterized the industry. After the first one, other waves of creative destruction followed in a situation of multiple potential standards and divided technical leadership. However, later on the potential waves of creative destruction in the PC industry set off by the widespread use of the Internet did not occur, because the openness and vertical disintegration of the PC industry had declined and established firms were able to block the threatened waves of creative destruction. Bresnahan concludes that vertical disintegration has been important for creative destruction because decentralized inventions of complements led to recombination and *ex post* flexibility. Established firms missed the waves of creative destructions because some waves were difficult to foresee – particularly in adequate details –, decentralization opened the innovation process to all firms, fundamental changes took place in the supply – demand match, and innovations by complementors created new opportunities and lowered barriers to entry.

Ashish Arora in his *Comments* points out that these two contributions highlight some essential elements in the innovation process. One refers to the fact that economists' focus on pecuniary incentives may be excessively narrow. How is it then possible to explain skunk works by engineers and scientists in firms, information trading by engineers, willingness to contribute to open-source software, and superior performance of firms that allow their researchers more freedom to publish? The other is regarding the link of the gales of creative destruction to the entire value chain in an industry and related markets, and to the presence of divided technical leadership, open standards, and modular knowledge. The last ones are conditions that allow for economic experiments and the introduction of variety in the economic system.

3.3   Innovation and industrial dynamics

In this section, Giovanni Dosi assesses the evidence reached so far in the statistical analyses of industrial dynamics and evolution and highlights the research challenges that need to be tackled. Steven Klepper and Peter Thomson, on the contrary, focus on a new area of research of industrial dynamics that has caught great attention recently: spin-offs, and their determinants.

Giovanni Dosi in "Statistical regularities in the evolution of industries. A guide through some evidence and challenges for the theory" examines the patterns of industrial evolution and the dynamics of the distribution of firms' size and performance and its relationship with the underlying technological characteristics. First of all, Dosi finds strong and persistent firm heterogeneity, not eliminated by aggregation, at the base of industrial dynamics. Then he examines the relationship between firms' characteristics and firms' relative success and the differences that exist in this respect across sectors. He links them to processes of technological and organizational learning on the one hand and market selection on the other. Dosi finds that different degrees of innovativeness yield persistent profitability differentials, and that there is an absence of any strong relationship between profitability and growth. Moreover, growth rates present a fat tail distribution. Market selection does not seem to work well, at least in the timescale in which statistics are reported. In sum, the empirical evidence on several sectors indicates various different processes of learning, competition, and growth, generating persistence over time and interindustry differences in firm heterogeneity.

Steven Klepper and *Peter Thompson* in "Spin-off entry in high-tech industries: motives and consequences" examine the origins of entrants in industries in which the rate of technological change is high. They focus on spin-offs and show that a good share of entrants in high-technology industries is constituted by spin-offs and that spin-offs are usually good performers. Existing theories of spin-off formation are based on a serendipitous discovery by an employee that is more valuable to the incumbent firm than to a start-up, but the presence of information asymmetries generates the spin-off; or on a discovery that is common knowledge and is less valuable to the incumbent than to a start-up; or finally on employees who learn from the employers how to compete. Klepper and Thompson propose a model of spin-off formation based on disagreement related to solipsism, which means that the asymmetric weighting of private and nonprivate signals is in favor of the former. Their model has some predictions in line with empirical findings: the hazard rates of spin-offs initially rises with the age of the parent firm and then declines; spin-offs are more likely in industries with considerable uncertainty; spin-offs have, on average, lower quality than their parents; spin-offs are more likely when there is a strong hierarchical structure in decision-making. This is a quite interesting avenue of research because it focuses on individuals' overweighting private information relative to public information rather than on the manipulation of individuals' pay-offs through deception or omission, or on firms trying to rip off their senior employees.

In his *Comments*, *Luigi Orsenigo* suggests that these two contributions highlight the relevance and growth of this sub field of industrial economics, with the introduction of new techniques of analysis, the development of new concepts, and the focus on new stylized facts and puzzles. He suggests that Dosi's piece identifies dynamics, regularities, and puzzles that are broadly in tune with evolutionary explanations of the dynamics of industrial structures and with competence-based theories of the firm. He also claims that Klepper and Thompson's chapter is an interesting contribution to the theory of entry in that it focuses on the process of new firm formation and on the endogenous definition of the population of potential entrants.

3.4   *Institutions for innovations and innovation in institutions*

The broad and relevant theme of institutions is tackled in two very different manners. One way is to look at institutions in a theoretical way; the other way is to examine the historical process through which institutions have been formed.

Masahiko Aoki in "Schumpeterian innovation in institutions" proposes a way to look at institutions in a game-theoretic, equilibrium view and focuses on major innovations in institutions. His unit of analysis is the primitive domain of the game people repeatedly play, and his focus is on the various modes of combinations, in terms of linkages between primitive domain. There are three primitive kinds of domain, with equilibria defining prototypes of institutions. Linkages between primitive domains offer some more realistic models of institutional arrangements. Aoki shows that there are three mechanisms of institutional change through recombinations, in terms of interlinks, de-links, and new links of games of different domains. In this way Aoki identifies overlapping social embeddedness, dynamic institutional complementarities, dis-bundling and new-bundling. Embeddedness and complementarities generate institutional change with inertial, path-dependent features, while bundling contributes to institutional discontinuity. Although conceptually distinct, these mechanisms of institutional change are likely to operate simultaneously and in an interactive manner. The evolution of Silicon Valley is an example of an innovation in an institution of innovation, in which the modularization of the initial design of IBM/system 360 and the emergence of Silicon Valley clustering is interlinked in a path-dependent way, with intense interaction and communication, reminiscent of the social embeddedness in the professional community.

Paul David in "Innovation in Europe's academic institutions" examines the role of universities in the knowledge economy, and discusses the changes undergoing the so-called "European Research Area." His starting general point is that universities are characterized by the integration of advanced education with research in the pursuit of scientific enquiry in a realm of open science. Thus, within the context of academic open-science norms and governance structures, the comparative advantage of university-based researchers lies in conducting inquiries that may provide the foundations for valuable commercial innovations. The best way to do that is not through closely managed, tightly coupled search for discoveries and inventions. The passage of the Bayh–Dole Act (1980) has been accepted as a model for emulation, but this seems to be too dangerously simplistic in several respects, and it does not offer an appropriate paradigm for the European Research Area. In the United States university patenting predates 1980, is concentrated mainly in biomedical knowledge, has brought significant gains only to very few U.S. institutions, and has imposed significant administrative and learning costs to the system by bringing into existence a new professional group- university technology managers. In the case of Europe, what is needed is not an emulation of the dubious paradigm of the Bayh–Dole regime, but a surge of institutional innovations complementing the universities and the institutions of higher education with novel organizations that populate the terrain situated between universities, state agencies, and the business corporations and that are suited for fostering the generation of commercially successful innovations based on the results of publicly supported research.

In his *Comments*, *Bengt-Ake Lundvall* notices that Masahiko Aoki develops the relevant point of institutions as rules of the game. Although critical of a view of institutional change seen only in terms of strategic behavior and centered on the equilibrium concept, Lundvall agrees that Silicon Valley is an example of the working of social capital, competence-building, sharing of tacit knowledge, and interactive learning. Lundvall shares Paul David's critical reflections on the possibility of mimicking the Bayh–Dole Act in Europe and on the attempt to make universities more market-oriented and "rational." If in the short run innovation may be speeded up through collaborations with companies, the long-term indirect effects may be quite serious in undermining scientific advances and the norms of open science.

3.5   Innovation, firms' organization, and business strategies

These two chapters, and related comments, focus on two largely understudied issues. First, both chapters emphasize the need for defining clear and shared criteria for evaluating performance. Unlike financial performance, the criteria through which technological and business strategies are evaluated are themselves unclear and emerging. Second, both chapters emphasize the fact that "technology," the hardware, is but one component of the problem, and possibly not the most complex one. The real challenges appear at the interfaces between technology, organization, and strategy.

Dan Levinthal in "Bringing selection back into our evolutionary theories of innovation" looks at selection as another dimension of firm-level heterogeneity. Not only do different firms rely on different selection criteria, but also – and more importantly – different selection criteria interact within the same firm. In fact, Levinthal argues that one of the key competitive challenges firms face is exactly maintaining such a variety of selection criteria. While we know a lot about the processes and outcomes of variety at the technological level, we know very little about heterogeneity in selection criteria. This issue is related to at least three key areas of current research. First, spin-offs. Failure in maintaining heterogeneity in terms of selection criteria might be, for example, one of the factors explaining why large firms tend to generate spin-offs: entrepreneurs may be led to set up their own firm "by their inability to convince their prior firm to pursue an opportunity that they feel has tremendous promise." Second, demand. Levinthal stresses that the key limit of large organizations is exactly this: they tend to rely on an established belief structure, and they are unwilling to change it. This is a major constraint when firms face rapidly changing demand conditions, as different market segments might require different selection criteria for allocating resources to competing projects. The third issue is about the object of selection. Looking at economics, the likelihood that selection pressures lead to equilibrium outcome crucially depends on the assumption of stability over the object of selection. Looking at the innovation-management literature, Levinthal's moving-target argument criticizes the nowadays popular applications of real-option theory to innovation-strategy issues. Intermediate selection, that is, the evaluation of R&D projects before they produce anything that can be tested in the market, cannot be simply linked to the evaluation of a number of financial options that can be easily evaluated in terms of their price variations. Whether you look at it from the viewpoint of economics or strategy, Levinthal's chapter identifies a clear gap in both literatures: what is being evaluated, and how?

Doz, Cuomo, and *Wrazel'* analysis in "From leadership to management: mobilizing knowledge for innovation in strategic alliances," resonates in many ways with Levinthal's theoretical piece. Starting from a practice-based perspective, Doz, Cuomo and Wrazel analyze how two large multinational firms, that is, STMicroelectronics and Hewlett-Packard, managed to forge a stable and successful alliance. This chapter analyzes the process by which two very different firms became progressively committed toward the development of a shared understanding of each other. The key point stressed by the authors is that the technological and market complementarities that led the two firms to initially join forces had to be transformed into what they call "substantial dependencies," that is, they had to commit through relation-specific asset investments, which generated significant risks. However, risk is only half of the story. At the heart of the successful transformation of complementarities into substantial dependencies lies the effort to design and manage the collaboration process over time. Such "engineering" effort in trust and mutual expectations is what really Doz et al. focus on. Technological design remains here in the background. It provides but the first step in a process of long, painstaking, and mutual learning about each other. Such a process is decomposed into three distinct phases: architecting – to match need and capabilities – organizing – to provide the basis for an ongoing operational collaboration – and structuring – to generate efficient routines. One of the key challenges throughout this process is developing a set of shared-performance-review criteria.

Sid Winter in his *Comments* on the Doz et al.'s chapter highlights the relevance of understanding the process by which successful alliances happen. While, as he stresses, it is clear that both firms had strong incentives to cooperate, a number of things might have gone wrong in the process of managing their relationship. They did not go wrong because managers in the companies involved struggled to buffer their project from "internal hazards" that might have hampered its progresses. Winter stresses how this buffering actually links this case study to the theoretical discussion by Dan Levinthal on heterogeneous selection criteria. Winter notices how Levinthal's discussion points to the trap in which most formal models fall, that is, the assumption that testing can be deferred once learning is complete. These two chapters together, Winter stresses, provide empirical evidence and theoretical grounding to the idea that firms themselves are selection mechanisms, and that such mechanisms are vital for determining the early survival of all innovative projects.

3.6   *Innovation and entrepreneurship*

There is little need to stress how wide and relevant the scientific and policy debate about entrepreneurship has become in recent years. Both these chapters intend to push further such debates developing a similar argument: we need to shift away from the analysis of "entrepreneurship" to look at "entrepreneurial action" or "entrepreneurial function." Entrepreneurship is inherently a process; thus we need to adopt a processual point of view to really come to grips with this concept.

Piera Morlacchi in "Schumpeterian legacies for entrepreneurship and networks: the social dimensions of entrepreneurial action" starts arguing that, despite the rather common association between Schumpeter's work and recent studies on entrepreneurship, relatively little of Schumpeter's early intuitions have been developed. In particular, she argues that much has been said about the relationship between Schumpeter and innovation, while less attention has been devoted to the analysis of entrepreneurial action. Morlacchi argues that entrepreneurship has been analyzed by specialized disciplines, which have collectively failed to deliver a general, integrated understanding of the key engine of capitalist development. She claims first that in going back to the original work by Schumpeter we can actually find elements of such integrated picture. Second, she argues that recent development in social network theory can push forward research inspired by Schumpeter's early work. Her starting point is that we know a lot about the *outcomes* of entrepreneurial action (e.g. new firms, new products, etc.). Yet, the very *process* by which entrepreneurial action unfolds is still unclear. Morlacchi notes that Schumpeter never implied that the entrepreneurial action should be embodied in a single physical person. Rather, he was quite open to the notion that entrepreneurial actions can take place within a social space within which different agents interact. Moreover, the entrepreneurial function can be "filled cooperatively." Hence, Morlacchi establishes a link between Schumpeter's intuition and recent development in network theory and analysis. However, Morlacchi also notes that network analysis has inherited from sociology and economics a strong emphasis on rational behavior. Instead, one of the fundamental drivers of entrepreneurial action is the "joy of creating." There is indeed a strong creative element implicit in Schumpeter's work on entrepreneurship that has been subsequently pushed to the background by approaches that emphasize rationality over intuition, computation over agency.

Ulrich Witt and *Christian Zellner* in "Knowledge-based entrepreneurship: the organizational side of technology commercialization" provide guidelines about the key dimensions that we need to study to explore the process by which entrepreneurial action is embedded in a specific social context. To understand this process, Witt and Zellner argue that we need to look at the constraints that entrepreneurship faces. Crucially, knowledge operates as both an enabling force and an obstacle to entrepreneurship. What kind of knowledge needs to be transferred explains to a large extent what kind of social context embodies the entrepreneurial action. One of the key functions played by entrepreneurs is what Witt and Zellner call the "self-sorting process", that is, the process that precedes the funding of a new firm or division, whose aim is to see a business opportunity generated by new scientific knowledge, to attract and coordinate resources, and to integrate new knowledge into the established organization. The outcome of such a process is the decision about the context within which to transfer newly generated scientific knowledge: a new firm or an established organization? It is hard not to see a parallel between this process and the "matching" process discussed by Doz et al., as well as the "selection" argument put forward by Levinthal. Witt and Zellner analyze then the different problems and constraints faced by entrepreneurs within established organizations and by entrepreneurs who found a new firm. Again, their key challenges are related to the extent to which the knowledge on which they build is tacit, and how rapidly it decays. Quite interestingly, in both cases they point to *career paths* as one of the fundamental channels through which organizations can provide options to their members: options to establish and maintain link also with other organizations, options to acquire and transfer new knowledge, and options to generate new business conceptions.

Maureen McKelvey in her *Comment*s stresses that more attention should be devoted to the analysis of how scientists and engineers who leave academia can succeed in renewing their knowledge and skills, thus slowing down that process of decaying highlighted by Witt and Zellner. Again, career paths and human resource issues are identified as central to reach a better understanding of technology transfer and, more generally, entrepreneurial actions. As far as Morlacchi's chapter is concerned, McKelvey stresses the need to use network analysis to foster our theoretical understanding of entrepreneurial action. She points to the need to look not only at entrepreneurial processes *per se*, but rather at how entrepreneurial actions cause further changes in the wider economic system.

3.7   Innovation and the evolution of the university system

The changing role of universities and the relationship between university and industry is nowadays of central relevance to the policy debate. The chapters by Bercovitz and Feldman and by Crespi and Geuna focus on the role of universities in modern economies. These two chapters present two different viewpoints on the subject.

Janet Bercovitz and *Maryann Feldman* in "Academic entrepreneurs and technology transfer: who participates and why?" deepen the discussion about knowledge-based entrepreneurship initiated by Witt and Zellner. Their analysis starts with the observation that institutional changes are not enough to explain the effectiveness of technology-transfer processes. Becoming an entrepreneurial university active in technology-transfer requires the participation and commitment of the faculty, because, in the end, it all depends on the propensity of faculty members to disclose their findings. And despite institutional changes about IPRs (an issue also discussed in Paul David's chapter), Bercovitz and Feldman highlight a number of reasons that might prevent disclosure. First, faculty specialized in basic research might perceive as a waste time the process of "applying" their findings to problems in the realm of business. Second, they might be worried about publication delays. Third, they might simply believe that commercial activities are not appropriate for scientist. Hence, the core empirical question of this chapter: given the institutional set up, what are the individual features of those faculty members who decide to disclose? They identify two key attributes that differentiate academics: inventive capacity and entrepreneurial propensity. The former refers to individuals' ability to generate new knowledge through bisociation, that is, the ability to relate two seemingly unrelated concepts. Such an ability is most likely found in individuals who hold multidisciplinary backgrounds and occupy boundary-spanning roles connecting different groups of people, and originate from countries other than that in which they work. Entrepreneurial propensity refers instead to the likelihood that an individual will disclose his or her ideas. There is both a "culture" and a "nurture" element here. As for the former, risk aversion plays a big role in explaining different disclosure behaviors. People have different propensity to risk, and such propensity can be proxied by migration decisions: people who leave their home country are likely to have a higher propensity to risk than those who stay. As for the latter, individuals trained in institutions where disclosure is a legitimate objective tend to disclose more willingly than individuals trained in "publish or perish" oriented institutions.

Gustavo Crespi and *Aldo Geuna* in "Modeling and measuring scientific production: a first estimation for a panel of OECD countries" perform a cross-country comparison of scientific productivity, whereby the notion of productivity is quite closely related to traditional output indicators for the university system: publications. From an analytical perspective, Crespi and Geuna explicitly rely on the often criticized "linear model." They argue that, despite its pitfalls, this model offers categories and constructs that can be operationalized and empirically assessed. Each category may be a "black box" (e.g. the "basic research" box, the "applied research" box), but as long as we have a fair estimate of what gets in and what gets out of each box, we can develop testable propositions (at least at the macro level). Crespi and Geuna analyze a panel that includes a sample of fourteen OECD countries over a period of twenty-one years. They explore the determinants of scientific research production. As proxies for output, they use both publications and citations, while ignoring the production of qualified personnel and new technologies. They focus on two key issues: the profile of time lag between investment in HERD (Higher Education Research and Development) and output, and the returns to national investments in science, with specific attention devoted to the possible role of cross-country spillovers. The aggregate results are quite interesting. First of all, they do find evidence of a positive, long-run relation between HERD and the output measures that they use. For both measures they also find evidence of decreasing returns at the national level, but not at the global level. The latter result points to the existence of positive spillovers, the main source of which is the United States. Also, they find a rather long lag structure in the relationship between HERD and output, that is, it takes at least two years to see a positive impact of an increase in HERD on publications and three years for citations. The maximum effect is reached, respectively, at years five and six. Because most evaluation exercises adopt a much shorter time frame, they undervalue the effects of changes in HERD levels on scientific production.

Ed Steinmueller's Comments to these two chapters set their results within the wider debate about the role of universities in modern economies. He stresses the fact that both chapters adopt a distinctively reductionist approach. Bercovitz and Feldman take the view that universities' productivity should be measured in terms of the money they generate through their patenting and licensing activities. Steinmueller notices how in the past universities have successfully contributed to wealth creation doing exactly the opposite, that is, disclosing their results without charging for the use of knowledge they had generated. Crespi and Geuna's chapter takes a similar view, although looking at universities in a more traditional way, that is, they analyze universities as producers of "science" as measured by publications. However, Steinmueller notices that they do not look at universities' output in terms of training and education. Either way, Steinmueller concludes that these two chapters contribute to the fast-growing literature about the role of universities in the knowledge-based economy, and also reflect on the narrowing of the contemporary discussion concerning university missions.

3.8   Innovation and public-policy

While most of the chapters discussed so far have public-policy, implications, no chapter has addressed directly the public-policy issues with respect to innovation or the policy-making process itself. The last three chapters of this book try to shed light on these issues. They approach this problem from three distinct, yet complementary, perspectives. Stan Metcalfe's chapter proposes the point of view of an academic whose work has had a major impact on policy issues. Paul Geroski's chapter provides the point of view of an economist who has – so to speak – jumped the fence to become a policy-maker as member and then Chairman of the U.K. Competition Commission. Finally, Paraskevas Caracostas is an economist who has spent most of his career as a policy-maker or, to use his own term, as a "policy-shaper."

   Stan Metcalfe's chapter, "Innovation systems, innovation policy, and restless capitalism", discusses what the fundamental characteristics of the economic system are that policy-makers need to know in order to affect its evolution. First, economies evolves because knowledge evolves. Second, changes in knowledge are embedded in the economic system. While knowledge is held by individuals, the growth of knowledge crucially depends on interactions among individuals. Third, the evolution of economies is inherently uncertain. That is precisely why we need markets: they provide a fundamental locus of experimentation through innovation and selection. These three features lead, according to Metcalfe, to a system-innovation policy perspective. As knowledge evolves in a distributed manner (across individuals, organizations, markets, sectors, and countries), it relies on an ecology of actors who need to interact. Interactions, however, may fail to occur: the ecology may not generate a "system." Policy-makers can play a fundamental role in facilitating and shaping the emergence of agents, and of connections among agents participating in an innovation ecology. What are then the key areas of policy concern? First, the generation through public investment of a large enough pool of educated minds that play a crucial role in experimenting to create new knowledge. Second, the support of firms' efforts to generate absorptive capacity through private R&D. Third, the development of bridges and connections among existing institutions, recognizing that there are fixed costs to be paid to establish a relationship. Stan Metcalfe also stresses that innovation policies need to be complemented by competition policies that keep the market system open to entrants and tolerate supernormal returns as the outcome of transient innovative superiority. This is exactly the topic of Paul Geroski's chapter. Second, Metcalfe stresses the necessity for policy-makers to be adaptive and ready to respond flexibly to a system that is necessarily open and continuously changing. This latter point is the focus of Caracostas' chapter.

Paul Geroski's chapter, "Intellectual property rights and competition policy" looks at the possible contradictions between the growing importance of intellectual property rights and competition policy. Its central argument is that the apparent contradiction between IPRs protection and competition policies rests on the misunderstanding of what role "monopoly" plays in each. Competition policy focuses on problems arising from the existence of monopoly *ex ante*. IPRs protection is instead concerned with monopoly *ex post*. The possibility of exploiting – *ex post* – a temporary monopoly provides strong incentives to firms to undertake innovative strategies. The existence of monopoly *ex ante* limits instead the strength of such incentives because monopoly pre-empts entry and thus makes the market closed, instead of being open to allow firms to experiment. So, generally speaking, there is no incongruence between competition policies and IPRs protection. Problems emerge, Geroski notes, only in two cases. First, when IPRs are abused. Restrictive practices can be fought through competition policies, though. Second, when IPRs prevent innovation, limiting the possibility of building upon prior knowledge. This may happen in two ways. One is the well-known problem of the anticommons. The other is the problem of hold-up in relation to complex innovations that build upon a variety of technological classes. In these cases, there is indeed a tension between IPRs defense and competition policy. This is the area of policy intervention that is attracting the most attention, with reference, in particular, to the issue of patent breadth.

Paraskevas Caracostas' chapter, "The policy-shaper's anxiety at the innovation kick: how far do innovation theories really help in the world of policy? claims that the knowledge bases on which "policy-shapers" rely is both highly malleable and broad. Old and new frameworks coexist at all times, and new policy measures are put in place within institutional contexts that were designed and modified according to different frames of reference. Within this context, policy-shapers are called upon to solve four different, yet highly overlapped, problems. First, what are the boundaries of the reality the policy-makers should try to affect? Second, what is the rational for intervening? Third, what are the specific problems to be solved? Fourth, what are the tools for solving those problems? Caracostas highlights the fact that policy-shapers may call upon different bodies of knowledge and different frameworks to answer these four questions. Yet, the answers to these four questions need to be integrated into a coherent whole! On the first issue, the complexity of the boundary issue is well reflected in the emergence of organizational innovations within governments. In this respect, however, the innovation-system approach provides useful representation of "innovation systems," but it does so only *ex post*. The second issue (i.e. legitimacy and rationale) leads Caracostas to claim (similarly to Crespi and Geuna) that the rhetoric of the linear model is still quite useful in providing the rationale for public interventions. The third issue (i.e. the problems) brings to the fore the advantages of evolutionary models to innovation and of the system of innovation approach, which is able to provide useful taxonomies of problems that policy-shapers can use to identify specific areas of intervention. Finally, the fourth issue (i.e. the tools) is best analyzed by looking at the analytical categories provided by the literature on the social shaping of technology, with its rich and deep analysis of the specific microlevel processes that lead to innovation. This approach emphasizes negotiation among subgroups and trade-offs among technological options, and challenges technological determinism and technocratic approaches to scientific and technological developments.

4.   As a way of conclusion . . .

We are convinced that the chapters in this book provide an detailed map of where research on innovation is going, where results have been obtained and – crucially – what challenges lie ahead of us. It is also clear from this book that it is our firm belief that understanding innovation in the economy requires intense interaction, open debate, and fruitful discussion among academics of different disciplines and different approaches, from economics to management, sociology, history, geography, and science and technology. Even within economics itself, the evolutionary and Schumpeterian traditions have to interact in a fruitful debate with the innovation system and institutionalist perspectives and with neoclassical theory. In conclusion, in spite of the differences in perspectives, approaches, and methodologies, one can identify a few major lessons.

First, openness and interdisciplinarity are necessary conditions for improving our understanding of innovation. However, the challenge here is to reconcile openness with new contributions from different fields, maintaining a high-level of scientific rigor and consistency – often difficult to obtain because scientific rigor is discipline-driven and discipline-bounded. Interdisciplinarity means interaction with different perspectives and visions, with one discipline trying to learn from another. It also means that we have to think hard about what methodologies we can deploy – and how – in order to produce research results that are both relevant and rigorous.

Second, besides the traditional distinction between static and dynamic approaches, we wish to stress another divide that most chapters touch – the divide between structural and processual analysis. Innovation is intrinsically a dynamic phenomenon and it has to be examined in that way. Looking at innovation requires an understanding of the processes through which it happens. Structural analysis has delivered much understanding about the proper configurations of complex systems (whether by system you mean an industry, an innovative network, an innovation system, or whatever else). However, if one looks at it from the point of view of industrial dynamics, entrepreneurship analysis, or strategy theory, we still know very little about the dynamic processes that lead to certain structures, and to their demise.

Third, and related to the above point, the richness and variety of approaches presented in this book remind us of the fundamental fact that research itself is a process. Different methodologies interact and complement each other. The search for the best methodology appears futile. A fruitful methodology in the analysis of innovation is one that identifies some empirical regularities and, within these, puzzles that need to be explained; and develops appreciative theorizing, possibly grounded in rigorous case studies that help in generating hypotheses, then builds formal models to test the rigor of the argument that is being developed, and feed back into empirical analysis in terms of tests, insights, and questions. Consistency between cases, appreciative theorizing, econometrics and modeling has to be present. Research on innovation should not be guided by techniques, and innovation theory should be driven by empirical questions and facts.

References

Rosenberg, N. 1976. *Perspectives on Technology*. Cambridge, UK: Cambridge University Press
Schumpeter, J. 1934. *The Theory of Economic Development*. Cambridge, MA: Harvard University Press
Schumpeter, J. 1939. *Business Cycles*. New York: MacGraw Hill
Schumpeter, J. 1950. *Capitalism, Socialism and Democracy*. New York: Harper Brothers

Part 1

Innovation and economic growth

1 Understanding economic growth as the central task of economic analysis

Richard R. Nelson

Introduction

This chapter has three parts. I begin by endorsing Schumpeter's argument that understanding economic growth ought to be the central focus of economic analysis, and proposing that modern evolutionary economic theory has its central focus just there. In Part II, I turn to the origins of modern evolutionary economic theory as an endeavor inspired by Schumpeter. However, I propose that, while significant progress has been made by proceeding along established paths, the endeavor now is running into diminishing returns. Part III offers my thoughts on new directions that I think are highly valuable to pursue, in order to develop a truly illuminating theory of economic growth.

1.1 Economic growth as the appropriate central focus of economic analysis

The vast increases in living standards and productivity experienced by a significant part of the world's population clearly is the most dramatic and beneficial achievement of the market-oriented economies that began to emerge in the late eighteenth and early nineteenth centuries. Surely, the primary task of economic theory should be to illuminate how this miracle was accomplished, and the determinants of economic growth in the future.

This is not simply my point of view. It certainly was Schumpeter's. And Schumpeter's position on this was not radical. Indeed it reflected the writings of many of the great classical economists whose work preceded his. Thus reflect on Adam Smith's *The Wealth of Nations*. (1937; First published 1776). This book is basically an analysis of the factors driving the economic growth that was occurring in the United Kingdom in the late eighteenth century, along with a diagnosis as to why it was not occurring so effectively elsewhere. The treatise starts out with the famous discussion of the dynamics that Smith believed had so dramatically improved productivity in pin-making. This central orientation to the phenomena of economic growth is present in many of the works of the nineteenth century "classical" economists. An analysis of the determinants of prices and wages also was an important issue in the classical economics writings, but, as in Smith's writing, tended to be treated after the sources of economic growth had been laid out.

However, this certainly is not the orientation of contemporary neo-classical economics, at least as the subject is laid out in general text-books. There the heart of modern economic science is presented as being the neoclassical theory of the determinants of the pattern of inputs, outputs, and prices, under conditions of a hypothetical equilibrium. The orientation is partly positive and partly normative, with the normative apparatus linked to the concept of Pareto optimality, and analysis of the conditions under which market equilibria meet, or deviate from, the necessary conditions.

This is not to say that economic growth is ignored in introductory texts. In many of them, analysis of economic growth is given high priority. However, generally economic growth is brought up as a subject of analysis only after the students are assumed to have standard microeconomic theory under control. And the tools of analysis of economic growth that are used are basically those of equilibrium micro-economics, augmented to take aboard the possibility of continuing technological advance. This is so not only in introductory treatments, but also on more advanced neoclassical treatises on growth. Solow's pioneering theoretical and empirical writings on growth (1956, 1957) were based exactly on neoclassical simple microeconomic theory, principally the theory of the firm in market equilibrium, that was the standard then and is now, augmented to include the possibility of technological advance over time. It is fair to say that the new neoclassical growth theory has stayed very much like the old, in these respects. (For a discussion, see Nelson, 1998)

Put more generally, contemporary neoclassical economics is basically about conditions of general equilibrium. Analysis of economic growth is largely a graft on that subject.

The shift in the orientation of the main line of economics away from a central focus on long-run economic growth, and toward a focus on conditions of economic equilibrium, comes with the rise of neoclassical economic theory. Marshall's reflections on this are interesting. In the preface to his *Principles* (1948, 8th edition, first published 1907) he says, in effect, that the important questions for economics lie in the dynamics, and that biological conceptions seemed the appropriate route into

economic dynamics. But then he goes on to say that the tools for analyzing equilibrium conditions were better honed, and so this is what his
book would largely be about. Marshall never got around to writing that
second volume on economic dynamics that he implicitly had promised.

Schumpeter's views here are very clear. In his writings from *The
Theory of Economic Development* (1934) through his *Capitalism, Socialism,
and Democracy* (1942) he is arguing against the prevailing trend among
economists to define the core of the discipline as about firm and
household behavior, prices and quantities, under conditions of equilibrium, whereas it was clear (to Schumpeter) that the main thing about
Capitalism was that it was an engine of progress.

This certainly does not mean a lack of interest in the question of what
lies behind the allocation of resources in an economy at any time, or the
pattern of output and prices. But Schumpeter's view on these matters
was dynamic and not static. He argued that one could not understand
the processes driving economic growth without consideration of what
was going on in different economic sectors that was leading to a changing pattern of prices and allocation of resources. That is, Schumpeter's
theory of the mix of outputs and inputs among industries, and product
and factor prices, was part of his theory of economic growth.

Nor does a central focus on economic growth play down the role of
market organization of economic activity, the activities of for-profit
firms, and competition as key elements behind the successful performance of Capitalism. Rather, it views successful performance in a different light, and sees the role of competition in a different way.

In any case, the central reason I am an evolutionary economist is that
evolutionary theory is, at its core, a theory of economic growth. It is
indeed concerned with illuminating the factors behind prevailing patterns of outputs, inputs, and prices, but sees these in a dynamic context.

1.2   The development of evolutionary growth theory,
         and diminishing returns

When Sidney Winter and I set out to write *An Evolutionary Theory of
Economic Change* 1982, much of our inspiration came from Schumpeter.
It may be somewhat ironic that I discovered Schumpeter, or rather came
to understand what he was arguing and the importance of that argument, in research motivated by Solow's neoclassical theory of economic
growth. Motivated by Solow's arguments and other empirical studies
documenting the importance of technological advance, I and other
young scholars set out to study the process empirically. A number of us
came to realize that the phenomena we were finding were completely

incompatible with those basic premises. While Schumpeter had made the argument long before, the economists studying technological advance, pointed in that direction by Solow, came on their own to see that innovation, technological or otherwise, could not be understood within the confines of a theory that assumed continuing equilibrium. Rather, one needed a theory that saw technology, and other aspects of the economic system, as undergoing continuing evolution.

The proposal that one should model technology as evolving, and that economic growth more broadly should be understood as proceeding through an evolutionary process, scarcely were new ideas. Thus in the early eighteenth century Mandeville(1924, first published 1714) argued that the basic design of the sophisticated naval fighting ships of his day, which he regarded as the pinnacle of technological accomplishment then, was the result of a multitude of cumulative advances made over a long period of time by many people, rather than something that was the result of a coherent worked-out plan. Adam Smith's discussion of the coevolution of advances in the technology of pin-making and the increasing division of labor in the operations, both driven by and interacting with a growing "extent of the market," has a similar evolutionary flavor. These early accounts, put forth well before Darwin, did not articulate a crisp theory of variation and selection as the cumulative mechanism at work, of the sort introduced in the new evolutionary growth theory. But something like that was implicit.

I confess that when Sidney Winter and I were developing *An Evolutionary Theory of Economic Change*, while we clearly recognized the intellectual base of our work in Schumpeter, I did not realize the extent to which what we were developing had been foreshadowed by an earlier pre-modern neoclassical tradition in economics. Of course, we had available to us a large body of technique and pieces of theory that were not there at the times of the earlier writings, like the theorizing of the Carnegie Tech crew – Simon, March, and Cyert – on bounded rationality, and their articulation of *A Behavioral Theory of the Firm* (1963). However, increasingly I am of the belief that modern economic evolutionary theory can be thought of as a renaissance of an older tradition in economics that got sandbagged.

What Winter and I did, of course, was to marry an evolutionary theory of technological change with a behavioral theory of the firm, augmented to include innovation as a central firm activity, and placed in a context of Schumpeterian competition. To attack the phenomena addressed by neoclassical growth theory, we treated technologies as activities that used labor and capital to produce output, and built in mechanisms regulating the change over time in supplies of labor and capital.

This formulation obviously struck a responsive chord. It has spawned a major research tradition. I want to express my particular enthusiasm for the fine mix of, and overlap between, empirical and theoretical research that has marked our research enterprise. The interaction between "appreciative" and "formal" theory has been strong, and I think very fruitful.

However, in my view, much, too much, of the research within this tradition has stayed too close to certain features of the early work, which I think is causing the endeavor to run into sharply diminishing returns. Here, I want to highlight three aspects of my early modeling with Winter that probably now are obstacles to further progress, and need to be got out of the way.

First, perhaps because we were so focused on showing that sophisticated effective practice could be explained without assuming that the individuals and organizations engaging in such practice had devised and chosen what they were doing from a large range of perceived alternatives, we played down the role of cognition, understanding, and conscious-problem-solving, in the evolution of practice. In so doing, in effect we were playing down the importance of human knowledge in the advance of know-how, and in particular were repressing the important roles that the advance of science had played in the evolution of practice in a number of areas. It is time, I believe, to build more closely into economic evolutionary models the nature and evolution of the knowledge that guides attempts to improve practice.

Second, the model that Winter and I developed to try to explain experienced economic growth focused on "technologies" as the body of practice that had experienced the most rapid evolution. While we stated that other aspects of business practice also went through evolutionary change, we did not do much with that proposition. A major reason, or at least my reason, was the conviction that it was the rapid and continuing evolution of technologies that was the basic driving force behind the growth that had been experienced.

I think the tack we took was the right one, then. But I think evolutionary growth theorizing has, until recently at least, neglected the evolution of business practice, organizational forms, and institutions more generally. Bhaven Sampat and I (2001) have proposed that these kinds of variables can be regarded as "social technologies," as contrasted with physical technologies, and that the evolution of social technologies is an important, and usually neglected, part of the economic growth story. In many cases social technologies have had to change in order that society be able to take advantage of the new physical technologies. At the same time the evolution of social technologies seems to be more sticky and less well oriented than the evolution of physical technologies. Getting a better grip on this set of issues ought to be high on the research agenda.

Third, we followed Solow and other neoclassical growth theorists in seeing economic growth as a macroeconomic phenomena. Solow's 1957 empirical article, while linked to his 1956 theoretical piece, also was in a tradition of empirical analysis of the factors behind economic growth that was being conducted by scholars at the National Bureau of Economic Research (NBER), which made use of the newly available time series of Gross National Product (GNP). The GNP series provided an aggregate measure of the total production and growth over time of an economy's output, which could be compared with aggregate measures of an economy's labor inputs and its capital stock and the changes in these over time. The evidence that aggregate output had increased at a significantly faster rate than had total inputs was reported in several publications before Solow's famous paper, and these earlier publications also put forth the proposition that the greater increase of output than inputs was evidence of the importance of technological advance.

I believe that it is highly useful to have an aggregate measure of economic production and of the rate of economic growth. However, a long time ago, particularly in his *Business Cycles*, Schumpeter insisted that viewing growth as a macroeconomic phenomena blinded the analyst to the fact that the real economy consists of many different economic sectors, and that economic growth involved in an essential way the rise of new industries and sectors and the decline of old ones. As Stanley Metcalfe has argued in several recent essays (2002), creative destruction is not simply about firms, but about industries. The current generation of evolutionary growth models has not recognized this adequately. I consider it an open question whether Schumpeter's long wave theory, the heart of which is the proposition that the driving force of growth at any time lies in the rapid advance of a small number of critical technologies, is basically correct or not. But I think it important that evolutionary growth theory be able to address that debate.

I believe that our common efforts to date on developing an evolutionary theory of economic growth have been very successful. But there are clear diminishing returns in continuing down the old paths. It is time, I would like to argue, for setting out in new directions.

1.3   Promising new directions

I focus here on the three limitations of the earlier evolutionary growth theory that I have identified above, and give my thoughts on promising new directions to take.

As I noted, the early versions of evolutionary economic theory perhaps leaned backwards too far in trying to demonstrate that the often very sophisticated and powerful human practices that were involved in economic activity could be, and should be, understood not as the result of human omniscience and global deliberation, but as the long-term achievements of an evolutionary process in which individual action and choice in any instance generally involved no more than ordinary sophistication and skill. The human and organizational "rationality" in evolutionary theories clearly is a "bounded rationality." The amazing progress achieved in many areas over the long run is the result of the power of the evolutionary processes at work.

While I am sure the basic perspective here is absolutely correct, it tends to repress the fact that, at least in modern times, the strength of human knowledge that is brought to search and problem-solving in a number of areas is extremely impressive. And while that knowledge itself needs to be understood as having been the result of an evolutionary process, the character and strength of knowledge at any time profoundly affects how the evolutionary processes at work at that time proceed. Joel Mokyr (2002, 2004) has argued that the development of strong scientific knowledge relevant to advancing technologies, which occurred during the nineteenth century, was the key factor enabling technological advance to become a sustained phenomena, rather than proceeding in fits and starts.

Economic evolution, human cultural evolution more generally, clearly differs from biological evolution in that the human and organizational actors are purposeful; they often make conscious efforts to find better ways of doing things, and their efforts to innovate are far from being completely "blind." I propose that when the knowledge that can be used to guide search (and problem-solving within search) is strong, it lends power to the effort in four different ways.

First, it enables the searcher to focus effectively; knowledge identifies certain potential pathways as likely dead ends, and identifies others as promising to pursue. Second, strong knowledge highlights markers that one can see if one goes down a particular path that indicate whether that path is going in a plausible broad direction or not, and also the kinds of change in direction that seem appropriate. Third, after a new practice is developed and actually employed, the strength of knowledge affects the ability to accurately evaluate that practice in a timely fashion.

Fourth, a strong knowledge base often permits a good deal of the searching and problem-solving to proceed "offline." In so doing, it changes the nature of the exploitation versus exploration conflict that

Jim March and others have highlighted, by permitting much of the latter to proceed offline, until strong evidence is accumulated that the practice being explored should be adopted. If one reflects on it, this is exactly what Research and Development is all about – offline exploration by doing theoretical calculations, constructing and testing "models," and working with pilot plants or test vehicles to learn more about their properties, without a commitment to actually put the new design or practice into operation until it is well tested.

Under this perspective, evolutionary processes are very much learning processes. A certain portion of the writing in evolutionary economics recognizes this, implicitly or explicitly. Of course, from a certain point of view, biological evolutionary processes can be interpreted as learning processes in which a species "learns" how better to survive and prosper. But what is going on in human cultural evolution is that knowledge is accumulating in the heads of human beings. Individuals, and individual organizations, are learning to do things better, and the society as a whole is learning.

A central part of that learning is simply learning about ways of doing things that had not been thought of before, or at least not seriously explored, and about the performance of these ways of doing things. However, it is clear that in the process of learning about and implementing new practices, like Mandeville's ship designs, what is learned transcends the details of particular practices, techniques, and designs, and a broad body of understanding thus evolves along with a body of practice. Mandeville's ship designers improve their general understanding of the principles of good ship design as they go about modifying their old designs, in most cases for the better, but occasionally for the worse.

However, while important parts of the knowledge base for search and problem-solving in a field develop almost as a by-product of actual experience, particularly over the last two centuries a large number of fields of applications-oriented science have been institutionalized. Today, virtually every field of human practice, from ship designing to the design of computers, to medical practice, to the practice of business management, has associated with it an applications-oriented field of research and training, like the engineering disciplines, or fields like pathology and bacteriology, managerial economics, and organization theory. But it is clear that some of these applications-oriented sciences are much more powerful than others.

More generally, the strength of the knowledge base to guide search and problem-solving, which has been achieved both by drawing the lessons of experience, and through the development of the background applications-oriented sciences, differs enormously across fields of human practice. In some areas, efforts at design and problem-solving work from a strong-enough base of understanding that theoretical and empirical calculation can relatively sharply identify highly promising directions, and evidence gained through offline experimentation and testing can provide quite reliable estimates of how a particular new design, or practice, will actually work. This powerful background knowledge does not eliminate the need for learning through actual doing and using, but it enables an enormous amount to be learned before the innovator actually has to go online, with the major commitments that that usually entails.

In other cases the knowledge base may be quite weak. Calculation and analysis of perceived alternatives may not take the venture very far, and the ability to learn through offline experimentation and testing may be highly limited. In this latter situation, about the only way to move forward is through actual trying, and learning through doing and using, and even that learning may be relatively unreliable and slow in coming. I propose that the rate of progress in the latter cases is going to be much slower than the rate of progress in the former.

I want to set this line of analysis aside for a moment and get into my second line of discussion, about the high priority of bringing organizational practice, organization form, laws and public policies, and institutions more broadly, explicitly into an evolutionary theory of economic growth. However, the connections that I will draw shortly between theme 1 and theme 2 might already be obvious.

The evidence is overwhelming that it is the advance of technology that has been the basic driving force behind the increase in productivity and living standards that has been achieved over the past two centuries. But changes in organizational practice and form, and institutional structures more broadly, are also an important part of the story. Adam Smith recognized this, in his discussion of pin-making. There he highlighted both the invention of many different kinds of machinery, and ·the increasing division of labor, associated with the dramatic increases in mechanization both as cause and effect.

Albert Chandler's great studies (particularly *Scale and Scope*, 1990) were focused on the changes in the structure of business firms, and business practice, that were needed to take full advantage of the development toward the middle of the nineteenth century of railroad and telegraph technologies, which opened the potentiality for firms to buy inputs and sell outputs over a much wider range of space than had been customary before, and the complementary advances in capital goods technologies, which together opened up the possibilities of great economies of "scale and scope." Chandler notes that these much larger firms required a larger and more sophisticated managerial team than that could be recruited by tapping family and friends, which had been the custom when companies were small. The concept of "professional management" came into existence, and shortly thereafter business schools arose to train professional managers. The very large financial requirements of the modern corporation led to changes in the organization of banking, and gradually to the emergence of the modern stock market. A wide range of new laws was needed to support, and control, these developments.

Samuel Beer (1959) and Peter Murmann (1993) have told a parallel story regarding the rise of the modern dyestuff industry during the last half of the nineteenth century. As with the Chandler story, advances in physical technology, in this particular case enabled by significant improvements in understanding and technique in organic chemistry, started the cascade of developments. The industrial research laboratory emerged as a structure enabling firms to hire and effectively employ "inventors" with advanced training in the relevant fields of science. The rapidly growing dyestuffs industry was the source of a large and rapidly growing demand for highly trained chemists. The German university system adapted to meet these demands helped by significant funding coming from governments.

Or consider developments in medical care over the last century. Again, the driving force has been significant improvements in scientific knowledge bearing on medicine, and the development of a wide range of chemical substances, physical devices and artifacts, and medical practice that are effective across a wide range of diseases. These advances greatly increased the skill requirements of physicians, and led to the development of the modern medical school. Hospitals changed their nature from places where the sick and dying were, in effect, simply kept, to places where sophisticated medicine was practiced. The new medicine was also very expensive; the institution of medical insurance began to arise. And a wide variety of new government policies came into place, both to provide financial support for the practice of medicine, and also for medical research. The modern research-based pharmaceutical company, drawing scientific understanding and trained people from the universities, and selling its products on a market dominated by third-party payment, is largely a post-World War II phenomena. And so are various forms of pharmaceuticals regulation.

In each of the cases above, while the advance of "physical technologies" was central in the story, development also involved new modes of organization and organizational practice, and new institutions more broadly. I have told these different stories in a certain amount of detail to make persuasive my argument that economic growth needs to be understood as a process driven by the coevolution of physical and social technologies, to use the terms Sampat and I proposed. It is fair to say that neither neoclassical nor evolutionary growth theory has taken the social technologies part of that story as seriously as it should.

Let me now link the discussion back to my earlier proposition about the significance of differences across areas of human practice in the extent to which the knowledge base permits sharp focus on promising pathways for improvement, ability to learn a lot by relatively low-cost offline experimentation, and quick reliable feedback of the efficacy of a new practice once it is put in place. Without denying significant intraclass variability, the apparent differences, on average, in these respects between efforts to advance physical technologies, and social technologies, are striking. Virtually all stories that I know about of significant physical invention in the twentieth century describe the calculation, the offline experimentation, and the deliberate and usually reliable testing that was involved in the efforts. In contrast, these aspects are strikingly missing from the accounts that I know about of efforts to advance social technologies, to implement a new business practice, or put in place a new public policy. Institutional learning seems to be just much more difficult than learning regarding physical technologies.

I want to turn now to the third area that I flagged. I think evolutionary growth theory needs to recognize more explicitly the multisector nature of economic activity. This would involve, first, recognizing and incorporating interindustry differences in the pattern of growth being experienced at any time, and second, coming to grips with interindustry coordination mechanisms. There are two building blocks I want to highlight here: the growing literature on industrial dynamics and the new writings on Schumpeter's theory of "long waves."

I do not know if the scholars who have been contributing to the advance of empirical and theoretical understanding of "industrial dynamics" (for example, Malerba and Orsenigo, 1997, and Malerba, 2002) would consider their work to be part of growth theory. But I would. A key characteristic of this work is that it recognizes, and attempts to explain, differences across industries. These differences have included the size of the firms who are most active in innovation, whether innovation is coming from firms in the industry or from upstream firms or both, and the links of technological advance in the industry with science. As a result of this work, we are now able to see significant differences across industries in these regards.

Also, technologies and industries change over time. Many (not all) seem to experience a more or less systematic product or technology "cycle," from infancy to maturity. To some extent, cross-industry variation at any time is associated with the different levels of "maturity" of different industries (see, for example, Klepper, 1996). A problem with the industry life cycle literature, at least in its early form, was that implicitly it saw industries as having a single cycle. However, as empirical research in this area has proceeded, it has become clear that many industries experience a succession of cycles, with a particular cycle being associated with the emergence of a promising technology, and then its maturation, followed by a renaissance of activity in the industry as a new technology emerges and replaces the older one, and so on (see, for example, Mowery and Nelson, 1999).

My own contribution to research in this area has been to propose that an industry or technology life cycle needs to be understood as involving the evolution of social technologies, as well as physical technologies, or rather the coevolution of both. Thus, organizational forms and practice, and the supporting institutional structures, change over the course of a technology or an industry life cycle. An extremely interesting question is whether the social technologies that are fruitful in one technological era are also the ones that need to be fruitful when a new technology succeeds the old. The considerable business school literature on competence-enhancing and competence-destroying technological advance is basically about this question. (For a survey and a collection of good studies, see, Dosi, Nelson, and Winter, 2000)

While there is little cross-referencing, the literature on technology life cycles and the rapidly growing literature on long waves of economic activity have a lot in common. The latter literature is, of course, motivated by Schumpeter's theory put forth in his *Business Cycles* (1939). Schumpeter's basic proposal was that economic growth in Europe and the United States had gone through a number of eras, with economic growth in each era largely driven by technological advance in a few key industries, whose effects fanned out to influence the economy as a whole. The "wave" aspect of the theory was very similar to the "life cycle" properties in the literature I have just discussed. In Schumpeter's case, a new cluster of technologies emerge, then advance rapidly, then slow down as they mature. The "successive cycles" phenomena in particular industries that I have described is very similar to Schumpeter's theory that growth more broadly proceeds in successive waves.

After a brief flurry of attention shortly after he put it forth, Schumpeter's long-wave theory received little continuing attention, perhaps because it seemed to have nothing to do with the neoclassical growth theory that soon emerged. Nor until recently have evolutionary theorists paid much attention to it. However, largely through the work of Carlotta Perez (1983) and Christopher Freeman (particularly in Freeman and Louca, 2001), in recent years there has been a surge of writing on growth oriented by that theory, but with a new twist.

What Perez and Freeman have done is to bring institutions and institutional evolution to the picture. The argument is that the forms of business organization and practice, legal structures, government policies, and institutions more generally that facilitate progress in one era often are not the same as those that facilitated in the preceding era. And institutional innovation or change, more generally, is difficult. Thus, the countries that led the world in one era often tend to fall back in the following era, where different countries are fortunate enough to have in place the bases for the institutions that have become appropriate, or somehow are able to create the right ones.

I find the broad outlines of this theory convincing. Thus far, its development has been exclusively through the vehicle of what Winter and I have called appreciative theorizing. But the time may be coming when some more formal theorizing can help sharpen and advance conceptualization.

It should be apparent that the basic theoretical ingredients needed to model industry product cycles also are needed to model broader economic development over a long wave, or a sequence of them. There is a need to explore the sources of diminishing returns to efforts to advance technology in a field, and the factors that renew opportunities. The effects of the pace and pattern of technological change on firm and industry structure needs to be modeled. There is need to incorporate social technologies in a model, in a way that captures the ways in which social technologies and their evolution both mold and reflect developments in physical technologies.

But there is also a need to deal explicitly with the multisectorial nature of economic activity. Under long-wave theory economic growth in any era is driven by rapid technological advance in a small number of industries. However, these rapidly advancing technologies are affecting a large number of industries, partly by providing new inputs, and partly because some industries are complements and others substitutes for the sectors where technological advance is most rapid. We need to learn to model these interactions, and their effects on relative prices, and in turn how changes in relative prices affect the allocation of resources across different industries.

I propose that we already have built into evolutionary economic theory the heart of an analysis of the factors causing changes in relative prices over time. To a first approximation, prices move with unit costs, although perhaps with a lag. Relative prices decline in industries experiencing the most rapid productivity growth and rise in those experiencing little progress.

To proceed further down this path, of course, requires that we develop a more explicit theory of how demand is influenced by prices than that contained in contemporary evolutionary models. Such a formulation would include specification within an evolutionary theory of concepts analogous to substitutes and complements in final consumption as well as in production. I suggest that this would involve both opening up the routine concept to incorporate variations tied to prices, and more elaborate treatment of how prices influence the direction of search, along the lines Winter and I sketched in Chapter 7 of our book *An Evolutionary Theory of Economic Change* (1982). These adaptations, together with more detailed treatment of the response of investments to differences in profits from pursuing different paths of expansion, would take evolutionary theory a long way forward.

I want to conclude this essay by observing that a successful development of evolutionary growth theory along these lines would do much more than simply improve its ability to illuminate economic growth as we have experienced it. It would enable evolutionary theory to encompass much of the subject matter treated in neoclassical economics as aspects of "general equilibrium" theory. But it would treat the prevailing pattern of inputs, outputs, and product and factor prices as a frame in the moving picture defined by the evolutionary processes driving economic growth. In my view, this would be an enormous accomplishment.

References

Beer, J. H. 1959. *The Emergence of the German Dye Industry*. Urbana: University of Illinois Press

Chandler, A. H. 1990. *Scale and Scope*. Cambridge: Harvard University Press

Cyert, R. and March, J. 1963. *A Behavioral Theory of the Firm*. New Jersey: Prentice Hall

Dosi, G., Nelson, R., and Winter, S. 2000. *The Nature and Dynamics of Organizational Capabilities*. Oxford: Oxford University Press

Freeman, C. and Louca, F. 2001. *As Time Goes By*. New York: Oxford University Press

Klepper, S. 1996. "Entry, Exit, Growth, and Innovation Over the Product Cycle," *American Economic Review* 86(3): 562–83

Malerba, F. 2002. "Sectoral Systems of Innovation and Production," *Research Policy* 31: 247–66

Malerba, F. and Orsenigo, L. 1997. "Technolgical Regimes and Secotral Patterns of Innovative Activities," *Industrial and Corporate Change*. Oxford: Oxford University Press, Vol. 6(1), pp. 83–117

Mandeville, B. 1924. (First published 1714.) *The Fable of the Bees*. Oxford: Oxford University Press, Vol. II, pp 141–2.

Marshall, A. 1948. (First published 1907.) *Principles of Economics*, 8th edn, London: Macmillan

Metcalfe, S. 2002. "Knowledge of Growth and the Growth of Knowledge," *Journal of Evolutionary Economics* 12: 3–15

Mokyr, J. 2002. *The Gift of Athena: Historical Origins of the Knowledge Economy*. Princeton: Princeton University Press

Mokyr, J. 2004. "Useful Knowledge as an Evolving System: A View From Economic History," paper presented at the Jena Workshop on Evolutionary Concepts in Economics and Biology, December 2–4

Mowery, D. and Nelson, R. 1999. *The Sources of Economic Growth*. Cambridge: Cambridge University Press

Murmann, P. 1993. *Knowledge and Competitive Advantage*. Cambridge: Cambridge University Press

Nelson, R. 1998. "The Agenda for Growth Theory, A Different Point of View," *The Cambridge Journal of Economics*, 22: 479–520

Nelson, R. and Sampat, B. 2001. "Making Sense of Institutions as a Factor Shaping Economic Performance," *Journal of Economic Behavior and Organization* 44: 31–54

Nelson, R. and Winter, S. 1982. *An Evolutionary Theory of Economic Change*. Cambridge: Harvard University Press

Perez, C. 1983. "Structural Change and the Assimilation of New Technology in the Economic and Social System," *Futures* 15(5): 357–75

Schumpeter, J. 1934. *The Theory of Economic Development*. Cambridge: Harvard University Press

Schumpeter, J. 1939. *Business Cycles*. New York: Prentice Hall

Schumpeter, J. 1942. *Capitalism, Socialism, and Democracy*. New York: Harper and Row

Smith, A. 1937. (First published 1776.) *The Wealth of Nations*. London: Henry G. Bohn

Solow, R. 1956. "A Contribution to the theory of Economic Growth," *Quarterly Journal of Economics* 70: 65–94

Solow, R. 1957. "Technical Change and the Aggregate Production Function," *Review of Economics and Statistics* 39: 312–20

2   Innovation and economic growth theory: a Schumpeterian legacy and agenda

Bart Verspagen

Introduction

Over the last twenty-five years, the analysis of economic growth has re-surfaced as one of the most important issues in economic theory. Schumpeter's intellectual legacy, as embodied in his *Theorie der wirtschaftlichen Entwickelung* (1912), *The Theory of Economic Development* (1934), *Business Cycles* (1939) and *Capitalism, Socialism and Democracy* (1943) is an important point of reference for the new growth theory of the last quarter of twentieth century. Different elements of Schumpeter's works have inspired the modern analysts of economic growth. Those in the "evolutionary" tradition stress mostly Schumpeter's interpretation of economic growth as an out-of-equilibrium process. In the neoclassical tradition, the notion of "creative destruction" is the most influential Schumpeterian concept. The idea of long business cycles driven by major innovations is now shared between evolutionary and neoclassical economists alike.

But Schumpeter's work on economic growth cannot easily be reduced to such conceptions with only limited depth and intellectual reach. For example, while the first two of the above-mentioned works certainly stress the out-of-equilibrium nature of economic growth, they also document Schumpeter's appreciation of a notion of equilibrium as a long-run attractor of the system. And while it is quite understandable that his *"Business Cycles"* would inspire modern economists working on "long waves" or "general-purpose technologies" *Capitalism, Socialism and Democracy* brings out more a tendency of ever-evolving history than one of "cyclical history."

The purpose of this chapter is to provide a broad survey of the Schumpeterian legacy in the modern field of economic growth, and to provide a critical discussion of its outlook for further progress. The survey, presented in Sections 2.1 and 2.2, is one of the main lines, rather than details of the various models and approaches found. Section 2.1 briefly reports on the work on economic growth that emerged in the immediate postwar period, and argues that this period essentially missed the point made by Schumpeter about economic growth and technology. Section 2.2 reports on economic growth theory in the 1980s and 1990s, and discusses the contrast that has emerged between the work on economic growth in the evolutionary tradition and that in the "endogenous growth theory" within the neoclassical framework.

The concluding Section 2.3 provides a critical analysis of both lines of investigation. Two main questions will dominate this analysis. The first one is whether or not a convergence between the two streams of thoughts is emerging, possibly leading to a common "Schumpeterian" view in modern economic growth theory. The second question is what, among other things from the point of view of Schumpeter's work on economic growth, is missing in the modern theory of economic growth.

2.1   Economic growth theory in the 1950s and 1960s: a legacy lost

Growth theory in the 1950s and 1960s was based on a simplistic view of technology as a "public good." Technological knowledge obviously has some characteristics of a public good, that is, more than one firm can use the same piece of knowledge at the same time (nonrivalry), and once knowledge is in the open, it is hard to exclude specific firms from using it (nonexcludability). In its extreme form, this view leads to the conclusion that all knowledge can be acquired externally as "general knowledge," and firms need not develop knowledge themselves.

On the basis of such a view of technology, the neoclassical growth models that appeared half a century ago (Solow, 1956) treated technological change as an exogenous phenomenon. These contributions were part of a debate on the stability of economic growth patterns, emerging out of the contributions by Harrod (1939) and Domar (1947), which, no doubt inspired by the historical event of the Great Depression, seemed to suggest that the long-run evolution of capitalism would fall victim to prolonged economic depressions and large-scale unemployment. The neoclassical answer of the Solow model to this pessimistic perspective was simple: factor substitution (between capital and labor) would ensure balanced growth. Technology was an explanatory factor "of least resort," in the sense that growth not explained by the variables included in the model was assumed to be the result of exogenous technological change. However, when empirical work – so called "growth-accounting" (Abramovitz, 1956; Solow 1957)[1] – indicated that

---

[1] Solow (1957) is often quoted as the standard reference on growth accounting, but the ancestry of the method lies earlier (e.g., Tinbergen, 1943).

the unexplained share of long-run economic growth tended to be very high, the interest in technological change and other possible explanatory factors not taken into account by the modelers increased.

Technology as an endogenous phenomenon was already present in the model presented in Kaldor (1957), taking the form of a so-called "technical progress function," which assumed a linear relation between growth of labor productivity and the growth of capital per worker. A central tenet of the post-Keynesian literature following the Kaldor model is the notion of technology as a private rather than a public good.[2]

Pure public goods do not require any special effort or special skills on the side of the consumer or receiver of the services of the good. This is obviously not the case for technological knowledge. Using technological knowledge, even if it stems from the public domain, requires considerable skills and efforts on the side of the receiver of this knowledge. The reason for this is that knowledge has a strongly cumulative and often tacit character. Every piece of new knowledge builds to a large extent on previous knowledge, and to apply that knowledge requires one to have command over the older knowledge on which the new knowledge is built.

From this perspective, the post-Keynesian tradition emphasizes the role of "cumulative causation" or "positive feedback." Contrary to the neoclassical idea of knowledge as a public good, these models assume that knowledge is specific to the agents that develop it and does not spill over easily to other agents or nations. This idea was applied to regional growth in Kaldor (1970), and goes back to Verdoorn (1949), Fabricant (1942), and Young (1928). In this view, generating knowledge is mainly a learning process deeply rooted in gaining experience with specific production processes and products; learning-by-doing and learning-by-using are key concepts. Only those engaged in the actual learning experiences will gain from it, and others, who do not profit from experience, will be left behind.

The consequence of this is a tendency for "success to breed success": those nations (or regions, or agents) that are growing rapidly accumulate experience and hence learn faster than others. This leads to a better competitive position for those already ahead and enables them to move further ahead. Hence, the crucial tendency here is one of divergence, in which some nations (regions) are able to grow rapidly while others are

---

[2] Besides the aspects of technology and innovation discussed here, the attention to demand as a factor in economic growth is typical of the post-Keynesian tradition, for example, Pasinetti (1993).

left behind. A model of regional growth along these lines was presented in Dixon and Thirlwall (1975).

Looking back at these topics that dominated the debate on economic growth in the 1950s and 1960s, it is evident that what can be learned about the role of technology in economic growth is only very limited. Compared with the central role of technology in Schumpeter's work, the (formal) theory of economic growth of the immediate postwar period has only very limited insight to offer. What is most prominent about technology, that is, the post-Keynesian view of cumulative learning, is much in line with Schumpeter's idea of "routine capitalism," as expressed in *Capitalism, Socialism and Democracy*, but obscures his ideas, expressed in earlier works, about entrepreneurs who introduce major epoch-making innovations, and the disruptive, nonequilibrium impact of these on the pattern of economic growth.

2.2   Competing Schumpeterian paradigms for explaining the relation between growth and technology

Two major approaches emerged during the 1980s and 1990s as the dominant approaches to the analysis of the relationship between technology and growth. These are the endogenous growth theory approach (in the neoclassical tradition), which is also dominant in other fields of economics, and the evolutionary approach. While the endogenous growth tradition consists of a relatively homogenous set of interrelated models, the field of evolutionary economics consists of a more loosely connected set of contributions. The evolutionary approach includes formal models as well as more "appreciative" or historical approaches, as will be explained in more detail below. Even the label used to describe this approach is not yet common understanding. Here, we will use, mainly for convenience, the short description of "evolutionary economics," even though this label may not be unanimously shared.

Both of these approaches agree on basic issues such as the importance of innovation and technology for economic growth, as well as the positive role that can be played by government policy for science and technology. Yet they disagree on the behavioral foundations underlying these respective theories. These differences can be characterized by saying that the endogenous growth tradition sacrifices a significant amount of realism in terms of describing the actual innovation process in return for a quantitative modeling approach that favors strong analytical consistency, while the evolutionary approach embraces the microcomplications of the innovative process and applies a more eclectic approach.

2.2.1   *The evolutionary approach to technology and growth*

Let us use Richard Dawkins' metaphor of the blind watchmaker to illustrate the general idea behind economic growth as an evolutionary process. Dawkins' story starts from the idea of William Paley, an eighteenth-century theologian. Paley argued that certain objects, like a watch, are by their nature obviously created by conscious design, whereas for others, like a rock, it is easy to believe that they "have always been around." His argument then went on to stress that nature contains many such objects that are obviously created by conscious design. The most famous of such objects discussed by Paley is the human eye. He then used this argument to offer the proposition that the world must have been created by a conscious being (God).

Dawkins uses Paley's examples to argue that the watch may look as if it was carefully designed, but it might just as well have been created by an evolutionary process that can be thought of as a *blind watchmaker*. This blind watchmaker is unable to design the watch by carefully planning it on a drawing board and then implementing it using precision instruments. Instead, he operates through the processes of random mutation and natural selection. His approach is to start with a simple device and add small and simple changes in a random way. These changes are subjected to a real-world test, that is, whether or not they lead to an improvement in keeping the time. Only if they do so are they kept; otherwise they are discarded. From a new design that incorporates such a successful small change, the process may start again, and step-by-step a more complicated design emerges. In the end, after a long and gradual process, a complicated artifact such as a watch may result. Although this artifact looks as if it were carefully designed, it was instead the blind watchmaker and his processes of random mutation and natural selection that created it.

Carrying the metaphor over to economic growth and technology, our watchmaker is blind because of the strong uncertainty facing the individual economic decision-maker. No businessman can foresee the huge potential of a new innovation when it first emerges, or even think of a probability distribution that describes the possible outcomes related to the innovation. It is through a process of incremental innovations, each one of which is implemented by an entrepreneur who sees some market for the newly resulting artifact, that the full potential of the technology unfolds. The incremental innovations are the economic counterpart of biological mutation. Natural selection has its counterpart in economic selection, that is, markets that decide whether or not certain innovations become successful. Just as in biology, many of the "mutations" (incremental innovations) are not successful, and the selection process erases them from history.

The metaphor is concluded by arguing that, although the individual entrepreneur has to cope with strong uncertainty and therefore cannot design a process that we may call a technological revolution, the capitalist system, working by means of a combination of the creation of novelty (innovation) and economic selection (markets), can create "objects" that seem as if they have been carefully designed. With hindsight, technological revolutions, such as the diffusion of steam power or Information and Communication Technologies (ICTs), may look as if they were planned from the very beginning to create a "new economy," but in reality, so it is argued by evolutionary theory, these technological systems were created by the trial-and-error method of the blind watchmaker.

The evolutionary approach to the analysis of economic growth is thus based on the axiom that individuals are unable to cope with the complexities of technological change in a complete manner. A single economic decision-maker, be it an entrepreneur from the early days of the Industrial Revolution or a large multinational corporation from the twenty-first century, simply cannot see all business opportunities that result from technological possibilities and/or manage them in a way that maximizes profits. These decision-makers thus operate under a scheme of bounded rationality, in which relatively simple and occasionally adaptive behavioral rules ("rules of thumb" or "routines") are used to make decisions. These are not fixed, but can be changed over time, especially so under the influence of feedback from economic performance.

Although these simple behavioral rules help economic decision-makers in a turbulent and complex world cope with strong uncertainty, their role sheds little light on the mechanisms through which complex modern economies remain on a path of constant technological improvement that we call economic growth. The explanation of aggregate economic performance in evolutionary economics relies on two forces: selection and the generation of novelty. Over time, the variety present in the system is reduced by selection, that is, the growth of those entities that are better adapted to circumstances, and the decline of those that are not. Novelty is constantly added to the system, however, and thus evolution is the outcome of a constant interaction between variety and selection. Innovation is an important novelty-generating process, and the market and other economic institutions are among the most important selection mechanisms in modern economies.

In biology, the generation of novelty (mutation) is purely random, and there is no way in which the mechanism of mutation itself can learn to generate "smarter" mutations. Each mutation is truly "blind" in the sense that there is no *ex ante* way of telling whether or not it will improve the performance of the organism. In economic evolution, however, decision-makers at the micro level are not "completely blind" – they plan their actions in order to generate potentially successful innovations in a process that more closely resembles the Lamarckian view of evolution. Thus, innovations introduced by profit-seeking, "satisficing" entrepreneurs will have at least some commercial potential; in other words, they are most likely biased in a "positive" direction. Nevertheless, uncertainty remains important, since it is difficult to foresee the cumulative effects of numerous, small, incremental improvements, and because of the systemic nature of knowledge that results from knowledge spillovers among fields. An actor operating in one field may invent something for which he does not see the full potential in other fields.

The evolutionary approach is particularly suited for analyzing historical processes. Evolution and history are both a complex mixture of random factors, or contingencies, and more systematic tendencies. It is a well-known error to think that the biological evolutionary process is goal-oriented, that is, it strives to achieve a predefined aim. Our discussion of the blind watchmaker metaphor may have misguided the reader into thinking that such a goal exists, that is, that it would be the aim of evolution to create a complex artifact such as a watch or a human eye. Instead, it is only the individual mutation that has a sense. The accumulation of incremental innovations may seem to have a purpose, but in fact there is no force in the system that has formulated or even tried to achieve such a goal. The same applies to economic evolution.

Such a view of the world as a mixture between chance and necessity is shared between the historical view of the world, the evolutionary view of the world, and the dialectic (Hegelian) view of the world. It is opposed to the Newtonian or Laplacean view that portrays the world as a clockwork in which future states of the system can be predicted with full accuracy if only enough information about the present state is known. We will argue below that the neoclassical economic growth theory is much more similar to the latter view.

Historical analysis is often used by evolutionary scholars to develop heuristic patterns that can be used to describe and categorize these developments in a more general way:

In the appreciative and applied evolutionary literature much has been made of the concepts of technological paradigm (Dosi, 1982) and natural trajectories (Nelson and Winter, 1977). This is indeed an attempt to impose additional structure on technology and differentiate discrete interrelationships in technological space from one another, if only *ex post* ... This should be contrasted with the smooth, substitutable, unbounded production possibility sets of neoclassical theory. (Silverberg, 2001, p. 1277).

Dosi (1982) defines a technological paradigm as a "model and pattern of solution of selected technological problems, based on selected principles from the natural science and on selected material technologies." The term is borrowed from Kuhn's philosophy of science (Kuhn, 1962), which posits that the normal development path of scientific knowledge relies heavily on a dominant framework jointly adhered to by the leading scientists in the field. The paradigm thus limits the possible directions technological development may take.

In the interpretation of Freeman and Louçā (2001), a small number of basic innovations set out a technological paradigm that may dominate techno-economic developments for a long time. Within the paradigm, the basic design of the innovation is constantly altered by incremental innovations, but the basic direction of technological development is limited by the paradigm. Still, there is some room for choice within the paradigm, and these choices are governed by the specific circumstances (e.g., scarcity of a particular resource) in which the technology develops. This development is termed a "technological trajectory."

Thus, in the paradigm/trajectory heuristic, a basic innovation can be thought of as setting out developments in the techno-economic domain for a number of years to come, but the success of the paradigm, and hence of the basic innovation, depends crucially on how well incremental innovation is able to adapt the paradigm to local (e.g., industry, geographical, and temporal) circumstances. These circumstances include the skills and capabilities of the workforce that has to work with new machinery, as well as factors such as cultural aspects of the society in which the paradigm develops.

Another set of heuristics developed in the historical part of evolutionary economics relates to the temporal clustering of innovations. This part of the literature starts from Schumpeter's observation that innovations "are not evenly distributed in time, but that on the contrary they tend to cluster, to come about in bunches, simply because first some, and then most firms follow in the wake of successful innovation" (Schumpeter, 1939, p. 75). Although Schumpeter was in fact referring to a tendency of incremental innovations to cluster following a large innovation (this is an idea not incompatible with the paradigm view summarized above), his idea has been interpreted in the literature as implying that large (or "basic") innovations cluster in time (e.g., Mensch, 1979, Kleinknecht, 1987). In this view, some historical periods are characterized by an above-average rate of (basic) innovations, while other periods show a relatively low rate of such activity.

Together, these two sets of heuristics have interesting implications for growth. They suggest that technological innovation can introduce an uneven temporal pattern into economic growth. In the early, exploratory stages of a paradigm, the technology progresses rapidly, but the pace of change slows down when the paradigm goes into its phase of "normal" development, and it slows still further when technological opportunities become less numerous (and the paradigm may start to break down as a result of this). The clustering-heuristic suggests variations over time in the rhythm of growth simply because the rate at which large, influential innovations occur differs over time.

One extreme interpretation of this temporal pattern of innovation is the idea of a "long wave" in economic growth, in which periodicity is bounded in a short range of fifty to sixty years (e.g., Kleinknecht, 1987; Freeman and Louça, 2001). Another view claims that growth patterns are inherently turbulent, but with little regularity in terms of strict cycles. In any case, the evolutionary view argues that the uneven temporal rates of technological change mean that the economy is almost always away from anything that could be characterized as a steady state.

Theories and historical analyses of this type propose a view of the interactions among technology, the economy, and the institutional context. The institutional environment is important because it is both a facilitator of and an impediment to technological change. Moreover, the institutional context is itself an endogenous factor that changes under the influence of technological and economic developments. Although it is sometimes claimed that theories of this type suffer from "technological determinism" (i.e., a tendency for one-way causality from technology to growth, see, e.g., Bijker et al., 1987), work such as that of Perez (1983) proposes an interactive relationship among institutions, the economy, and technology that emphasizes mutual causality.

Evolutionary ideas have also been used to formulate formal models of economic growth and technology. The starting point of this tradition is the model described in Nelson and Winter (1982), which defines heterogeneity in terms of firms, using production techniques that employ a fixed ratio of labor and capital (so-called Leontief-technology). The generation of novelty (new fixed-proportion techniques) occurs as a result of search activities by firms, but search is initiated only when the firm's rate of return falls below a certain (arbitrarily set) value. Search may take two different forms: local search or imitation. In the first case, firms search for new, yet-undiscovered techniques, each of which has a probability of being discovered, which linearly declines with technological distance from their current technology (hence the term *local* search). In the second search process, imitation, a firm searches for techniques currently employed by other firms but not yet used in its own production process.

Like most models in this tradition, the Nelson and Winter model has to be simulated on a computer to obtain an impression of its implications. The model, which is calibrated with the Solow (1957) data on total factor productivity for the United States in the first half of the century, yields an aggregate time path for capital, labor input, output (GDP), and wages (or labor share in output) that corresponds in a qualitative sense to those observed by Solow. On the basis of these results, Nelson and Winter argue that "it is not reasonable to dismiss an evolutionary theory on the grounds that it fails to provide a coherent explanation of . . . macro phenomena" (p. 226). More specifically, they argue that although both the neoclassical explanation of economic growth offered by Solow and the Nelson and Winter model seem to explain the same empirical trends, the causal mechanisms underlying the two perspectives differ greatly:

The neoclassical interpretation of long-run productivity change . . . is based upon a clean distinction between "moving along" an existing production function and shifting to a new one. In the evolutionary theory . . . there was no production function. . . . We argue . . . that the sharp "growth accounting" split made within the neoclassical paradigm is bothersome empirically and conceptually. (Nelson and Winter, 1982, p. 227)

Evolutionary models following Nelson and Winter (1982), such as Chiaromonte and Dosi (1993) and Silverberg and Verspagen (1994), extend these conclusions. A more complete overview is in Silverberg and Verspagen (2005). The model by Chiaromonte and Dosi shows how growth rates in a cross section of nations may differ. The models by Silverberg and Verspagen show how "routines" of R&D investment may arise endogenously in a population of firms, and how growth patterns vary along the history of an economy that learns in such a "collective" way.

One of the rare models in this tradition that is solved analytically rather than by numerical simulation is that of Conlisk (1989). Under the assumption that technology advances are random, Conlisk constructs a model in which the growth rate of the aggregate economy is a function of three variables: the standard error of the productivity distribution of new plants (which can be interpreted as the average innovation size), the savings rate (which is defined somewhat unconventionally), and the speed of diffusion of new knowledge. Moreover, by changing some of the assumptions about the specification of technical change, the model emulates three standard specifications of technical change found in growth models in the neoclassical tradition. In this case, the first and third factors no longer have an impact on growth (they are specific to the "evolutionary" technical change specification of the model). However, the impact of the savings rate can be compared between the various model set-ups. Conlisk finds that using purely exogenous technical change (as in the Solow model), or learning-by-doing specifications (as in Arrow, 1962), the savings rate does not have an impact on (long-run) economic growth. This result, which is in fact also well known from standard neoclassical growth theory, marks an important difference between these models and his more evolutionarily inspired specification.

The recent so-called "history-friendly models" (Malerba et al., 2000) aim to bring evolutionary models closer to empirical reality by reproducing the historical evolution of a particular industry, for example the computer industry. To this end, they start with a descriptive analysis of industry variables such as growth, concentration, and employment, and incorporate the insights from this analysis into a model, the behavioral foundations of which are consistent with the evolutionary view. This model is calibrated and simulated to reproduce real-world trends as closely as possible. While this approach generates empirically relevant models, the simulations employ a relatively narrow set of parameter values. The work devotes little attention to a more open-ended investigation of which *minimal* set of assumptions is necessary to generate certain aspects of the structural evolution of specific industries.

2.2.2   *Neoclassical views of economic growth and technology*

How has mainstream economic theory coped with the complexity of technological change? The literature on neoclassical models of endogenous technology grew rapidly in the 1980s and 1990s following the publication of Romer (1986). Romer's model and others in this tradition were motivated by the apparent flaws associated with the possibility in the Solow model of decreasing marginal returns to capital: holding all other production factors (labor, land, infrastructure, and buildings) fixed, the productivity of an extra (marginal) unit of investment would fall with growth in the existing capital stock. Decreasing marginal returns to investment could cause growth to slow down or even cease in the long run. As growth proceeds, capital accumulates, that is, the capital stock increases, and hence an extra unit of investment generates less and less growth. Exogenous growth or productivity (knowledge) had been the traditional answer, but Romer (1990) and Grossman and Helpman (1991) proposed to make technology endogenous by modeling the R&D process. Abstracting from technicalities (a survey is provided by Verspagen 1992), this can be summarized as follows.

All the models assume that R&D is essentially a lottery in which the prize is a successful innovation. In the model by Aghion and Howitt (1992), this innovation-prize buys the firm a temporary monopoly of supplying the best-practice capital good used for production of consumption goods. The temporary monopoly vanishes when the next firm makes an innovation. Hence, the innovation process is modeled as a "quality ladder" of innovations, in which each new innovation supersedes the old one. In the industrial organization literature, this is called "vertical differentiation" of products.

In the model by Romer (1990), the innovation prize buys the successful firm a new variety of capital that will be demanded by producers of consumption goods forever, but has to compete with all other varieties (invented in the past, with the range continuing to expand in the future as a result of R&D). In this model, varieties of goods (innovations) do not go out of the market. Substitution between variations of goods is governed by a utility function or production function (depending on whether innovation takes place in consumer goods or intermediate goods) with a "constant elasticity of substitution." This is called "horizontal differentiation."

More tickets for the R&D lottery can be bought by doing more R&D, which is of course a costly process. Relative to the evolutionary models considered above, the crucial assumption is that the outcomes of the R&D process can be realistically characterized by weak uncertainty, that is, the firm is able to estimate the probability that it will get the innovation prize, given its level of R&D spending. With expected benefits and costs of R&D known, the firm may make a cost–benefit analysis and derive an optimal level of R&D spending. This will, on average, correspond to a given amount of innovation, and produce a given growth rate. Although additional assumptions are necessary (e.g., with regard to the working of capital markets in which R&D expenditures have to be financed), this mechanism is the key to generating endogenous growth.

Before endogenous growth is possible in these models, there is one essential assumption about the nature of technology that needs to be made. This is related to the (partly) public-good nature of technology. In the new growth models, this is represented by the assumption that there are technology spillovers between firms in the R&D process. The assumption takes two forms, depending on which flavor of model is used. In the horizontal differentiation-type models (also called "love-of-variety" models), each innovation increases the level of general knowledge available in the economy, and this increases the productivity of the R&D

process itself (Romer, 1990). This assumption is necessary because of the ever more severe competition between the varieties of capital goods, and the falling profit rates that this causes. A tendency for R&D to be more productive (i.e., the costs of R&D to fall) offsets this falling profit rate, and keeps R&D feasible in the long run (Grossman and Helpman, 1991).

In the quality-ladder models (vertical differentiation), each new innovation destroys the monopoly of the old innovator. However, the new innovator also builds on the previous innovation, because the quality of the new capital good is a fixed increase over the previous one. In other words, each new innovator is "standing on the shoulders of giants," and knowledge spills over intertemporally from one innovator to the next one. Without this spillover, endogenous growth would not be possible.

It is in the quality-ladder model that the Schumpeterian notion of creative destruction is most referred to. In this model, the incumbent monopolist is periodically replaced by a successful challenger. Obviously, this replacement process is a rather limited representation of Schumpeter's original notion, which went further to include, for example, infrastructures and related investment. But the quality-ladder model does bring out the crucial notion that technological change is often a process of sharp competition, in which negative as well as positive externalities exist.

The technological spillovers in endogenous growth models lead to increasing returns to scale at the aggregate level. Even though the production functions of firms at the micro level are characterized by constant returns to scale, the R&D spillovers that flow from one firm to the rest of the economy imply increasing returns at the aggregate level. In terms of the expression for the aggregate growth rate of the economy, this feature of the endogenous growth models implies that growth at the country level depends (*ceteris paribus*) on the size of the country. Taken literally, this means that (*ceteris paribus*) larger countries will grow more rapidly. Related to this issue is the fact that the basic endogenous growth models are quite sensitive to small changes in the model specification with regard to technology spillovers. A slightly different specification of the impact of "general knowledge" on R&D productivity will lead to either zero growth in the long run or to increasing growth rates in time (Grossman and Helpman, 1991).

Technological spillovers make endogenous growth possible, but pose a challenge for policy-makers. When technology generates positive externalities, the social benefits of R&D are larger than the private benefits (a rational firm investing in R&D does not consider the benefits of its R&D for its competitors). Hence the amount of R&D investment "generated by the market" will be too low from a social point of view.

Technology policy in the form of R&D subsidies may bring the economy to a higher, socially optimal growth path. A similar conclusion is reached in a model of human capital and growth in Lucas (1988). In Aghion and Howitt (1992), there is also a negative externality: each new innovator destroys the rents of the existing monopolist (this is called "business stealing," or, in line with Schumpeter, 1939, "creative destruction"). In this model, private R&D investment also can be too high from a social-welfare perspective, depending on which of the two forms of externalities (creative destruction or standing on the shoulders of giants) is stronger.

The development of the endogenous growth models raises promise and problems. On the positive side, it can be argued that this new growth theory takes seriously a number of arguments about technological change previously championed by evolutionary theorists but ignored by main-stream economists. These include the notion that R&D and technology are essentially stochastic phenomena (although evolutionary theory would argue that the type of uncertainty, that is, weak uncertainty in which the probability distribution is known, is still not very adequate), and the importance of technology flows between agents (spillovers) for growth in the long run. The implication in many of these models that technology policy matters for growth also is relatively consistent with evolutionary theory, but may be less easily accommodated by mainstream economic theories that emphasize the efficiency of market forces.

On the negative side, these new growth models still propose a view of the interaction between economic growth and technology similar to the Newtonian clockwork world, although a certain degree of "weak" uncertainty is added to this. In other words, the new growth theory still portrays the relationship between technology and growth as one of a steady-state growth pattern, which can be "tweaked" relatively easily by turning the knobs of the R&D process. The evolutionary inclination, on the other hand, is that the nature of the growth process is more complex and variable over time. While the importance attached to the technology factor is shared with the new growth models, the belief that the relation between technology and growth is easily tweaked is not. In the evolutionary view, it is hard to predict exactly the impact of a policy measure, because it impacts on a complex range of interrelated factors. Moreover, while relations between a number of factors may have been revealed by careful research for a specific instance in time, it is to be expected that the nature of this relationship will change over time, exactly because of the (co-)evolutionary nature of the process.

A more recent branch of new growth theory is the group of models that comes under the heading of "general-purpose technologies" (GPT, Helpman, 1998). A GPT is defined in essentially the same way as a basic innovation or paradigm in the evolutionary tradition, and thus builds to an important extent on Schumpeter's ideas in *Business Cycles*. It consists of a basic technology (radical breakthrough), but this needs to be developed in the form of a range of intermediate (capital) goods. Within each GPT, the determinants of productivity are essentially the same as in one of the variants of the new growth models discussed earlier. Technological change takes the form of an ever-expanding range of capital goods, but this is time-specific to the GPT. Thus, we see that at least two ideas from the evolutionary tradition are captured: the idea of differences in innovation size, and the idea that incremental innovations are responsible for the diffusion of a basic technology.

The GPT model generates cyclical growth. In its simplest form, the cycle consists of two phases. In the "low growth phase," the new GPT has been discovered, but is not yet in operation. New capital goods are being developed for it, and this activity has been halted for the old GPT. Thus, economic growth is low, because the main technology in use is no longer being developed. Once enough capital goods are available for the new GPT, its productivity outperforms that of the old GPT, the old GPT vanishes, and the economy shifts into a "high growth phase."

The GPT model resembles the evolutionary, Schumpeterian idea of long waves in economic growth. But scholars in the latter tradition have moved away from the fixed and deterministic cycle that characterizes the GPT model. Its clockwork view of economic growth has been dominant in the neoclassical tradition since the Solow model. One illustration of the limitations resulting from this view is the fact that, in the GPT view of the world, there is only room for substitution between subsequent paradigms. But economic and technological histories are filled with examples of the adaptation and survival, often in modified form, of old paradigms. For example, although the automobile is typical of the mass-production paradigm, it still plays a crucial role in the modern "Information Economy," although ICT has indeed been applied in the production of cars.

The new growth models led to a tidal wave of empirical work on growth. Temple (1999) provides a detailed overview of this literature. A crucial notion in the empirical debate following the endogenous growth models is the respective roles of steady state growth-rates and convergence toward them. While the Solow model predicts that countries will converge to identical steady states (dependent on the exogenous rate of technological progress available to everyone), endogenous growth models predict that steady states will generally differ between countries. Empirical work on this issue has used a wide range of variables in regressions of growth-rate differentials between countries, in order to examine cross-national differences in steady state growth rates.

Unfortunately, this approach is data-driven rather than theory-driven: an overall framework that governs and justifies the selection of factors is lacking. Also, many of the estimation results are sensitive to a small number of observations in the large sample (Levine and Renelt, 1992). Nonetheless, this work leads to the conclusion that steady state growth rates differ between nations. Growth rates may converge toward a country-specific steady state growth path at best (so-called conditional convergence), leading to the divergence of growth paths among countries.

Jones (1995a, b) has argued that the observed empirical record on R&D and growth is inconsistent with the theoretical predictions of endogenous growth models. He observes that the postwar empirical evidence does not confirm the relationship proposed by R&D-based endogenous growth models that an increase in the number of R&D workers leads to higher rates of economic growth, and notes that the number of R&D workers has increased since the 1960s, but growth rates (of total factor productivity) have been either constant or declining during the same period. The so-called "Jones critique" has led to still more work in the endogenous growth tradition since its publication. Jones suggests an alternative model, which differs from the R&D-based endogenous growth models by Romer, Grossman, and Helpman, and Aghion and Howitt by a different specification of the invention process. Whereas these original R&D-based growth models assumed that the growth rate of knowledge depends on the number of R&D workers in a linear way, Jones assumes that there are decreasing returns to R&D labor. This assumption is based on the idea that "the most obvious ideas are discovered first, so that the probability that a person engaged in R&D discovers a new idea is decreasing in the level of knowledge ... [and] the possibility that at a point in time the duplication and overlap of research reduce the total number of innovations" (Jones, 1995a, p. 765). In this so-called "semi-endogenous growth model," endogenous growth is only possible when the population grows.

2.3 The Schumpeterian legacy assessed: an outlook on the theory of innovation and growth

Neoclassical work in "new growth" or "endogenous growth" has recently shifted toward more "realistic" models that can accommodate a range of phenomena previously of interest only in the evolutionary tradition. Heertje (1993) described this convergence as follows: "neo-Schumpeterians [i.e., the evolutionary tradition] have been productive in their criticism of the neoclassical scheme on the basis of an evolutionary approach, but the questions they have raised have been addressed more or less successfully by many scholars, who have close links with the neoclassical tradition (...) I would not be surprised to see the present Schumpeterian mood to be part of mainstream economics before the end of this century" (p. 273–5).

It is certainly true that the ideas that were introduced by evolutionary writers such as Nelson and Winter (1982), Pavitt and Soete (1982), Freeman, Clark, and Soete (1982), and Fagerberg (1988) are recognizable in the endogenous growth models that emerged in the late 1980s and 1990s. The notion of technology as a driver of economic growth, the importance of (business) R&D and the stochastic nature of technology are all ideas that can be traced back to Schumpeter, and that were introduced into the modern literature by the evolutionary "school," and finally became central in the endogenous growth theory. The idea of a long wave driven by basic innovations has been the subject of fifty to sixty years of intense debate in the European evolutionary tradition, before it became incorporated into the neoclassical guise of general-purpose technologies. But does this sequence of inheritance of ideas point to a real convergence between the evolutionary school and the mainstream neoclassical approach, as Heertje argues?

Despite the similarities that exist, it is certainly also possible to point to major remaining differences between the two approaches (e.g., Nelson, 1994). Here, we have mostly stressed the differences in underlying assumptions about the microeconomic behavioral patterns, as well as the consequences of this for differences in the general view about the nature of the process of economic growth. Whereas the endogenous growth literature adheres to the standard economic model of optimizing agents and thus requires more or less perfect foresight with regard to the implications of R&D investment, the evolutionary view starts from bounded rationality. In this view, rules of thumb dominate the behavior of individual firms, but the "grand direction" of technological change is the result of an evolutionary process of selection in which those innovations that are best adapted to market circumstances survive. Whereas the full rationality view generally leads to steady state equilibrium growth patterns that are essentially a-historical, the evolutionary view embraces (nonperiodic) fluctuations, sudden and unexpected trend reversals, and sustained growth rate differentials between countries. In short, the world of the endogenous growth theory works like a Newtonian clockworks system in which each action has a predictable effect, whereas the evolutionary world is full of historical contingencies that may change the world forever, and in which economic growth is more a process of transformation and development than one of pure increase of living standards.

It thus seems fair to conclude that convergence has taken place, at least in a limited sense, but that this is far from complete. Is further convergence of the two traditions likely? One avenue on which this may happen is the further analysis of the intertemporal variability of growth patterns. At least some new growth models (e.g., Aghion and Howitt, 1992) argue that time series of economic growth show variability, and this is a main topic in evolutionary models. The application of Pareto-type probability distributions, in which very large innovations have non-negligible probability, may bring the two approaches closer together, since they provide an intuitive way of modeling "strong uncertainty" (see, e.g., Kortum, 1997, Sornette and Zajdenweber, 1999).

Each of the two approaches also contains a range of important and interesting lines of research to be pursued. In the endogenous growth tradition, the returns to purely theoretical work seem to have slowed down, but important empirical challenges remain open. The most fruitful avenue of research here seems to be further theoretical refinement induced by empirical work on technology and growth, with the explicit aim of developing empirically relevant models instead of new explorations motivated by technical problems with the existing models (Eaton and Kortum, 1999, is an example of such a research strategy). For a long time, empirical research has led the way in the mainstream analysis of technology and growth, and this approach still seems to be the way forward.

A main challenge still confronts the evolutionary tradition. This is to develop a research program that goes beyond just emulating, although with a more plausible microfoundation, the results of neoclassical analysis. The strategy of criticizing neoclassical assumptions about rationality and technological foresight may have worked well for evolutionary theory in the past, but it seems that models based on this alone are now running into decreasing returns. The models by Nelson and Winter (1982) and the followers that were reviewed earlier have shown that economic growth driven by profit-seeking, but boundedly rational, agents can be plausibly modeled. This demonstration may be argued to come a long way towards the process that Schumpeter described in *"Theorie der wirtschaftlichen Entwickelung"* (1912) and *"The Theory of Economic Development"* (1934). Economic growth in the boundedly rational evolutionary world is a far-from-equilibrium process, much as Schumpeter described it. Innovations are the force that drives the economy away from (neoclassical) equilibrium, as they provide a source of above-normal profits and hence keep the economy away from the zero-profit equilibrium of perfect competition.

But such a state of nonequilibrium growth is still essentially a-historical. The salient features of growth in these models are captured by the growth rate of GDP and its variability over time. Historical transformations, such as the Industrial Revolution and the like, are not the subject of these models. But it is exactly this type of transformation that occupied Schumpeter in the later periods of his career. *Business Cycles* (1939) already brings out fully the historical interests of Schumpeter, and lays out the argument that every epoch in modern capitalism has its own peculiarities that need to be taken into account when attempting to explain economic growth.

The argument is even grander in *Capitalism, Socialism and Democracy* (1943), when Schumpeter links the transformation of business organization to the dynamics of innovation, and uses the combination of the two to speculate about the future of capitalism. In his view, the demise of the true entrepreneur, pictured as a heroic individual by the early Schumpeter, in favor of routine innovations produced in R&D labs of large firms operating according to the managerial model, would inevitably lead to a lack of creativity which would transform the dynamic capitalist system, which he admired so much, into a socialist system, which he tried to appreciate as a neutral intellectual. One may argue that his notion of "routine" innovation has proven to be much less destructive for capitalism than Schumpeter expected: even in the age of huge multinational firms, major innovations such as ICTs were still able to transform the world economy, and the Schumpeterian entrepreneur lived on in cultural icons such as the Apple II computer that was invented in Steve Jobs' garage.

Such a complex argument about the history and transformation of the capitalist economic system is admittedly not easy to capture in any formal model. But unless the evolutionary approach takes on this challenge, it might not become a serious alternative to the more mainstream endogenous growth theory and remain marginalized as a provider of ideas for these models. The proposed extension of the evolutionary research agenda could benefit from closer interaction with the nonformal work in the evolutionary tradition and greater reliance on historical research, as well as work on contemporary transformations operating at a large scale, such as that provided, for example, by Castells (1996).

Such a research strategy could start from identifying observed historical regularities ("stylized facts") in the relation between growth and technology, or in the generation of technological change *per se*. A new class of models that employ relatively simple, evolutionary, microeconomic foundations to generate a broader range of phenomena in the evolutionary interpretation of technology and growth could be the alternative to the often observed trend of increasing the sophistication and level of realism of the microfoundations of evolutionary models. In such a new class of models, the focus would be on the range of phenomena that can be explained, rather than on the range of behavioral foundations that can be taken on board. A much clearer focus on the salient macro features and what really drives them at the micro level may result from this approach, which is necessary to close the gap between the historical, evolutionary view and model building.

That such an approach may indeed lead to models that show elements of patterns that bear broad resemblance to historical transformations such as the Industrial Revolution is shown in the model by Silverberg and Verspagen (1994). In this model, economic growth shows a regime shift as a result of endogenous changes in the collective R&D behavior of firms. Whereas the old regime is characterized by high market concentration and low growth, the gradual increase of R&D efforts leads to a new regime in which market concentration is suddenly much lower and economic growth much higher. Such a regime shift is admittedly a long way from the historical Industrial Revolution, or from Schumpeter's argument on the evolution of the capitalist system, but it provides a useful starting point for a new direction in the evolutionary modeling of economic growth, and a viable alternative to the endogenous growth models in the neoclassical tradition.

Once again, Schumpeter points out the way forward for the theory of economic growth.

References

Abramovitz, M. A. 1956. "Resources and Output Trends in the United States since 1870," *American Economic Review* 46: 5–23

Aghion, P. and Howitt, P. 1992. "A Model of Growth Through Creative Destruction," *Econometrica* 60: 323–51

Arrow, K. J. 1962. "The Economic Implications of Learning by Doing," *Review of Economic Studies* 29: 155–73

Bijker, W. E., Hughes, T. P., and Pinch, T. (eds.) 1987. *The Social Construction of Technological Systems*. Cambridge, MA: MIT Press

Castells, M. 1996. *The Rise of the Network Society. The Information Age: Economy, Society and Culture*. Oxford: Blackwell, Vol. 1.

Chiaromonte, F. and Dosi, G. 1993. "Heterogeneity, Competition, and Macroeconomic Dynamics," *Structural Change and Economic Dynamics* 4: 39–63

Conlisk, J. 1989. "An Aggregate Model of Technical Change," *Quarterly Journal of Economics* 104: 787–821

Dixon, R. J. and Thirlwall, A. P. 1975. "A Model of Regional Growth-Rate Differences on Kaldorian Lines," *Oxford Economic Papers* 11: 201–14

Domar, E. D. 1947. "Expansion and Employment," *American Economic Review* 37: 34–55

Dosi, G. 1982. "Technological Paradigms and Technological Trajectories," *Research Policy* 11: 147–62

Eaton, J. and Kortum, S. 1999. "International Technology Diffusion: Theory and Measurement," *International Economic Review* 40: 537–70

Fabricant, S. 1942. *Employment in Manufacturing 1899–1939.* New York: NBER.

Fagerberg, J. 1988. "Why Growth Rates Differ," in G. Dosi, C. Freeman, R. Nelson, G. Silverberg, and L. Soete (eds.), *Technical Change and Economic Theory.* London: Pinter, pp. 432–57

Freeman, C. and Louçà, F. 2001. *As Time Goes by: From the Industrial Revolutions to the Information Revolution.* Oxford: Oxford University Press

Freeman, C., Clark, J., and Soete, L. 1982. *Unemployment and Technical Innovation.* London: Pinter

Grossman, G. M. and Helpman, E. 1991. *Innovation and Growth in the Global Economy.* Cambridge, MA: MIT Press

Harrod, R. F. 1939. "An Essay in Dynamic Theory," *Economic Journal* 49: 14–33

Heertje, A. 1993. "Neo-Schumpeterians and Economic Theory," in Magnusson (ed.), *Evolutionary Approaches to Economic Theory.* Dordrecht, Kluwer: 265–76.

Helpman, E. 1998. *General Purpose Technologies and Economic Growth.* Cambridge, MA: MIT Press

Jones, C. 1995a. "R&D Based Models of Economic Growth," *Journal of Political Economy* 103: 759–84

Jones, C. 1995b. "Time Series Tests of Endogenous Growth Models," *Quarterly Journal of Economics* 110: 495–525

Kaldor, N. 1957. "A Model of Economic Growth," *Economic Journal* 67: 591–624

Kaldor, N. 1970. "The Case for Regional Policies," *Scottish Journal of Political Economy* 17: 337–48

Kleinknecht, A. 1987. *Innovation Patterns in Crisis and Prosperity. Schumpeter's Long Cycle Reconsidered.* London: Macmillan

Kortum, S. A. 1997. "Research, Patenting and Technological Change," *Econometrica* 65: 1389–419

Kuhn, T. 1962. *The Structure of Scientific Revolutions.* Chicago and London: The University of Chicago Press

Levine, R. and Renelt, D. 1992. "A Sensitivity Analysis of Cross-Country Growth Regressions," *American Economic Review* 82: 942–63

Lucas, R. E. B. 1988. "On the Mechanics of Economic Development," *Journal of Monetary Economics* 22: 3–42

Malerba, F., Nelson, R., Orsenigo, L., and Winter, S. 2000. "'History-Friendly' Models of Industry Evolution: The Computer Industry," *Industrial and Corporate Change* 8: 3–40

Mensch, G. 1979. *Stalemate in Technology. Innovations Overcome Depression.* Cambridge: Ballinger

Nelson, R. R. 1994. "What Has Been the Matter with Neoclassical Growth Theory?," in Silverberg and Soete (eds.), *The Economics of Growth and Technical Change.* Aldershot: Edward Elgar

Nelson, R. R. and Winter, S. G. 1977. "In Search of a Useful Theory of Innovation." *Research Policy* 6: 36–76.

Nelson, R. R. and Winter, S. G. (eds.) 1982. *An Evolutionary Theory of Economic Change*. Cambridge, MA: Harvard University Press

Pasinetti, L. L. 1993. *Structural Economic Dynamics*. Cambridge: Cambridge University Press

Pavitt, K. and Soete, L. (1982). "International Differences in Economic Growth and the International Location of Innovation," in H.Giersch (ed.), *Emerging Technologies: The Consequences for Economic Growth, Structural Change and Employment*, Mohr, Tuebingen, pp. 105–33

Perez, C. 1983. "Structural Change and the Assimilation of New Technologies in the Economic and Social Systems," *Futures* 15: 357–75

Romer, P. 1986. "Increasing Returns and Long Run Growth," *Journal of Political Economy* 94: 1002–37

Romer, P. M. 1990. "Endogenous Technological Change," *Journal of Political Economy* 98: 71–102

Schumpeter, J. A. 1912. *Theorie der wirtschaftlichen Entwickelung*. Leipzig: Duncker & Humboldt

Schumpeter, J. A. 1934. *The Theory of Economic Development*. Cambridge, MA: Harvard University Press

Schumpeter, J. A. 1939. *Business Cycles: A Theoretical, Historical and Statistical Analysis of the Capitalist Process*. New York: McGraw-Hill

Schumpeter, J. A. 1943. *Capitalism, Socialism and Democracy*: Harper and Row

Silverberg, G. 2001. "The Discrete Charm of the Bourgeoisie: Quantum and Continuous Perspectives on Innovation and Growth," *Research Policy* 31: 1275–89

Silverberg, G. and Verspagen, B. 1994. "Learning, Innovation and Economic Growth: A Long-Run Model of Industrial Dynamics," *Industrial and Corporate Change* 3: 199–223

Silverberg, G. and Verspagen, B. 2005, 'Evolutionary Theorizing on Economic Growth', in: Dopfer (ed.), *The Evolutionary Foundations of Economics*, Cambridge University Press, pp. 506–39

Solow, R. M. 1956. "A Contribution to the Theory of Economic Growth," *Quarterly Journal of Economics* 70: 65–94

Solow, R. M. 1957. "Technical Progress and the Aggregate Production Function," *Review of Economics and Statistics* 39: 312–20

Sornette, D. and Zajdenweber, D., 1999. "The Economic Return of Research: The Pareto Law and Its Implications," *European Physical Journal* B 8(4): 653–64

Temple, J. 1999. "The New Growth Evidence," *Journal of Economic Literature* 37: 112–56

Tinbergen, J. 1943. "Zur Theorie der Langfristigen Wirtschaftsentwicklung," *Weltwirtschaftliches Archiv* 55: 511–49

Verdoorn, P. J. 1949. "Fattori che Regolano lo Sviluppo della Produttivitá del Lavoro," *L'Industria* 1: 45–53

Verspagen, B. 1992. "Endogenous Innovation in Neo-Classical Growth Models: A Survey," *Journal of Macroeconomics* 14: 631–62

Young, A. 1928. "Increasing Returns and Economic Progress," *Economic Journal* 38: 527–42

Comments to Chapters 1 and 2:

Understanding economic growth

Jan Fagerberg

What could be more important than, say, understanding why some countries excel in terms of income levels and productivity, with healthy increases year after year, while living conditions in others are so harsh that large parts of the population do not get enough to eat? And finding out what can be done about it? As Richard Nelson rightly points out in his contribution to this book, questions like these – which have to do with understanding economic growth – ought to be central to economist's agenda.

And at times they have been. For instance, a few centuries ago such distinguished authors as Adam Smith and David Ricardo were busy arguing that stagnant growth and generally poor living conditions were not at all necessary, if only institutions and policies were geared toward allowing the capitalist machine to work at full speed. They argued forcefully that such changes in institutions and policies, although detrimental to the narrow interests of some stakeholders in the existing system, would be enormously beneficial to society as a whole. For various reasons, toward the end of the nineteenth century this dynamic, political–economy perspective gave way to a more static approach, focusing on the equilibrating character of market forces rather than the economic growth that these might (or might not) give rise to. But with the global economic downturn and the massive unemployment of the 1930s, economists again started to search for a more adequate understanding of economic growth (or rather the lack of such), and what policy could contribute to the solution of the pressing economic and social problems of the time. The resulting "Keynesian revolution" in economic thinking led to the formulation of a new perspective on economic growth, the so-called "post-Keynesian growth theory" associated with the work of Evsey Domar and Roy Harrod among others, which showed that long-run growth with full employment was indeed possible but depended on extensive intervention by the government (especially with respect to income distribution).[1]

I tell this story,[2] which ends where Bart Verspagen starts his, to illustrate one central point. Growth theory at its best is deeply political. Right or wrong, it shows, in a simple and transparent way, which factors underpin the current dynamics, what social and economic problems this might lead to and how policy can have a say in combating them. Admittedly, not all growth theories are of this sort. But not all contributions to growth theory are very influential either. Arguably, there are not that many contributions that really have been influential in shaping others' views of what causes growth and what can be done to influence this dynamics. Apart from the grand masters of the past, such as Adam Smith, David Ricardo, Karl Marx, Joseph Schumpeter, and some post-Keynesians (among which Nicholas Kaldor perhaps is the best example), who deserve mentioning?

Bart Verspagen, in his contribution to this volume, provides a highly readable guide to the development of growth theories during the last fifty years, with the major emphasis on the contrast between neoclassical and evolutionary theories. It is customary to date the advent of neoclassical growth theory to Robert Solow's celebrated article from 1956, although as common in the history of economic thought, there were other contributors that advocated similar ideas at roughly the same time. As Verspagen shows, the Solowian theory essentially embraced the central tenet of neoclassical economics, namely that the economy can be seen as composed of a high number of selfish actors, endowed with so-called "perfect information" (about everything worth knowing), who exchange goods and services with the purpose of maximizing their own individual gains. This process, Solow showed, leads to an equilibrium, a constant state of affairs, in which labor productivity is constant as well. Although capital accumulation – substitution of labor by capital – may increase productivity in the short run, on the path towards equilibrium, in the long run such accumulation will not be economical to undertake. Thus, following this theory, in the long run any growth in productivity must be exogenous, related – for instance – to (exogenous) advances in science and technology. As for policy, the most important finding was that long-run growth with full employment was indeed possible as long as market forces were allowed to operate freely. Hence the central argument of the post-Keynesians that sustainable economic growth required extensive intervention by the state was effectively put to rest (at least for a while?).

[1] For an overview, see Pasinetti (1974).
[2] The interested reader will find it presented in some more detail in Fagerberg (2000).

Although celebrated at first, the conclusion that long-run growth had to be explained by "noneconomic" forces was hard to accept for many neoclassical economists. The central prediction of the theory that under otherwise identical circumstances everyone would be equally well off in the long run was also hard to reconcile with the available empirical evidence. As a result of these concerns, and advancements in formal mathematical techniques, new neoclassical theories of economic growth – so-called "new growth theory" – were invented during the 1980s and 1990s, and Verspagen presents these in quite some detail. Although, as he points out, there are several versions of "new growth theory," the central argument is that individual investments in new technology have positive effects beyond that of the individual investor, through so-called "spillovers," and that this counteracts the tendency towards decreasing returns to such investments, which would otherwise have led these to halt in the long run (as in the Solow model). Hence instead of Solow's stationary equilibrium, the theories predict a "moving equilibrium," in which growth is driven by the incentives to investment in new technology (knowledge) and the "spillovers" from such investments. Compared with the Solowian approach this clearly implies a change of agenda, both with respect to policy implications and applied research, and so one might perhaps see this as an example of a theory that has succeeded in addressing central issues in a novel way. On the other hand, it might be argued that these changes in the agenda occurred before the theory, induced by other types of contribution, of a much less formal but not less influential nature. In fact, there is by now a lot of work on issues such as R&D and innovation activities at the level of the firm, sector, or industry, and diffusions of innovations, published in books and specialized journals such as, for instance, *Research Policy*. Most of this has been going on for a long time and is not specifically related to the new growth theories.

In one respect the new growth theories resemble the old, however, and this is in how the agents are perceived. These are still seen as selfish, optimizing agents endowed with perfect information and so on. This is where, according to Verspagen, neoclassical and evolutionary theories of economic growth part. There are two, mutually supporting reasons for this that evolutionary theorists emphasize.[3] The first is the uncertain character of technological advance. Successful innovations defy planning.

---

[3] The classic work on the subject is Nelson and Winter's book on *An Evolutionary Theory of Economic Change* (1982). For a discussion of how this relates to Schumpeter and work inspired by him, see Fagerberg (2003).

The second has to do with the limited cognitive abilities of humans, individually as well as collectively. Humans, it is argued, are simply not able to calculate the consequences of all possible actions and choose between them in the way neoclassical economists usually assume. The world is too complex and the mass of information too large to allow for this type of decision-making. What humans actually do, following this view,[4] is to practice a simpler and less demanding type of decision-making called "bounded" or "procedural" rationality. At the firm level this leads, according to Nelson and Winter, to the formulation of "routines" that guide firm action. Such routines, although relatively stable, may be changed through search processes (that are also guided by "routines"). Routines may differ across firms (or sectors), and taking this into account leads to complex dynamics that, as explained by Verspagen, is most often explored through simulations.[5]

Verspagen illustrates the difference between neoclassical and evolutionary economics with the help of the fable of "the blind watchmaker." Basically, the idea is that trial and error, combined with selection processes, drive the evolutionary dynamics, just as random mutation and selection drive biological evolution (in the Darwinian interpretation). Verspagen obviously sees this explanatory framework as a major strength of evolutionary economics when compared with the magnificent albeit completely unrealistic optimisation framework adhered to by neoclassical economists. But is this Darwinian dynamics a valid presentation of how economies, or technologies for that sake, change? Verspagen concedes that in the real world watchmakers are not "completely blind," but leaves it there.

Interestingly, Richard Nelson, in his contribution to this volume, also considers the role of intentionality in human evolution. Nelson points out that "human and organizational actors are purposeful, they often make conscious efforts to find better way of doing things and their efforts to innovate are far from completely blind." Commenting on his own work with Sidney Winter more than two decades ago, he argues that much subsequent work in this area has stayed too close to their original framework, and that this represents an obstacle to further progress. In particular, he notes that they "played down the role of

---

[4] Nelson and Winter's views on how economic agents operate were inspired by prior work by Herbert Simon and other writers in the so-called "behavioralist" tradition in organization theory (Simon 1959, 1965; Cyert and March 1963). See Andersen (1994) for a good discussion of the relation between Nelson and Winter's work and the work of Simon and the behavioralists.

[5] See Fagerberg (2003) and Silverberg and Verspagen (1998) for extended accounts.

cognition, understanding, conscious problem-solving" and with this "the importance advance of human knowledge." He now believes that it is high time to make up for this. Related to this he suggests that "the evolution of business practice, organizational forms and institutions," what he calls "social technologies," deserve a high rank on the research agenda in this area. Nelson also makes the very interesting point that progress seems to be much easier in "physical" than "social" technologies. Why is this so? And what are the implications for economic growth, and the future research agenda? No lack of exciting issues for further research here, as the reader may have realized already.

I leave the further exploration of these issues to the reader. Instead I will get back to some of the questions raised earlier about what we can – and should – expect from growth theory. Both Nelson and Verspagen discuss this to some extent. Nelson clearly has a very broad concept of growth theory, ranging from formal models to systematic explorative work into "industrial dynamics." Verspagen talks about the need for closer interaction with "nonformal work" and historical research, and suggests creating a new class of models based on observed "stylized facts" that employ "relatively simple, evolutionary microeconomic foundations." Although it is not entirely clear where the latter suggestion might lead us, it is in my view promising that both authors recognize the need for several different methodological approaches and levels of analysis in explorations of economic growth.

Arguably, in spite of the drive towards mathematization that has characterized economic growth theory for at least a century, it is not the case that mathematical models necessarily have been more influential than more informal reasoning. Historically, it seems to be the case that the theories that have been most influential, such as those of Smith, Ricardo, Marx, and Schumpeter, were put forward verbally. Attempts to put their contributions into a mathematical language, such as Solow (1956) of Adam Smith, Tugan Baranowsky (1905) of Karl Marx, and Sraffa (1960) of David Ricardo have often tended to base themselves on very narrow interpretations of what the authors meant (perhaps reflecting what the available mathematical methods were suited for, rather than what the authors had to say?). As a consequence such mathematical formalizations have commonly ended up much less rich than the original accounts and often directly misleading. An interesting question is to what extent this also is the case for Schumpeter's work? Compare, for instance, Schumpeter's perspective as expressed in his works with the so-called "Schumpeterian" models that have emerged within "new growth theory." What seems to be the case is that only certain aspects of Schumpeter's work are taken into account, while other elements, arguably equally essential, are disregarded without any justification.[6] In particular, Schumpeter's analysis of what shape innovation decisions tends to be neglected in these formalization attempts.

Thus, there are reasons to be skeptical toward the strong tendency toward formalism that dominates growth theory. This is not because there is anything wrong with mathematics in itself, but that the application of mathematical methods to growth theory has tended to change the questions that are addressed in a less-interesting and relevant direction. Perhaps it is time to be more truthful to issues and arguments, and let these determine the choice of methods and forms of exposition, rather than the other way around?

I argued above – and illustrated it with some examples – that growth theory at its best is deeply political. How are evolutionary growth theories doing on this account? Not too well, one might be tempted to say. Perhaps the most important challenge for economists with an evolutionary leaning is not to construct clever models, but to be able to address, on the basis of evolutionary reasoning, some of the most important challenges facing mankind today.

References

Andersen, E. S. 1994. *Evolutionary Economics, Post-Schumpeterian Contributions.* London: Pinter

Cyert, R. M. and March, J. G. 1963. *A Behavioural Theory of the Firm.* Englewood Cliff, NJ: Prentice-Hall

Fagerberg, J. 2000. "Vision and Fact: A Critical Essay on the Growth Literature", in Madrick (ed.), pp. 299–330, 350–54 (reprinted as Chapter 6 in Fagerberg, J. 2002. *Technology, Growth and Competitiveness: Selected Essays.* Cheltenham: Edward Elgar)

Fagerberg, J. 2003. "Schumpeter and the revival of evolutionary economics: An appraisal of the literature," *Journal of Evolutionary Economics* 13: 125–59

Nelson, R. R. and Winter, S. G. (eds.) 1982. *An Evolutionary Theory of Economic Change.* Cambridge, MA: Harvard University Press

Pasinetti, L. 1974. *Growth and Income Distribution.* Cambridge: Cambridge University Press

Silverberg, G. and Verspagen, B. 1998. "Economic Growth and Economic Evolution: A Modeling Perspective," in F. Schweitzer and G. Silverberg (eds.), *Evolution and Self-organization in Economics, Jahrbuch für Komplexität in den Natur-, Sozial- und Geisteswissenschaften,* vol. 9. Berlin: Dunker & Humblot

Simon, H. A. 1959. "Theories of Decision Making in Economics," *American Economic Review* 49: 253–83

---

[6] See Fagerberg (2002), Introduction, and Fagerberg (2003) for extended discussion of these issues.

Simon, H. A. 1965 *Administrative Behaviour*, 2nd edn, New York: Free Press

Solow, R. M. 1956. "A contribution to the theory of economic growth," *Quarterly Journal of Economics* 70: 65–94

Sraffa, P. 1960. *Production of Commodities by Means of Commodities: Prelude to a Critique of Economic Theory*. Cambridge: Cambridge University Press

Tugan-Baranowsky, M. 1905. *Theoretische Grundlagen des Marxismus*. Leipzig: Duncker & Humbolt

Part 2

The microdynamics of the innovation process

3 Schumpeter's prophecy and individual incentives as a driver of innovation

Wesley M. Cohen and Henry Sauermann

Introduction

Over sixty years ago, in his *Capitalism, Socialism and Democracy* [*CSD*], Joseph Schumpeter (1962) predicted the demise of capitalism. He argued that the growth of the large business enterprise was frustrating the entrepreneurial incentives vital to capitalism's vitality and growth. Accordingly, the very success of capitalism's flagship institution – the large modern business enterprise – was to be its undoing.

In *CSD*, Schumpeter argues that the locus of innovation and technological progress had shifted to the large modern firm. That shift would, however, undermine capitalism by embedding innovation within large corporate bureaucracies, leading to its routinization, with a consequent replacement of entrepreneurial incentives – attendant upon the prospect of individual gains and losses from entrepreneurial initiative – with those associated with either the salaried employee or the shareholder. Neither status, in his view, confers the "substance of property," which entails a "sense of personal responsibility for success" (1962, p. 133) – that is, a strong personal stake in the creation and subsequent performance of the enterprise.[1] In his view, neither top management, salaried personnel

We wish to thank Ashish Arora, Steven Klepper, and Franco Malerba for comments on this chapter, and the Ewing Marion Kauffman Foundation for its support.

[1] Schumpeter states: "The capitalist process, by substituting a mere parcel of shares for the walls of and the machines in a factory, takes the life out of the idea of property. It loosens the grip that once was so strong – the grip in the sense of the legal right and the actual ability to do as one pleases with one's own; the grip also in the sense that the holder of the title loses the will to fight, economically, physically, politically, for 'his' factory and his control over it, to die if necessary on its steps. And this evaporation of what we may term the material substance of property – its visible and touchable reality – affects not only the attitude of holders but also that of the workmen and of the public in general. Dematerialized, defunctionalized and absentee ownership does not impress and call forth moral allegiance as the vital form of property did. Eventually there will be nobody left who really cares to stand for it nobody within and nobody without the precincts of the big concerns." (1962, p. 142)

nor shareholders, nor, in turn, the modern business enterprise itself, can sustain what he believed to be the economically critical function of the entrepreneur in "exploiting an invention or, more generally, an untried technological possibility for producing a new commodity or producing an old one in a new way, by opening up a new source of supply of materials or a new outlet for products, by reorganizing an industry and so on." (1962, p. 132). Thus, in achieving its very dominance, the large modern enterprise was sowing the seeds of the decline of the system for which it had become the bedrock institution.[2]

As we know, Schumpeter was wrong about capitalism's demise. He did not foresee the enduring capacity of the large business enterprise to innovate, nor the startling growth of the entrepreneurial sector in the United States. So, why was Schumpeter wrong? Rosenberg (1994) provides part of the answer when he argues that Schumpeter both overestimates the degree to which capitalism "automatizes" innovation, and underestimates the degree to which commercial success depends upon the more mundane and rationalized processes associated with downstream R&D and related activities. But there is more, and we suggest that some of the answers reside in Schumpeter's own analysis.

First, he focuses our attention on incentives *at the level of the individual*, whether that individual be an entrepreneur, a salaried manager, or even a shareholder. Although Schumpeter was mainly concerned with the effect on economic performance of the incentives of the principals and leaders in large firms versus more entrepreneurial ventures, his discussion points toward a consideration of the importance of individuals' incentives more generally. Second, Schumpeter's discussion expands our notion of incentives beyond a focus on the pecuniary gains from innovation. He suggests that the benefits that motivate innovation and economic initiative more generally on the part of individuals may be non-pecuniary, including, for example, an individual's satisfaction of realizing a "vision," or "the opportunity to fling himself into the fray," (p. 133) or the opportunity to exercise autonomy and responsibility – "the legal

---

Regarding the group of large shareholders in particular, Schumpeter states: "even if it considers its connection with the concern as permanent and even if it actually behaves as financial theory would have stockholders behave, it is at one remove from both the functions and the attitudes of an owner." (p. 141).

[2] Schumpeter's discussion suggests that his views of the role of the small, entrepreneurial firms versus that of the large monopolistic firm in driving technical change did not change as much as often supposed between his earlier, *The Theory of Economic Development*, and his later, *Capitalism, Socialism and Democracy*. On one theme he is consistent. He retains the belief that the entrepreneurial incentive is key to technological progress and economic growth in the long run, and that that incentive is best preserved at a much smaller scale of enterprise than that provided by the large industrial firm.

right and the ability to do as one pleases with one's own" (p. 142).[3] Thus, his notion of incentives is psychologically richer than simply the garnering of pecuniary rents. Indeed, in *CSD*, he even derides the power of salary or equity shares alone to motivate entrepreneurial incentives (p. 156).[4]

Schumpeter's false prophecy of capitalism's demise, therefore, offers two suggestions that may enrich our understanding of innovation. First, consider the impact on innovation and economic performance of the incentives of the individuals involved – not simply the firms. We will extend this logic to suggest that, in Schumpeter's terms, it may be illuminating to consider the individual incentives of not simply those concerned with innovation and commercialization, but those concerned with invention and the generation of new ideas as well. Second, he understands that individual incentives are broader and psychologically richer than the pecuniary incentives that are typically the focus of economists' attention.

In this chapter, we argue that studying individual-level incentives – which we define as desired benefits that motivate individual behavior – will help us understand the determinants of innovative activity and performance. Second, we will suggest that to understand technical advance, scholars' notions of incentives should transcend the pecuniary. Finally, we suggest that a more complete conception of the drivers of technological change that integrates individual-level incentives

---

[3] This attention to nonpecuniary goals in *CSD* is consistent with Schumpeter's earlier characterization of the motives of entrepreneurs in his *Theory of Economic Development* where he argued that entrepreneurs work too hard and long to believe that they are driven by the "prospective consumption that wealth affords." Rather, he suggests that entrepreneurs are driven by a range of goals, including the "joy of creating a private kingdom," to prove oneself superior to others, the joy of accomplishment, or the pleasure derived from the exercise of one's energy or ingenuity (Schumpeter, 1934, pp. 94, 95). He goes on to suggest that entrepreneurs enjoy not the fruits of success, but success itself.

[4] Schumpeter was not the only economist of his time to highlight the importance of individual, nonpecuniary incentives for innovation. Shortly after the publication of Schumpeter's *Theory of Economic Development*, F.W. Taussig (1915) argued that humans benefit from "an instinct of contrivance" originating from a process of Darwinian natural selection: "There is abundant evidence that the human animal follows an instinct of contrivance. The utilitarians would indeed explain it in their familiar fashion: men contrive and invent *because* they find it advantageous to do so. But the matter is not so simple as this . . . . The instinct of contrivance is widespread in the animal world – among insects, birds, mammals . . . In origin doubtless it goes back to the fact of having been at some stage advantageous. For its evolution, as for that of every other instinct, one turns almost as a matter of course to the Darwinian organon. Those creatures which were disposed to contrive had a better chance in the struggle than their fellows; they survived, and their nervous structure was transmitted to their descendents" (Taussig, 1915, pp. 12, 13). Veblen (1914) makes a similar argument, upon which Taussig builds, in *The Instinct of Workmanship and the State of Industrial Arts*.

will yield dividends for both policy and management. For policy-makers, a consideration of the interests of those individuals responsible for advancing technology may contribute to the formulation of tax policies, intellectual property policy, and policies toward universities. For management, the question is how can the incentives of employees as well as others upon whom firms also depend for new ideas and expertise be structured to increase firms' innovative performance. A better understanding of the nature of individual-level incentives as well as of the ability of firms to manage and exploit these incentives may allow us to understand the drivers of firms' innovative performance as well as differences in such performance across firms.

3.1   Technological change and the incentives of individuals

Dating from the 1950s and early 1960s, economists such as Schmookler (1962), Griliches (1957), and Nelson (1959), among others, set out to show that the rate and direction of technological change could be understood as the outcome of rational, profit-driven behavior. These arguments were posed at a time when the study of invention and technological change was largely the province of sociologists, historians, and even psychologists who understood the direction of technical advance as reflecting either the logic of technology itself that compelled the direction of subsequent advance, the work of "hero-genius" inventors such as Pasteur and Edison, or the outcome of the curiosity or other predispositions of individual technologists. None of these explanations appealed principally to profit incentives. In making the case for the primacy of profit as a driving force behind technical change, economists sensibly focused scholars' attention on firms (or, in the case of Griliches' seminal work on the diffusion of hybrid corn technology, on farms) since it is indeed firms that are responsible for both a good deal of innovation and particularly its commercialization. In doing so, however, they subordinated consideration of the impact of individuals, and, in turn, their incentives, on technical advance.

Why should the incentives of individual employee-technologists – as opposed to those of the firm as a whole – matter? As is true of any employee, a firm's technologists will invariably be able to exercise some degree of autonomy. Moreover, since there is typical uncertainty about how to tackle a technical challenge, and the technologists themselves will have considerable expertise about the technology in question, top management would typically prefer that the technologists retain some degree of that autonomy. In addition to information asymmetry with respect to substantive expertise, there is also information asymmetry with respect to the level and quality of effort expended. And, given the uncertainty endemic to the outcomes of R&D projects, observable outcomes are not necessarily informative of the level or quality of effort expended by the employee-technologists. Thus, R&D labs are settings where there is some significant delegation of authority to the individual employee-technologists. As a consequence, the innovative performance of firms will be affected by the motives of its employee-technologists, especially to the degree that those motives or individuals' responses to the firms' incentive system are inconsistent with the firm's interests.

Although applied economists have neglected the empirical study of the impact of individual incentives on technical advance in particular, economic theorists have considered implications of at least a stylized view of individual incentives for firm performance generally. Assuming that employees' incentives are pecuniary, that individuals prefer leisure over work, that individuals' incentives may be contractible, and that there is information asymmetry between the employee and the employer, economic theorists have considered how firms should structure contracts with individual employees (i.e., agents) to align their behavior as much as possible with that of the firm (i.e., the principal) (e.g., Gibbons, 1998; Prendergast, 1999). We would suggest, however, that this representation of individual incentives may depart importantly from key features of the incentives of individuals engaged in innovative activity.

For a more comprehensive characterization of incentives than what economists typically consider, we can look to research in social psychology and organizational behavior. These literatures discuss several types of individual-level incentives, which can be classified as extrinsic, intrinsic, or social incentives.[5]

Extrinsic incentives or rewards do not result directly from either task engagement or outcome. Rather, some environmental condition, such as a market, or actor such as an employer, a superior, a judging body, or a customer, provides such a reward conditional upon an evaluation of the task outcome. Extrinsic incentives are those considered by economists, and within this class of incentives, economists typically focus on

---

[5] Kreps (1997) highlights, however, some of the challenges in distinguishing different classes of incentives from one another. In particular, he suggests that some behaviors that are attributed to intrinsic motivation because no explicit extrinsic incentives are visible in fact reflect indirect and more subtle extrinsic incentives.

those which are pecuniary, or those, such as promotions, that are closely related to pecuniary benefits. Other examples of extrinsic rewards include monetary or other tangible rewards such as raises, royalty income from patents, performance-contingent pay, bonuses, research funding, or a paid vacation.

Intrinsic incentives. The motivation to work on something because it is interesting, involving, exciting, satisfying, or personally challenging is called intrinsic motivation (Amabile, 1993; Deci and Ryan, 1985). Some intrinsic rewards, such as task enjoyment and intellectual challenge, are realized directly from the process of engaging in certain behaviors. Others, such as a feeling of achievement, mastery, or self-competence, result directly from task performance and outcomes. Intrinsic incentives therefore originate within the individual or the task – not the environment – and are often a function of the interaction between characteristics of the task (e.g., challenge of the task) and of the individual (e.g., interest in the task). While an organization cannot directly satisfy intrinsic incentives, organizations can provide conditions under which such incentives are more likely to be satisfied, such as the provision of autonomy, or the assignment of challenging, engaging work (Deci and Ryan, 1985; Hackman and Oldham, 1976).

Social incentives. Social psychologists typically think of intrinsic and extrinsic incentives as covering the full range of task incentives. In related work, Sauermann (2004) suggests that one might usefully identify another class, namely *social incentives*, that encompass intangible rewards originating from outside the task itself and originate from the individual's perceived social relations. Fehr and Falk (2002), for example, review a growing body of literature showing that people act either to gain social approval or to avoid social disapproval. Similarly, the desire to fulfill a psychological contract with peers or the desire to reciprocate others' contributions can be a powerful incentive to engage in certain behaviors (Fehr and Falk, 2002). Such social incentives may be particularly important in teams or organizations to the extent that members develop a high degree of cohesion and mutual commitment, possibly to the point where team or organizational goals become internalized (Gagne and Deci, 2005; Ouchi, 1979). While social rewards may be tied to pecuniary rewards (e.g., via the economic payoff to reputation or to the approval of a superior), there is considerable evidence that social incentives operate also in the absence of any pecuniary incentives (Fehr and Falk, 2002). Although social incentives overlap conceptually with extrinsic incentives with regard to their external origin, we suggest that they are worth distinguishing, given their potentially different impact on behavior, and the different degree to which they are susceptible to managerial influence.

Individuals are generally motivated by a combination of extrinsic, intrinsic, and social incentives, and different types of incentives are often provided simultaneously. For example, a publicly announced year-end bonus involves both pecuniary elements (the monetary value) and social elements (recognition). Individuals respond to all three kinds of incentives to some degree, but there are individual differences with respect to the strength of the response (cf. Amabile et al., 1994; Deci, Koestner and Ryan, 1999). To some extent, incentives can also be substituted for one other: a given level of activity or performance can be achieved by different incentive structures or "mixes" of incentives. Little is known, however, about how and the degree to which incentives may substitute for one another.[6] Second, some types of incentives are simply hard to provide for certain behaviors. Gottschalg and Zollo (2004) suggest, for example, that intrinsic incentives may not be sufficient to induce people to process animal cadavers in a slaughter house.

The literature provides some evidence on how different classes of rewards may affect different kinds of incentives and associated behaviors. A finding that has received increased attention recently in the economics literature is that extrinsic rewards (i.e., appealing to extrinsic incentives) may actually crowd out intrinsic motivation under certain conditions (Amabile, 1996; Deci et al., 1999; Frey and Jegen, 2001). Different explanations concerning the psychological mechanisms underlying crowding out have been offered. Some authors suggest that salient contingent rewards reduce perceived self-determination and autonomy, which in turn are important facilitating factors for intrinsic motivation. Others suggest that the presence of extrinsic rewards may be construed as a signal that the task "cannot be fun," leading to lower task interest and task enjoyment (Benabou and Tirole, 2003; Deci and Ryan, 1985; Lepper and Greene, 1978). Although the psychological mechanism behind the crowding-out effect remains unclear, it is conjectured to apply especially to tasks offering greater potential for intrinsic motivation, making the effect especially relevant for the R&D function within firms.

The social psychology literature also suggests that some types of rewards may be more effective for motivating particular behaviors than others, implying that types of incentives should be matched to desired behaviors when possible. For example, empirical research has shown

---

[6] One exception is Stern (2004) whose work suggests a precise trade-off between salary and the nonpecuniary rewards of doing science.

that creativity is enhanced by intrinsic motivation and can be stifled by pecuniary rewards (Amabile, 1993, 1996; Hennessey and Amabile, 1998). Amabile (1993) argues that intrinsic incentives should, therefore, dominate in the problem-presentation and idea-generation stages of R&D projects, which require much creativity. As the project enters the idea validation and implementation stages, creativity is less important and more weight on pecuniary incentives is appropriate. As a more practical example, consider the evolution of most academic papers. Truly creative ideas rarely emerge "on command" and under external pressure. However, once conceived, ideas rarely get written up without some external pressure such as editors' deadlines.

Cooperation in teams is another behavior that may be more effectively induced by some types of rewards than others. Consistent with agency theory, pecuniary rewards to individuals may undermine cooperative behaviors, whereas pecuniary rewards given to teams are problematic because of free-rider problems (Alchian and Demsetz, 1972; Prendergast, 1999). Perhaps teamwork is best supported by intrinsic or social rewards where the team members strive for some collective goal and feel mutual obligation (Fehr and Falk, 2002; Osterloh and Frey, 2000).[7]

At this point, one might ask what benefit for the study of innovation may be had from paying attention to different types of incentives at the level of the individual. In the next two sections, we review evidence that individuals are motivated by a range of incentives, especially in settings where innovative activity occurs. Second, we provide examples to suggest that paying attention to individual-level incentives and their diversity may enrich our understanding of the sources of innovation and technical advance.

3.2   Incentives of researchers, scientists, and inventors

A large body of research in organizational behavior has continued to investigate what motivates people and has consistently found that extrinsic, intrinsic and social incentives can matter a great deal (Baron and Kreps, 1999; Jenkins et al., 1998; Stajkovic and Luthans, 2001; Wood and LeBold, 1970). While most of these studies were not specifically conducted with samples of scientists and engineers, recent surveys sponsored by the National Science Foundation (NSF) indicate that scientists and engineers also care about nonpecuniary rewards, and probably

---

[7] Intrinsic incentives based on task enjoyment, task interest, and challenge are nonrivalrous; they may even be "contagious." Incentives based on a feeling of achievement or based on social recognition are performance-contingent and may therefore be rivalrous.

more so than people in other settings. In the "2001 Survey of Doctorate Recipients" over 30,000 engineers and scientists with doctorate degrees indicated the importance of a range of factors when thinking about a job. The results (Figure 3.1) show that people perceive nonpecuniary factors as extremely important when thinking about jobs. For example, seventy-nine percent of doctorate recipients indicated that intellectual challenge is a very important aspect of a job and fifty-four percent stated that contributions to society are very important. In contrast, only forty-three percent of the respondents assigned that level of importance to salary.

To compare the results for scientists and engineers with that of production workers, we analyzed data from the General Social Survey (GSS), which is conducted regularly by the University of Chicago National Opinion Research Center. Among other questions, the survey asks respondents to rank a number of job characteristics according to their importance.[8] Figure 3.2 shows the results for two types of occupations, scientific- and engineering-related specialty occupations, and production.[9]

According to these data, the importance of nonpecuniary incentives is larger for scientific and engineering professionals than for production workers, as reflected in the relative importance of the characteristics "High Income" and "Work is important and gives a feeling of accomplishment".[10] Thus, the incentives of engineering and scientific professionals appear to differ from those of production workers. There are also undoubtedly differences in incentives within the group of scientists and engineers, and possibly across different industries. More research is needed to better understand why different types of incentive correspond to technologists distinguished by training, task assignments, and so on, and how such incentives may be conditioned by socialization, norms, feedback, and selection.[11]

---

[8] The exact question is "Would you please look at this card and tell me which one thing on this list you would most prefer in a job? Which comes next? Which is third most important? Which is fourth most important?"

[9] The former group encompasses the 1980 Census Occupational Codes 044 to 083; the second group encompasses the codes 703–799. For more information, see http://webapp.icpsr.umich.edu/GSS/.

[10] While it is common practice to use direct questions to elicit respondents' preferences for job attributes, the responses may suffer from social desirability bias (Rynes, Gerhart, and Minette, 2004). In particular, respondents may overstate the importance of socially desirable attributes such as "contribution to society" and understate their preferences for extrinsic rewards. Research using indirect measures of preferences is needed to replicate findings such as those reported here.

[11] In another paper (Sauermann and Cohen, 2006), we begin to examine these questions using data from the 2003 Scientists and Engineers Statistical Data System (SESTAT).

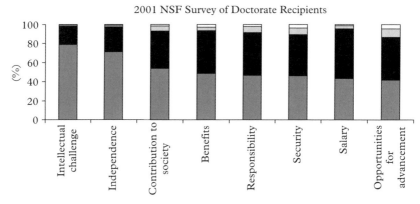

Figure 3.1 Importance of job characteristics for S&E doctorate recipients (n = 31.366) *Source* Based on 2001 Survey of Doctorate Recipients, National Science Foundation, Division of Science Resources Statistics

The results highlighting the importance of nonpecuniary incentives for science and engineering professionals from both the NSF and the GSS are consistent with the findings from over seventy years ago of Joseph Rossman (1964), who asked 710 inventors (whose inventions had been granted patents) "What motives or incentives cause you to invent?" The top three answers, with associated percentages reported in parentheses (answers were not mutually exclusive) were: (1) Love of inventing (27.2%); (2) Desire to improve existing devices (26.6%), and (3) Financial gain (23.5%). Underscoring the role of intrinsic incentives, Rossman states: "The sheer joy of inventing, resulting from an irrepressible urge to invent has been felt as the greatest urge by the inventors of this study. The pleasure resulting from manipulation and experimentation, the satisfaction of solving problems and the desire to create were considered sufficient in themselves as objectives by the inventors." (Rossman, 1964, pp. 152–153). Rossman suggests, however, that, although intrinsic incentives are key to invention, they are not sufficient, and that another necessary condition is the prospect of economic gain.[12] Nonetheless, Rossman's early findings, like the more

---

[12] Reiterating a position first articulated by Taussig (1915), Rossman (1964, p. 153) states: "We must bear in mind that the mere love of inventing would in itself often be insufficient to be an incentive to invention. Before any invention is perfected and marketed a great deal of money must be spent in developing and perfecting the original mental conception. The inventor either must spend his own money or interest business-men in his invention.

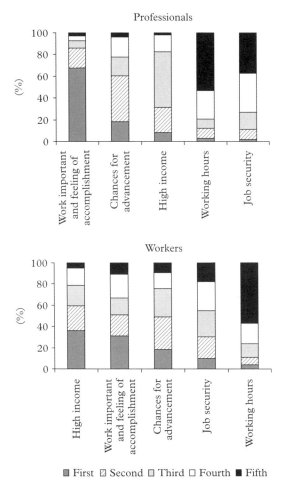

Figure 3.2   Rankings of importance of job characteristics; engineering and science professionals versus workers (n = 477) *Source* Based on GSS 1972–2000 Cumulative Datafile, University of Chicago National Opinion Research Center

recent survey results reported above, raise the question of just how importantly nonpecuniary rewards affect the quality and quantity of effort expended by individuals tasked with innovation.

In either case, unless there was a prospect of gain, the chances are that no money would be spent in developing the invention to a practical basis.''

Although empirical scholars concerned with the determinants of the rate and direction of technical advance have largely ignored the role of the incentives of individual technologists, more practice-oriented writers have not. Manners Steger, and Zimmerer (1997), for instance, emphasize the importance of creating "excitement" about innovative work and argue that performance *per se* is immediately exciting and that, while money rarely motivates people, the quest for outstanding results can (p. 33). Other authors provide page-long lists of incentives and rewards that can be used in the R&D setting, ranging from challenging research and time for personal projects over public praise and plaques to special parking and paid education (Koning, 1993; Mower and Wilemon, 1989). That practitioners commonly highlight the importance of nonpecuniary incentives for innovative performance does not, however, mean that they are indeed important.

3.3   Do individual incentives matter for innovation?

It should not be news that individuals have different incentives. It does not necessarily follow, however, that individual-level incentives need to be considered in our efforts to explain innovation since innovation and its commercialization is typically carried out not by individuals, but by firms. It is conceivable that, relative to firm profit, the incentives of individual technologists matter little as determinants of firms' and industries' innovative activity and performance. In this section, we provide a number of examples to show how attention to individual-level incentives may enrich our understanding of R&D productivity and innovation, and can provide alternative explanations for observed phenomena.

First, consider Cockburn and Henderson's (1998) study of the determinants of research productivity among pharmaceutical firms. Their key finding was that those firms that permitted their research scientists to publish and participate in "open science" were more productive in drug discovery. The reason offered is that, by virtue of publication and participation in scientific conferences, these scientists were better connected to relevant research networks and were thus more effective conduits of scientific knowledge to the firm. We would suggest another, though not mutually exclusive, explanation, which is closely related to that proposed by Stern (2004). By permitting their scientists to publish and participate in their respective scientific communities, the firms are allowing their scientists to satisfy their curiosity and desire for eminence, indulging their "taste for Science."

As a consequence, these firms attract more research-oriented personnel, and the personnel who work for them may be more motivated, and may work harder and more creatively than the R&D employees of firms where the scientific ambitions of their staff are not so encouraged. As Stern (2004) characterizes it, Cockburn and Henderson's result may thus reflect a "preference effect" on the part of the firms' scientists, as opposed to the R&D "productivity effect" associated with information flows. In addition, Stern finds that the scientists who were allowed to publish realized twenty-five percent less salary on average, suggesting the substitution of nonpecuniary for pecuniary incentives.

Business histories illustrate how the management of technologists' nonpecuniary incentives could substantially affect the innovative performance of firms. Consider, for example, that DuPont could only recruit Wallace Carrothers from Harvard by offering to indulge his personal goals of advancing fundamental understandings of polymer science and publication (Hounshell and Smith, 1988). Carrothers subsequently invented nylon, one of the most successful polymers of all time. Just as attention to technologists' incentives can make a firm more innovative, inattention can destroy it. Shockley – the coinventor of the transistor and the founder of the Shockley Corporation – wanted to keep focusing his firm's R&D largely on foundational breakthroughs in semiconductor technology (Holbrook et al., 2000). In pursuing this goal, Shockley disregarded the motives of his gifted engineers who, manifesting a mix of pecuniary and intrinsic incentives, wanted to solve concrete, practical problems, see those solutions implemented in practice and make money. He consequently lost the allegiance of eight key employees (including Robert Noyce and Gordon Moore), who left the firm to found the Fairchild Semiconductor Corporation, which subsequently spawned Intel, National Semiconductor, and numerous other firms that populate the semiconductor industry today.

Nonpecuniary incentives may also drive R&D employees to pursue research programs contrary to the desires of top management. Consider, for example, the history of Digital's Alpha chip (Katz, 1993). In the early 1980s, an R&D team within Digital was working on a new microprocessor architecture, code-named PRISM. By late 1987, however, believing that internal development was too slow, Digital canceled the project and decided to adopt an existing design from an external provider in order to quickly gain a foothold in the emerging market. The team continued to "discreetly" work on the design, however, driven by the vision of building a chip twice as fast as anything available in the industry. Over time, the team convinced Digital's management of the promise of the PRISM chip and eventually brought it to life as the Alpha chip. Katz (1993) described the Alpha team's motivation: "They were not preoccupied with their individual careers; they were more interested in having their peers within the engineering community see them as being one of the world's best design teams. Ambition, promotion, and monetary rewards were not the principal driving forces. Recognition and acceptance of their work output by their technical peers, and by society was, for them, the true test of their creative abilities" (1993, p. 17).

The Alpha-chip case highlights the importance of intrinsic and social incentives and how group dynamics within firms can reinforce these incentives. It also suggests that technologists within large firms may indeed possess sufficient autonomy to depart rather substantially from the directions set by top management.[13]

Another instance where individual incentives induce R&D employee behavior unsanctioned by top management, in this case with pervasive effect upon R&D spillovers across rivals (cf. Cohen, Nelson, and Walsh, 2002), is the pattern of "tit-for-tat" information exchange first identified by von Hippel (1987). In the semiconductor and steel industries, von Hippel (1987) and Schrader (1991) observed that engineers employed in firms' R&D labs routinely received solutions to R&D challenges from friends and acquaintances employed in rival firms with the loose expectation that the favor would be reciprocated at some point in the future. The incentives here would seem to combine both social incentives to assist professional colleagues, but to do so in a way that prospectively enhances your performance in the lab – and perhaps even extrinsic rewards – even though such an exchange may undermine the competitive position of the firm.

Attention to individual-level incentives may also help us understand not only firm-level innovative performance, but regional innovative performance as well. Saxenian (1994), for example, argues that the more rapid growth of Silicon Valley relative to that of Route 128 area surrounding Boston is due to the more vertically disintegrated structures of Silicon Valley firms that rely more heavily upon outsourcing and specialized, R&D-intensive firms for their components and systems, and benefit from greater mobility of technical personnel that makes for a richer information environment.[14] One implication is that the more

---

[13] Examples of such cases of "civil disobedience" abound and they even form an explicit part of the culture in innovative companies such as 3M (Bartlett and Mohammed, 1995). However, it is not apparent that such departures from the desires of top management benefit firms on average.

[14] Fallick, Fleischmann, and Rebitzer (2005) show that there is indeed greater job mobility across computer firms in California relative to Massachusetts. Gilson (1999)

rapid growth of Silicon Valley relative to the Boston Route 128 region is at least partly due to the greater R&D spillovers with which such mobility is associated. Again, there is an alternative, though not mutually exclusive, explanation that appeals to individual incentives rather than spillovers affecting R&D efficiency. One might conjecture that such mobility benefits predominantly the most capable technologists who are most scarce relative to demand, and can consequently command greater pay in the form of salary, stock options, and so on. As a consequence, such "star" employee-technologists are able to appropriate a greater share of the pecuniary returns to their innovative efforts. Such mobility may also enhance nonpecuniary benefits such as perceived autonomy and self-determination. Such greater individual-level appropriability of pecuniary returns and additional nonpecuniary incentives may induce Silicon Valley technologists to work harder, expend greater cognitive effort, and so on.

Another example where individual incentives matter at a market-wide level, transcending the employment relationship, is open-source software. In this case, most contributors to open-source projects are operating as individuals, even when they work for firms. The puzzle here was initially why programmers would essentially work for free. Although there is not yet a consensus about precisely which individual incentives are most important for open-source software, a range of individual-level goals have been proposed, illustrating every form of task incentive identified above. These include the intrinsic pleasure of creating, the social incentive of fulfilling perceived obligations to the community, and reputational gains that can yield higher salaries in the future (Lakhani and von Hippel, 2003; Lerner and Tirole, 2002).

These examples suggest that both pecuniary and nonpecuniary individual-level incentives may matter for innovation. However, to consider the importance and role of these different types of incentive requires an assessment of where the literature stands and the questions that need to be addressed.

3.4   Individual incentives and innovation: an agenda

Prior literature on the determinants of within-industry differences in innovative activity and performance have focused on two classes of capabilities associated with innovation and its commercialization: procedural

---

further argues that the greater job mobility in Silicon Valley probably reflects the absence of enforceable noncompete agreements in California.

capabilities that reflect organizational structures and processes, and substantive capabilities that represent a firm's knowledge and expertise with respect to specific technologies (Cohen, 1995). In this chapter, we suggest that, to understand the drivers of firms' R&D performance, we also need to know something about the motivations of the firms' individual technologists: what those motives are, and how they interact with the firm's goals, organizational processes and structures.

Recent theoretical work in strategy has considered the role of incentives and of firms' ability to manage incentives for strategy and competitive advantage (Gottschalg and Zollo, forthcoming; Sauermann, 2004; Wright, Dunford, and Snell, 2001). Adopting a dynamic perspective, Kaplan and Henderson (2005) suggest that rigid individual-level incentives might constitute an important cause of established firms' difficulty in responding to radical or discontinuous technological change. Empirical work on the relationship between incentives and individual and firm performance is usually limited to non-R&D functions or experimental tasks (Jenkins et al., 1998; Lazear, 2000). Others have examined related issues by looking at the implications of human resource management practices for innovative performance (e.g., Mumford, 2000) or firm performance more generally (Huselid, 1995; Ichniowski, Shaw, and Prennushi, 1997). However, there is little empirical study of the impact of individual-level incentives on innovation. In this section, we will briefly outline questions that might be considered in examining the impact of individual incentives on the innovative activity and performance of firms.[15]

First, other than limited evidence cited above, there is no quantitative evidence on what incentives may apply to technologists within firms. It is important to go beyond the broad categories of intrinsic, extrinsic, and social incentives and identify more precisely the motives that apply and their intensity. For example, just how important are peer recognition, the satisfaction of achievement, or even financial rewards in affecting innovative performance. Given that more than one incentive is often operative at any given moment, more work on the effect of the interaction across different types of incentive and the impact of a given type of benefit (e.g., pecuniary) on different classes of incentives (e.g., intrinsic, social) is warranted. Moreover, it would be important to know what happens when incentives conflict. For example, how do people make choices about actions, and how strong

---

[15] A related and important issue, which we will not consider, is the impact on innovation of the incentives of individuals who are not employed by the firm, but who are nevertheless important for the firm's innovative performance (e.g., academic scientists).

are different types of incentives relative to each other? The empirical work on the crowding-out effect and other work in social psychology and human resource management should provide some foundation for such inquiries.

One would also want to know about differences in the incentives across technologists engaged in different fields of science or engineering, or in different types of R&D activity such as basic and applied research, and development. For example, while the motivations of molecular biologists may conform to the norms of science, electrical engineers in the computer industry may be more motivated by the satisfaction of solving concrete problems and seeing their solutions implemented in practice. Once we have some idea of the incentives that matter, research on their distribution across technologists within firms should provide a foundation for studying their impact on firms' innovative performance.

From the firm's perspective, a fundamental issue is then how well individual incentives are aligned with the firm's objective, profit. The likelihood that individual incentives are not aligned with the interests of the firm raises the question of the degree to which the firm will be able to reconcile those incentives with its own. This, in turn, depends upon what mechanisms the firm can employ to directly manage its technologists' incentives to advance its interests. Some of the levers available to management, such as pay, praise, or recognition, can operate directly on individual incentives. The firm's ability to affect or address other goals is, however, limited and indirect. The firm may be able to provide conditions necessary for the satisfaction of some goals, such as providing the time or resources to permit their researchers to pursue scientific eminence, but they will not be able to satisfy them directly. The ability of firms to manage incentives also depends on the degree to which management is aware of the particular preferences of their personnel. Research by Heath (1999) and others suggests that this is a nontrivial issue. Heath found that managers tend to believe that their employees are driven much more by extrinsic incentives such as money than the employees themselves report.[16]

For firms to maximize their innovative performance, it is not simply a matter of aligning the incentives of their employee-technologists with

---

[16] In a follow-up study, DeVoe and Iyengar (2004) explored this effect in a cross-national study and found that North American managers perceived their employees as being more extrinsically than intrinsically motivated, whereas Asian managers perceived that their subordinates were equally motivated by intrinsic and extrinsic incentives. Latin American managers perceived employees as primarily intrinsically motivated. In each of the three regions, however, the employees reported themselves as being more motivated by intrinsic than extrinsic incentives, a finding that is consistent with the data we reported above.

those of the firm. In many industries, firms will need to reconcile and integrate different incentives across individuals within the R&D function as well as across different business functions such as R&D, manufacturing, and marketing. As we have seen in the data discussed above, individuals in different business functions may exhibit marked differences in their preferences and goals. For example, at Steve Jobs' computer animation firm, Pixar, programmers and software engineers had to work closely with those responsible for creating content. In some instances, the goals of creators of content, emphasizing novelty of stories and graphics, could run afoul of the interests of the programmers who are concerned with receiving material that is programmable and can be effectively engineered for the medium.

The use by firms of the various incentive-related instruments and policies entails costs and benefits, neither of which are well understood, especially in the R&D context. Expected gross benefits to the firm turn partly on the questions posed above regarding the effect of different types of benefit or reward upon individuals' innovative activities and performance. Apart from such a direct effect of firm policies on current employees, another important effect is on the recruitment of talent. A firm's policies toward prospective employees' varied motives will constitute a part of the employment negotiation, and will affect the firm's reputation with respect to future applicants, with implications for some future distribution of recruits' preferences for different types of reward. To the extent that those preferences are correlated with innovative ability and performance, such selection effects may impact the firm's payoff to its policies. It would be worth considering whether, for example, scientists whose motivations are largely intrinsic tend to be more capable researchers than others whose motivations are more pecuniary. Another illustration of the relationship between selection and ability is highlighted by agency theorists' observation that individuals who believe themselves to be strong performers will be attracted to firms where pay is tied to performance.

The costs of managing employee-technologists' incentives are not well understood. The cost of appealing to or managing technologists' various incentives may be direct and out-of-pocket, as in the cases of pay and bonuses. They may be indirect as well. For example, there are costs associated with the development of a system of controls, monitoring, and allocation of managerial attention. There are also the less tangible opportunity costs associated with allowing technologist-employees to pursue agendas and activities that do not contribute to the firms' profits. For example, Dupont's support of Carrothers' pursuit of his own research and writing agenda represented a cost to Dupont.[17]

The cost of a firm's management of technologist incentive also depend importantly on what these incentives are, whether the firm and the technologist goals are both pecuniary, or whether the technologist is more interested in nonpecuniary goals. If pecuniary, it is basically a question of the wage bargain and the sharing of rents due to innovation between the employer and employee. Ultimately, that is even true when the firm satisfies its technologists' nonpecuniary motives, but how much a firm's accommodation of a nonpecuniary motives may cost turns on the degree to which the pursuit of that nonpecuniary objective may conflict with, or perhaps augment, firm earnings.

Again consider Stern's (2004) finding that the salary of scientists employed in the life sciences is significantly lower if they are allowed to publish. Stern argues that because of researcher's "taste for Science," the firms are able to extract a wage discount, estimated to be about twenty-five percent, that researchers "pay" for being able to publish. Thus, the cost of appealing to the scientists' intrinsic and social incentives may be negative. Stern may have underestimated the long-run cost, however, because he did not consider the competitive advantage foregone owing to disclosure of information to the firm's competition. Indeed, a common point of conflict between academically oriented scientists and firms is that researchers want to publish while the firms would like to keep findings secret to protect their competitive advantage.[18] More generally, while the costs of satisfying some nonpecuniary goals may be straightforward, the consequences of appealing to others may be multifaceted, with offsetting cost and benefits for the firm. In any event, firms that want to innovate have to ensure that individuals can expect sufficient benefits of the sort they care about.

To estimate firms' costs of managing and responding to the incentives of their technologists, it would be helpful to identify their exogenous determinants. Presumably, much of those costs will depend upon those factors that condition individual technologists' bargaining power, a key determinant of what we called "individual-level appropriability", or the share of the firm's rents due to innovation that individual technologists

---

[17] As top management at DuPont changed, tensions ultimately arose between management and Carrothers' academic research orientation, with management feeling that Carrothers' research should become more commercially rewarding.

[18] Even that calculus is further complicated, however, by the degree to which the firm's productivity may benefit by allowing its scientists to interact freely with the broader scientific community, as highlighted by Henderson and Cockburn (1994).

can command. Such individual-level appropriability depends upon the demand and supply for any given technologist's ability. This market for technologists' skills will depend partly upon the mobility of technologists, which can be importantly affected by legal conventions governing the rights of employees to work for rivals. Gilson (1999), for example, argues that the unenforceability of noncompete covenants in California has increased the ability of technologists to switch firms there, which, as we suggest above, may have increased the ability of at least the most talented technologists to command higher pay. Another factor affecting individual-level appropriability is the portability of knowledge or disembodied innovation across firms (Becker, 1964; Topel, 1991). That portability should depend upon the strength of intellectual-property protection afforded to the firm (the stronger is such protection, the less portable is the innovation) as well as the degree to which the further development and commercialization of any given innovation requires collocation with other personnel or assets possessed by the originating firm (i.e., the degree to which the innovation is firm-specific or "sticky" (von Hippel, 1994)).

Whatever it costs to manage and satisfy employee-technologists' incentives, an important feature of such costs is that, *ceteris paribus*, the higher they are as a share of the private returns to innovation, the less incentive the firm has to conduct R&D to begin with. At the same time, a firm's performance will depend upon the quality and manufacturing costs of its products, which in turn is affected by its investment in R&D. Thus, there is a trade-off between the firm's innovative performance and its employees' share of rents due to innovation (i.e., individual level appropriability). The interesting question is what factors condition the terms of that trade-off.

As another step toward understanding the impact of individual incentives on the rate and direction of technical advance across firms, or for entire industries, we should examine differences in such incentives across firms within industries, distinguished by size, specialization, R&D capabilities, and so on. We would also want to explore the impact of cross-firm, intraindustry differences in the ways that firms manage their technologists' incentives.

A focus on the link between firm characteristics and the sorts of incentive that might characterize their employee-technologists brings us back to Schumpeter's contention that the small, entrepreneurial firm is the best vehicle for supporting the incentives that permit the achievement of superior innovative performance.

There is some evidence that individual-level incentives in entrepreneurial and small firms may differ systematically from those in large firms. Hamilton (2000), for example, appeals to the nonpecuniary returns to entrepreneurship as a possible reason why the self-employed less income than wage employees, controlling for ability and other factors. Little, however, of the empirical research comparing the incentives of individuals in larger incumbents versus smaller ventures or startups has been conducted in the context of R&D and innovation. It is conceivable that large firms create particular "R&D cultures" that are distinct from that of the larger organization; the individual incentives within these subunits might actually resemble more closely those of individuals in small organizations than is the case in other business functions such as manufacturing.[19] More empirical research is clearly needed to compare and contrast the incentives of research personnel in small versus large organizations, including the antecedents of such differences in incentive structures, the importance of self-selection, socialization, and managerial reward systems. Even if we find that different incentive structures apply to small, entrepreneurial firms and larger incumbents, we cannot infer that small, entrepreneurial firms' R&D efficiency is greater than that of larger firms because of any particular ability to capitalize on the individual-level incentives. Economists do not even know whether the R&D efficiency of small firms is greater than that of large firms (Cohen and Klepper, 1996), nor whether any such advantage might be attributable to the incentives of the technologists employed by each.

Although Schumpeter's prediction about the frustration of entrepreneurial incentives and the consequent demise of capitalism was wrong, it would not surprise us to learn that there indeed exist systematic differences in individual incentives between technologists who are founders or employees of new ventures versus employees in the R&D labs of larger incumbent firms. Even if nonpecuniary incentives differ between the entrepreneur and the employee-technologist, however, there may be ample opportunity for the latter to realize considerable nonpecuniary payoffs from his work, albeit perhaps of a different sort. He can feel proud about having come up with an exciting idea. He can enjoy the recognition of peers. And he can still feel that he has contributed to the advancement of the field and the betterment of society. Indeed, deeper study of the full range of individual incentives that motivate innovation may well be the key to understanding why individuals continue to innovate in larger firms where the R&D function is more rationalized and bureaucratized.

---

[19] With a focus on firm strategy, Tushman and O'Reilly (1996) develop a similar idea in their work on ambidextrous organizations.

Having largely embraced Jewkes, Sawers, and Stillerman's (1969) argument that innovation within an industry is likely to be best supported by a mix of firms distinguished by size, capabilities, specialization, and so on (Arora, Fosfuri, and Gambardella, 2002; Cohen, 1995; Scherer and Ross, 1990), scholarship on innovation has moved away from trying to find some ideal firm type best suited for innovation, distinguished by size and other characteristics. Similarly, one might entertain the possibility that the issue is not whether smaller, entrepreneurial firms versus large incumbents best support the incentives of those who pursue innovation. Rather, one might explore whether different types of firms foster different incentives and associated behaviors, and whether such differences at least partly underpin the distinctive and sometimes complementary innovative activities observed across firms within industries.

3.5   Implications

Apart from contributing to our understanding of the determinants of technical advance, what might a systematic understanding of the role of individual incentives in affecting innovation do for us? Does it help managers? Does it allow government to devise more effective policies for advancing technology?

3.5.1   Management

In a co-authored discussion of the early history of Silicon Valley, Gordon Moore, one of the founders of Fairchild Semiconductor and Intel, highlights the importance of the management of individual incentives, stating: "Aligning the goals and incentives of the firm with those of the talented individuals whose efforts build a successful firm takes on greater importance in highly technical, skill-intensive firms." (Moore and Davis, 2001). Effective "incentive management" for innovation, however, is far from straightforward and managers may face a number of challenges.

First, pecuniary incentives might be harder to administer in innovative than in noninnovative contexts. Agency-theory literature discusses the factors conditioning the cost and effectiveness of pecuniary incentive systems (e.g., Gibbons, 1998; Prendergast, 1999) and suggests that problems arise from imperfect output measurement, weak links between individual effort and performance, technological nonseparabilities and teamwork, as well as multidimensional and heterogeneous tasks.

It should be apparent that all these challenges apply to the kind of work with which we are concerned – that of innovation.

Second, to the extent that technologists have particularly strong preferences for heterogeneous sets of nonpecuniary rewards, and to the extent that pecuniary incentives cannot be administered effectively, management might have to focus on a large variety of intrinsic and social incentives. This could create distinct challenges; while it is relatively easy to hand a check to a high-performing individual, it will be more difficult to give someone a certain amount of "fun," "feeling of achievement," or "commitment to the organization." Owing to differences in the preferences of research and other personnel as well as the particular difficulties involved in providing nonpecuniary incentives, we also suspect that centralized corporate incentive systems may have to be complemented with decentralized informal and formal reward management focusing on nonpecuniary incentives. Moreover, to the extent that both pecuniary and nonpecuniary incentives are hard to manipulate once an individual has joined an organization, *ex ante* self-selection of individuals into innovative organizations may be more important relative to *ex-post* managerial interventions (cf. Sauermann, 2005a). Interestingly, research typically focuses on personnel selection and the management of incentives as distinct processes rather than related antecedents of the same outcome – individual motivation.

A third factor that makes individual-level incentives hard to manage is that it is not sufficient to manage pecuniary and nonpecuniary incentives independently if these types of incentives affect each other. While field studies showing such interactions are still rare (e.g., Frey and Jegen, 2001), there is a large body of experimental evidence that extrinsic rewards can undermine intrinsic and social motivation under certain conditions (Deci et al., 1999; Fehr and Falk, 2002). To the extent that such "motivation crowding out" occurs, management has to consider the potentially problematic "side effects" of rewards that otherwise seem to have strictly positive effects on motivation. For example, might strong pecuniary incentives for patenting or publishing negatively affect scientists' intrinsic and social motivation, and have negative net effects on the quality or quantity of their effort? Ideally, of course, firms would create conditions where different rewards do not undermine each other but are additive or even reinforcing. Recent work by Sauermann, Massey, and Larrick (2006), for example, suggests that providing individuals with explicit choices about their tasks may eliminate the otherwise negative effect of contingent rewards on intrinsic motivation. Furthermore, to the extent that different types of incentive and motivation are more conducive to certain innovative behaviors and processes (e.g., intrinsic motivation may be more conducive to creativity than extrinsic motivation), pecuniary, social, and intrinsic incentives cannot be considered perfect substitutes. As a result, firms may be able to gain from creating a "fit" between the desired innovative processes or outcomes and the incentives of their research personnel (Amabile, 1993; Sauermann, 2005b). Thus, the management of the mix of extrinsic, intrinsic and social incentives may be particularly important.

This discussion has outlined some of the key challenges associated with managing individual-level incentives in innovative contexts. It also suggests, however, that firms are likely to differ in their abilities to manage individual-level incentives for innovation, and such an ability may therefore represent a competitive advantage. In a similar vein, strategy research has recently started to examine to what extent human resource management capabilities – including reward systems – might provide a basis for competitive advantage (Colbert, 2004; Wright et al., 2001).[20]

3.5.2   *Policy*

Assumptions about individual incentives lie at the heart of a range of policies designed to support technological progress. Consider patent policy, for example. It presumes that the pecuniary gain, and specifically that which accompanies the right to exclude others from imitating or using your invention, motivates innovative effort. While this logic applies well to firms (depending on the industry), it may not apply so clearly to individuals. If, however, some other, nonpecuniary motive is sufficient to elicit the requisite effort – such as the joy of achievement or advancing knowledge, the satisfaction of curiosity, recognition, or in Taussig's terms, the "instinct of contrivance" – then patents may not be necessary. This argument has been applied with greatest force to academic research where, as Merton (1973) suggests, academics are motivated to do research by the nonpecuniary rewards of the pleasure that comes from advancing knowledge and the achievement of eminence. Yet, since the passage of the Bayh–Dole Act in 1980 and related legislation, academic institutions are now eligible for patents for their publicly funded research. The rationale for Bayh–Dole was not to provide a commercial incentive to academic institutions or individual

---

[20] While empirical tests involving a comprehensive set of pecuniary and nonpecuniary incentives are still lacking, there is considerable evidence that certain human resource management practices are systematically linked to higher firm performance (Huselid, 1995; Ichniowski et al., 1997; Lazear, 2000).

academics, but to firms to invest in the commercialization of academic research. Yet, academic institutions now typically provide a share of licensing revenues to the academics whose inventions yield commercially successful inventions. Moreover, as a consequence of Bayh–Dole, academics are now more likely to start their own firms. The prospect of either licensing income or the acquisition of equity stakes in firms built around their research raises the question of the impact of commercial rewards on the intrinsic incentives of academics. Amabile's work suggests that pecuniary rewards can diminish creativity that tends to benefit from intrinsic motivation. Dasgupta and David (1994) suggest that, stimulated by the prospect of commercial rewards, academics will move away from foundational to more applied, more readily commercialized research. The fundamental question is how mutable academics' intrinsic incentives are in the face of the prospect of pecuniary benefits.

Although, some have expressed concern that academics may now have too much of an incentive to allocate their attention to commercial applications of their research, others are concerned that faculty are still not sufficiently motivated to participate in technology-transfer efforts. Arguably, Bayh–Dole has had its greatest effect on technology transfer in the life sciences. One reason may be that many academics in the life sciences do not have to change what they would normally do in the course of their academic research to get their discoveries to the point where firms can use them; their academic incentives are not that inconsistent with developing commercially important technology. Academic research in the life science fields almost uniquely generates commercially valuable findings, partly owing to the closeness of the field to commercial application, and partly owing to the strength of patent protection that permits its sale in disembodied firm. This is quite different from most academic research in the sciences, or even in many fields of engineering that tend to be more distant from commercial application than, for example, molecular biology (Cohen, Nelson, and Walsh, 2000). And once a field is further away, it becomes all the more difficult or costly to align academic with commercial incentives, assuming that is desirable to begin with. The broader point is that innovative performance of many types does not depend on simply whether the "*organization*" has the requisite incentive to innovate – The "*individuals*" responsible for the innovative activity must be appropriately motivated as well.[21]

---

[21] This is consistent with Link and Siegel's (2005) finding that those universities that allocate a higher percentage of royalty payments to faculty members tend to be more efficient in technology-transfer activities.

The challenges of technology transfer apply to the industry side of the relationship as well, notwithstanding the intent of Bayh–Dole to provide firms with the incentive to invest in the commercialization of public research. Even if the interests of a firm to which technology is being transferred are clear, if the firm does not provide sufficient incentive to the technologists within the firm to devote the effort necessary to understand and adapt the technology in question, the technology transfer may not succeed. Reflecting, for example, on the technology transfer from a university–industry R&D center dedicated to semi-conductor research, Randazzese (1996) concluded on the basis of twenty-seven interviews that the structure of the transfer processes mattered little as compared with the provision of incentives to the responsible technologists within the firm.

In addition to affecting technology transfer between universities and firms, public policy can also affect the incentives of individual technologists within firms. We see this in the way the rights of employee-inventors to their inventions are treated in the United States, Germany, and Japan. In the United States, employees contractually cede owner-ship to any intellectual property that they may produce. In contrast, in Germany and Japan, employees cannot do that. The ownership of any patent is granted to the inventor by statute, even if it is an employee of a firm. Although patent policy in these countries also requires that the inventor-employee license the patent to their employers, they must be "adequately" compensated. This rule has recently become a point of controversy in both Japan and Germany, and the policies are currently under review in both nations. In Germany, the financial compensation provided to inventors is relatively modest, and is subject to an explicit schedule (Harhoff and Hoisl, 2004). In Japan, the compensation had been only nominal until a number of recent court decisions on high-profile suits. In January 2004, for example, Hitachi was ordered to pay a former employee about $1.5 million for three patented inventions cov-ering technology for reading data from optical disks. In the most pro-minent case of this sort, the Japanese firm Nichia was ordered to pay a former employee, Shuji Nakamura, over eight million dollars (compared with the $182 million initially awarded by a lower court) for his patented invention of the blue-light-emitting diode (LED).

Whether policy should directly influence the incentives of individual technologists is an interesting and potentially complex issue that would benefit from further research. For example, what is Japanese patent policy's effect on pecuniary rewards doing to the intrinsic motivation and creativity of firms' scientists? Or, given that typically many more individuals than the initial inventor must expend effort within a firm to bring an innovation to market, what happens to the performance of the team as a whole in Japan when only the inventor whose name appears on a patent receives a substantial financial reward? On the other hand, perhaps policy should begin to address the incentives of individuals even more aggressively. For example, returning to the issue of technology transfer from universities to industry, what might be the effect of providing both individual faculty and firms' employee-technologists a tax credit for whatever effort they dedicated to technology transfer?

3.6   Conclusion

In his *Capitalism, Society and Democracy*, Joseph Schumpeter predicted that large business organizations would be unable to provide the individuals with sufficient incentives to overcome the competitive challenge represented by socialist economies. Although this prediction has not been realized, Schumpeter's discussion highlighted the importance of the incentives of individuals within firms for economic outcomes. For those of us concerned with innovation, his argument highlights, first, that individual-level incentives may matter for innovative performance, in addition to the firm-level incentives that are typically at the center of economists' attention. Second, these individual-level incentives are of a diverse nature and include not only pecuniary incentives, but also intrinsic incentives (e.g., the joy to invent) and social incentives (e.g., peer recognition). We have outlined in this chapter how an understanding of individual-level incentives may not only explain why large capitalist organizations have maintained the ability to generate significant innovations, but also inform us about the drivers of technical change more generally. There are a multitude of examples and case studies suggesting an important role of individual-level incentives in motivating innovative behaviors by technologists employed in large firms as well as by entrepreneurs. However, systematic empirical studies of the importance and role of individual incentives for innovation are lacking. We have outlined a research agenda that may lead us to important insights about this role. In addition to more knowledge about the structure of incentives operating for individuals engaged in innovation, research is also needed on organizations' ability to manage and control these incentives. Moreover, we need to gain a better understanding of the concrete mechanisms by which incentives affect innovative outcomes, and how they interact with organizational processes and knowledge stocks – which have thus far been the focus of innovation researchers – in doing so. We suggest that, if individual incentives are indeed important, such research will not only advance our understanding of the drivers of innovation and technological change, but will also have important implications for public policy and management.

References

Alchian, A. A. and Demsetz, H. 1972. "Production, Information Costs, and Economic Organization," *American Economic Review* 62(5): 777–92

Amabile, T. 1996. *Creativity in Context*. Boulder, CO: Westview Press

Amabile, T. M. 1993. "Motivational Synergy: Toward New Conceptualizations of Intrinsic and Extrinsic Motivation in the Workplace," *Human Resource Management Review* 3: 185–201

Amabile, T. M., Hill, K. G., Hennessey, B. A., and Tighe, E. M. 1994. "The Work Preference Inventory – Assessing Intrinsic and Extrinsic Motivational Orientations," *Journal of Personality and Social Psychology* 66(5): 950–67

Arora, A., Fosfuri, A., and Gambardella, A. 2002. *Markets for Technology: The Economics of Innovation and Corporate Strategy*. Cambridge, MA: MIT Press

Baron, J. N. and Kreps, D. M. 1999. *Strategic Human Resources: Frameworks for General Managers*. New York: John Wiley

Bartlett, C. A. and Mohammed, A. 1995. "*3M: Profile of an Innovating Company*," Case Study 9–395–016, MA: Harvard Business School, Boston

Becker, G. S. 1964. *Human Capital; a Theoretical and Empirical Analysis, with Special Reference to Education*. New York: National Bureau of Economic Research; distributed by Columbia University Press

Benabou, R. and Tirole, J. 2003. "Intrinsic and Extrinsic Motivation," *Review of Economic Studies* 70(3): 489–520

Cockburn, I. M. and Henderson, R. M. 1998. "Absorptive Capacity, Coauthoring Behavior, and the Organization of Research in Drug Discovery," *Journal of Industrial Economics* 46(2): 157–82

Cohen, W. M. 1995. "Empirical Studies of Innovative Activity," in Stoneman (ed.), *Handbook of the Economics of Innovation and Technological Change*: 182–264 Cambridge, MA: Blackwell Publishers, Inc.

Cohen, W. M. and Klepper, S. 1996. "A Reprise of Size and R&D," *Economic Journal* 106(437): 925–51

Cohen, W. M., Nelson, R. R., and Walsh, J. P. 2000. "Protecting Their Intellectual Assets: Appropriability Conditions and Why U.S. Manufacturing Firms Patent (or Not)" NBER Working Paper 7552

Cohen, W. M., Nelson, R. R., and Walsh, J. P. 2002. "Links and Impacts: The Influence of Public Research on Industrial R&D," *Management Science* 48 (1): 1–23

Colbert, B. A. 2004. "The Complex Resource-Based View: Implications for Theory and Practice in Strategic Human Resource Management," *Academy of Management Review* 29(3): 341–58

Dasgupta, P. and David, P. A. 1994. "Toward a New Economics of Science," *Research Policy* 23(5): 487–521

Deci, E. L. and Ryan, R. M. 1985. *Intrinsic Motivation and Self-Determination in Human Behavior.* New York: Plenum

Deci, E. L., Koestner, R., and Ryan, R. M. 1999. "A Meta-Analytic Review of Experiments Examining the Effects of Extrinsic Rewards on Intrinsic Motivation," *Psychological Bulletin* 125(6): 627–68

DeVoe, S. E. and Iyengar, S. S. 2004. "Managers' Theories of Subordinates: A Cross-Cultural Examinations of Manager Perceptions of Motivation and Appraisal of Performance," *Organizational Behavior and Human Decision Processes* 93(1): 47–61

Fallick, B., Fleischmann, C. A., and Rebitzer, J. B. 2005. "Job Hopping in Silicon Valley: Some Evidence Concerning the Micro-Foundations of a High Technology Cluster" NBER Working Paper 11710

Fehr, E. and Falk, A. 2002. "Psychological Foundations of Incentives," *European Economic Review* 46(4–5): 687–724

Frey, B. S. and Jegen, R. 2001. "Motivation Crowding Theory," *Journal of Economic Surveys* 15(5): 589–611

Gagne, M. and Deci, E. L. 2005. "Self-Determination Theory and Work Motivation," *Journal of Organizational Behavior* 26(4): 331–62

Gibbons, R. 1998. "Incentives in Organizations," *Journal of Economic Perspectives* 12(4): 115–32

Gilson, R. J. 1999. "The Legal Infrastructure of High Technology Industrial Districts: Silicon Valley, Route 128, and Covenants Not to Compete," Paper presented at the University of Chicago Law School, Law and Economics Workshop

Gottschalg, O. and Zollo, M. 2004. "Towards a Motivation-Based Theory of the Firm" INSEAD Working Paper

Gottschalg, O. and Zollo, M. forthcoming. "Interest Alignment Rents and Competitive Advantage", forthcoming in Academy of Management Review

Griliches, Z. 1957. "Hybrid Corn: An Exploration in the Economics of Technological Change," *Econometrica* 25(4): 501–22

Hackman, J. R. and Oldham, G. R. 1976. "Motivation through the Design of Work – Test of a Theory," *Organizational Behavior and Human Performance* 16(2): 250–79

Hamilton, B. H. 2000. "Does Entrepreneurship Pay? An Empirical Analysis of the Returns to Self-Employment," *Journal of Political Economy* 108(3): 604–31

Harhoff, D. and Hoisl, K. 2004. "Institutionalized Incentives for Ingenuity – Patent Value and the German Employees" Invention Act," *EPIP Conference*, Pisa, Italy

Heath, C. 1999. "On the Social Psychology of Agency Relationships: Lay Theories of Motivation Overemphasize Extrinsic Incentives," *Organizational Behavior and Human Decision Processes* 78(1): 25–62

Henderson, R. and Cockburn, I. 1994. "Measuring Competence – Exploring Firm Effects in Pharmaceutical Research," *Strategic Management Journal* 15: 63–84

Hennessey, B. A. and Amabile, T. M. 1998. "Reward, Intrinsic Motivation, and Creativity," *American Psychologist* 53(6): 674–75

Holbrook, D., Cohen, W. M., Hounshell, D. A., and Klepper, S. 2000. "The Nature, Sources, and Consequences of Firm Differences in the Early History of the Semiconductor Industry," *Strategic Management Journal* 21 (10–11): 1017–41

Hounshell, D. A. and Smith, J. K. 1988. *Science and Corporate Strategy: Du Pont R&D, 1902–1980.* Cambridge and New York: Cambridge University Press

Huselid, M. A. 1995. "The Impact of Human-Resource Management-Practices on Turnover, Productivity, and Corporate Financial Performance," *Academy of Management Journal* 38(3): 635–72

Ichniowski, C., Shaw, K., and Prennushi, G. 1997. "The Effects of Human Resource Management Practices on Productivity: A Study of Steel Finishing Lines," *American Economic Review* 87(3): 291–313

Jenkins, G. D., Mitra, A., Gupta, N., and Shaw, J. D. 1998. "Are Financial Incentives Related to Performance? A Meta-Analytic Review of Empirical Research," *Journal of Applied Psychology* 83(5): 777–87

Jewkes, J., Sawers, D., and Stillerman, R. 1969. *The Sources of Invention,* 2nd edn, New York: W. W. Norton

Kaplan, S. and Henderson, R. 2005. "Organizational Rigidity, Incentives and Technological Change: Insights from Organizational Economics?," Working Paper

Katz, R. 1993. "How a Band of Technical Renegades Designed the Alpha-Chip," *Research–Technology Management* 36(6): 13–20

Koning, J. W., Jr. 1993. "Three Other R's: Recognition, Reward and Resentment," *Research Technology Management* 36(4): 19–29

Kreps, D. M. 1997. "Intrinsic Motivation and Extrinsic Incentives," *American Economic Review* 87(2): 359–64

Lakhani, K. R. and von Hippel, E. 2003. "How Open Source Software Works: 'Free' User-to-User Assistance," *Research Policy* 32(6): 923–43

Lazear, E. P. 2000. "Performance Pay and Productivity," *American Economic Review* 90(5): 1346–61

Lepper, M. R. and Greene, D. (Eds.) 1978. *The Hidden Cost of Rewards: New Perspectives on the Psychology of Human Motivation.* Hillsdale, NY: Erlbaum

Lerner, J. and Tirole, J. 2002. "Some Simple Economics of Open Source," *Journal of Industrial Economics* 50(2): 197–234

Link, A. and Siegel, D. 2005. "Generating science-based growth: an econometric analysis of the impact of organizational incentives on university-industry technology transfer," *European Journal of Finance* 11(3): 169–81

Manners, G., Steger, J. A., and Zimmerer, T. W. 1997. "Motivating Your R&D Staff," *Research Technology Management* 40(6): 29

Merton, R. K. 1973. *The Sociology of Science: Theoretical and Empirical Investigations.* Chicago: University of Chicago Press

Moore, G. and Davis, K. 2001. "Learning the Silicon Valley Way" SIEPR Discussion Paper No. 00–45, Stanford University

Mower, J. C. and Wilemon, D. 1989. "Rewarding Technical Teamwork," *Research-Technology Management* 32(5): 24–29

Mumford, M. D. 2000. "Managing Creative People: Strategies and Tactics for Innovation," *Human Resource Management Review* 10(3): 313–51

Nelson, R. R. 1959. "The Simple Economics of Basic Scientific Research," *Journal of Political Economy* 67(3): 297–306

Osterloh, M. and Frey, B. S. 2000. "Motivation, Knowledge Transfer, and Organizational Forms," *Organization Science* 11(5): 538–50

Ouchi, W. G. 1979. "A Conceptual Framework for the Design of Organizational Control Mechanisms," *Management Science* 25(9): 833–48

Prendergast, C. 1999. "The Provision of Incentives in Firms," *Journal of Economic Literature* 37(1): 7–63

Randazzese, L. P. 1996. "Exploring University-Industry Technology Transfer of Cad Technology," *IEEE Transactions on Engineering Management* 43(4): 393

Rosenberg, N. 1994. *Exploring the Black Box: Technology, Economics, and History.* Cambridge; New York: Cambridge University Press

Rossman, J. 1964. *Industrial Creativity; the Psychology of the Inventor,* 3rd edn, New Hyde Park, NY: University Books

Rynes, S. L., Gerhart, B., and Minette, K. A. 2004. "The Importance of Pay in Employee Motivation: Discrepancies between What People Say and What They Do," *Human Resource Management* 43(4): 381–94

Sauermann, H. 2004. "Competitive Advantage through People?," A Fuqua School of Business Working Paper

Sauermann, H. 2005a. "Vocational Choice: A Decision Making Perspective," *Journal of Vocational Behavior* 66: 273–303

Sauermann, H. 2005b. "The Role of Incentives in Organizational Capabilities", *Conference on Microfoundations of Organizational Capabilities.* Copenhagen Business School

Sauermann, H. and Cohen, W. M. 2006. "What Makes Them Tick? – Employee Motives and Industrial Innovation," Fuqua School of Business Working Paper

Sauermann, H., Massey, C., and Larrick, R. 2006. "Crowding-out of Intrinsic Motivation by Extrinsic Rewards – a Matter of Choice?," Fuqua School of Business Working Paper

Saxenian, A. 1994. *Regional Advantage: Culture and Competition in Silicon Valley and Route 128.* Cambridge, MA.: Harvard University Press

Scherer, F. M. and Ross, D. 1990. *Industrial Market Structure and Economic Performance,* 3rd edn, Boston, MA: Houghton Mifflin

Schmookler, J. 1962. "Economic Sources of Inventive Activity," *Journal of Economic History* 22: 1–20

Schrader, S. 1991. "Informal Technology-Transfer between Firms– Cooperation through Information Trading," *Research Policy* 20(2): 153–70

Schumpeter, J. A. 1934. *The Theory of Economic Development; an Inquiry into Profits, Capital, Credit, Interest, and the Business Cycle.* Cambridge, MA.: Harvard University Press

Schumpeter, J. A. 1962. *Capitalism, Socialism and Democracy,* 3rd edn, Cambridge, MA.: Harvard University Press

Stajkovic, A. D. and Luthans, F. 2001. "Differential Effects of Incentive Motivators on Work Performance," *Academy of Management Journal* 44(3): 580–90

Stern, S. 2004. "Do Scientists Pay to Be Scientists?," *Management Science* 50 (6): 835–53

Taussig, F. W. 1915. *Inventors and Money Makers.* New York: Macmillan

Topel, R. 1991. "Specific Capital, Mobility, and Wages – Wages Rise with Job Seniority," *Journal of Political Economy* 99(1): 145–76

Tushman, M. L. and OReilly, C. A. 1996. "Ambidextrous Organizations: Managing Evolutionary and Revolutionary Change," *California Management Review* 38(4): 8

Veblen, T. 1914. *The Instinct of Workmanship, and the State of Industrial Arts.* New York: The Macmillan Company

Von Hippel, E. 1987. "Cooperation between Rivals: Informal Know-How Trading," *Research Policy* 16(6): 291

Von Hippel, E. 1994. "Sticky Information and the Locus of Problem Solving: Implications for Innovation," *Management Science* 40(4): 429–39

Wood, D. A. and LeBold, W. K. 1970. "The Multivariate Nature of Professional Job Satisfaction", *Personnel Psychology* 23: 173–89

Wright, P. M., Dunford, B. B., and Snell, S. A. 2001. "Human Resources and the Resource Based View of the Firm," *Journal of Management* 27(6): 701–21

4   Creative destruction in the PC industry

Timothy Bresnahan

4.1   Introduction: because it is there! Why?

The personal computer (PC) industry offers a marvelous opportunity to study creative destruction. Over its first twenty years, the industry experienced a number of Schumpeterian waves of creative destruction. Each wave involved many of the distinct markets in the industry. Waves struck established dominant firms in hardware, in software, and networking; in general-purpose technologies; and in applications. While not numerous enough for systematic statistical analysis, the many instances of waves in many markets present an opportunity to think analytically about the causes and consequences of creative destruction. The waves of PC industry creative destruction stopped ten years ago. Even though occasions for waves continued, creative destruction of established firms' positions in the most strategic PC markets ceased.

The PC industry offers us the chance to see how creative destruction occurs. It also lets us to distinguish analytically between circumstances permitting creative destruction and other circumstances blocking it.

4.1.1   Technology and demand

Any analytical enquiry into creative destruction must answer a series of "Why?" queries, and this is no exception. Schumpeter observed that competition from new commodities, new technologies, new sources of supply, and new types of organization is particularly important for long-run growth.[1] That normative observation addresses the social value of creative destruction.

I would like to thank Shane Greenstein, Rebecca Henderson, Franco Malerba, and Manuel Trajtenberg for many helpful comments. My collaboration with Greenstein and Henderson has been particularly useful in creating the analytical frame used in this industry study.

[1] Schumpeter (1942), pp. 82–85 (page cite to 1975 Harper edition.) In addition to "waves" Schumpeter called such times "revolutions." Computer-industry leaders

Sometimes we interpret the normative observation as also providing a positive theory of creative destruction. Why are there Schumpeterian waves? Because they are crucial to growth! Yet that is seriously incomplete. It does not answer what changes occur in technology or demand to make a wave of creative destruction part of the socially desirable innovation path. Even more important, the normative observation does not answer critical positive questions. Why do waves occur at particular times in particular markets? What moves an industry from a regime of repeated waves of creative destruction to a regime of persistent dominant firms?

To address these questions in the PC industry entails the key aspects of technology and demand. At the heart of PC technical progress is Moore's law, a quantitative engineering prediction about the rate of improvement in microelectronic components. Moore's law has been a driver of change since the beginning of the PC industry. Many PC industry participants correctly see it as driving opportunities for ongoing improvement in things like software (not itself subject to Moore's law). There must, however, be more to the story. Moore's law suggests a continuous stream of innovation, not a series of waves of creative destruction. And careful measurements suggest that Moore's law increased in speed about a decade ago, just as the waves of creative destruction ceased. Market analysis, not technical determinism, is needed to explain the early waves and the later cessation. Other consideration of demand and technology in the PC industry raise more questions about creative destruction. Demanders and inventors in the PC industry make sunk investments that add costs to radical change. Demanders and inventors prefer new products or technologies with "backward compatibility," that is, the ones that can be used without abandoning already sunk investments. A consumer who bought a new word processor, for example, would prefer it to read old files, run on existing computers, and accept old commands. Similarly, an applications developer would prefer that new computers run existing programs. Backward compatibility is a conservative force that leads quickly to the question, why did the PC industry have so many waves of creative destruction?

More generally, any enquiry into creative destruction in the PC industry should understand why and how technology and demand change over time to call forth waves of creative destruction.

---

sometimes say "strategic inflection point" or "paradigm shift," wanting to suggest something revolutionary and technical.

4.1.2   Organizations

Schumpeterian enquiry links waves of creative destruction to organizational capabilities and incentives.[2] An essential feature of a wave is that new firms create and existing dominant firms are destroyed. This observation, too, leads to a chorus of "why?" Why did existing established firms not invent the new technology? Why, instead, must new firms be the innovators who set off a wave of creative destruction? The answer to this query could arise because of limitations on the abilities or knowledge of existing firms. They may not see opportunities for advance that are, instead, seen by entrants. Satisfactory discussion of this answer by scholars calls for first learning precisely what it was that existing firms did not see, and second for adopting an *ex ante* invention perspective in order to understand why they did not see it. The answer to this query could also arise in incentives. Established firms may have seen the technological opportunity but not had an incentive to take it up, while entrants' incentives are the reverse. Satisfactory discussion of this answer calls for careful statement of why and how the incentives varied at the decision moment(s).

One simple organizational theory posits that entrepreneurs are innovative, while established dominant firms are less so. Another theory posits (exogenous?) technical and market eras; firms that are strong in one era are weak after the changes that usher in the new era. Of course, it is logically possible that the market is selecting the most suitable firm both before and after each wave of creative destruction. Then the timing of arrival of new and better firms explains the timing of waves. But that is not the only logical explanation. Another is that entry barriers keep out more suitable firms before a wave; creative destruction arises when entry barriers fall.

These organizational questions have to be answered with care and precision in the PC industry, for there are many different phenomena. In Figure 4.1, I sketch the history of dominant firms in a number of important PC markets.

The figure makes it clear that we cannot use only the simple theory that entrepreneurial firms see new opportunities while established firms see only existing ones. Consider the history of the spreadsheet and word-processor markets. Each has had three dominant firms over three distinct eras. All of the dominant firms shown in the figure are entrepreneurial firms that think of themselves as forward-looking and

---

[2] See Henderson (1993), Henderson and Kim (1990), and Christensen (1997). Incentives-based theories related to creative destruction are reviewed in Reinganum (1989).

Market / Era	1970s	1980s		1990s	Today
Word processor	WordStar	WordPerfect		MS Word	MS Word
Spreadsheet	VisiCalc	Lotus 1-2-3		MS Excel	MS Excel
Programming tools	MS	MS + Borland		MS	MS
Browser					MS IE
Operating system	CP/M	IBM PC-DOS / MS-DOS	MS-DOS	MS Windows	MS Windows
Box	Apple + Many	IBM	IBM + Clones	Many	Many
Microprocessor	Intel + Zilog	Intel + AMD	Intel + AMD	Intel + AMD	Intel + AMD

Figure 4.1 Leading products and firms in widely used PC markets

innovative. How should we explain the role of the "middle" firms, Lotus (spreadsheets) and WordPerfect, which were successes in creative destruction just a few years before they were swept away by it?

Another problem can be identified by looking at the PC market row in the figure. Most of the firms successful in the early PC industry or, indeed, successful in it today, are young entrepreneurial firms that think of themselves as more forward-looking than the "dinosaurs" of the traditional computer industry, like IBM. In light of that, how should we understand the core event of the industry's first wave of creative destruction, the entry of IBM with the IBM PC in 1981?

I am not singling out explanations of organizational capabilities here. Similar problems apply to explanations of incentives. The Office suite of personal productivity programs is the most valuable monopoly in history.[3] That casts real doubt on a simple incentives theory of the most recent wave of creative destruction. Why did Microsoft have an incentive to introduce Word and Excel, the components of Office, while Lotus and WordPerfect, makers of the predecessor applications, did not? To assume they did seems an implausible theory. My point here is not to discard either organizational theories or incentives ones, and certainly not both. My point is that any answer to "Why did established firms miss the waves?" needs to be stated with precision.

4.1.3   Threats and policies

Schumpeter observed that the threat of creative destruction can give powerful incentives to incumbents. Instead of waiting to be destroyed, an incumbent should act as soon as the threat appears; in that manner, the incumbent is disciplined by the threat of creative destruction. This leads to the logical possibility that actually completed creative destruction is not, strictly speaking, necessary. Perhaps entrants can play their role as creators and changed incumbent incentives can rid society of wasteful destruction. This argument seems to have at least two shortcomings. One is, why entrants have the incentive to take costly actions to attempt creative destruction? If they will not succeed, but instead merely serve as an example and a threat, how much resources should go into creation? Similarly, if an entrant has created valuable new technology, why go into competition with well-positioned incumbents instead of selling out to them? Many of our colleagues

---

[3] Obviously, this is undiscounted. If the Bourbon monarchy had cashed out in, say, 1760, and invested the money at reasonable rates of interest, it would today be worth more than Office.

think that such incentives are very limited, and that the supply of entrant or outsider technologies is largely by volunteers, such as entrepreneurs in garages, scientists, or the excessively optimistic. A modern example in the PC industry might be the open-source movement. The same logic would also provide an explanation of why creative destruction is rare – if not, why it was so common for so long in the PC industry.[4] A second problem with the "mere threat" approach arises with incumbent incentives or with organizational heterogeneity. Clearly the threat of creative destruction is an incentive for incumbents. If, however, incumbents can always respond to threats and evade destruction, is that incentive weakened? And if incumbents and entrants are fundamentally different organizations, is the mere threat of creative destruction sufficient? Will there sometimes not be advantages of replacing one organization with another, better suited to new technological or market circumstances?

Schumpeter also argued that large established firms can be important engines of growth, perhaps more important than creative destruction by outsiders.[5] This argument would interact positively with the "mere threats" theory. We might have a market in which large established firms, with all their resources and market connection, successfully innovate themselves and also successfully respond to the threat of creative destruction.

The frequency of creative destruction is sometimes explained by analysis of government support of innovation, for example, through patent policy. An entrant might have a patent on a superior technology, for example, spurring a wave. Yet government protections for innovation have been unimportant in the PC industry, both at times when there has been a great deal of creative destruction and at times when there has been less. Like incumbents, entrants have had little protection from patents, copyrights, or the like. Firms have relied far more on trade secrets.

Another Schumpeterian policy debate swirled in the PC industry recently. In its landmark antitrust case, the U.S. government accused Microsoft of blocking creative destruction in order to avoid competition. Critics of the case, such as Richard Schmalensee, suggested that the government simply did not understand "Antitrust Issues in Schumpeterian

---

[4] Gans, Hsu, and Stern (2002) provide an argument, on the basis of the opportunities for contracting between entrants and incumbents, why creative destruction is rare, and considerable evidence that it is in general rare. I shall return to the reasons the PC industry was exceptional below.

[5] Schumpeter (1911).

Industries."[6] Liebowitz and Margolis (1999) offer a radically different view of the sources of Schumpeterian competition in the PC industry than you will find here. However you see it, creative destruction is not the stuff of empty academic debate, but of immediate importance to growth and to policy.

The phenomenon of creative destruction is there. In the PC industry, the key positive questions are "Why was the pace of creative destruction so fast?" and "Why did creative destruction cease?" I begin to answer them by looking at the first wave of creative destruction in the industry.

4.2   A wave, preceded and proceeding

The PC industry was founded in 1975. A wave of creative destruction began in 1981 with the introduction of the IBM PC. While the consequences of the creative destruction took longer to play out in some PC industry markets than in others, they were far-reaching. The dominant sellers of word-processing programs, spreadsheets, operating systems, and computers, among others, were all replaced within a few years. All of these were entrepreneurial firms, and most had won difficult competitive races to gain their dominant position. In this section, I first discuss the conditions that preceded the wave, then analyze the creative destruction itself.

4.2.1   Preceding a wave

The story of the PC industry shortly after its founding is familiar, and we need only to retell it from an economic perspective to see the essential features of the time before a wave of creative destruction.[7]

The founding of the PC industry was based on the entrepreneur innovation of firms like *Intel* (invented the microprocessor) *MITS* (founded the industry with a PC kit) and *Microsoft* (wrote a programming language, BASIC, for the first PCs). The initial industry sold primarily to hobbyists, that is, very technically fluent users.

Entrepreneurial *Apple* introduced fully assembled personal computers and sold them primarily to home and hobbyist users. Apple encouraged outsiders, developers of applications software, to write programs for its computer. Commercially oriented Apple focused particularly on

---

[6] That is the title of Schmalense (2000).

[7] Throughout this chapter I draw heavily on histories of the computer industry that treat the PC industry in detail, such as Campbell-Kelley and Aspray (1996) and Chandler (2001). I also draw heavily on works of very careful journalism, such as Freiberger and Swaine (2000) and Manes and Andrews (1993).

encouraging the supply of computer games and home applications. Supply was vertically disintegrated, and the well-documented and open interface between the Apple and applications meant that anyone could write an application and attempt to gain widespread distribution for it.[8]

Similarly, the entrepreneurial sellers of the *CP/M* operating system (OS) encouraged a wide variety of complementary inventions to go with their product.[9] Their OS ran on many different brands of computer. They encouraged innovation by computer makers and by developers of applications. Since much of the demand was from hobbyists or from hobbyist-entrepreneurs seeking to make small business computers, CP/M's makers encouraged the development of hardware and software that would be useful to such demanders. Like Apple, one of their most important tools of encouragement was an open and well-documented interface between CP/M and applications programs.

One factor that makes the PC industry easier to study is the extensive public discussion of interface standards, extensive technical discussion that crosses firm boundaries. Information that might be inside the firm in another industry is publicly discussed in this one. One reason is network effects, which give PC firms selling GPTs (such as the Apple II or CP/M) an incentive to collaborate with many other firms.[10]

While there were other kinds of personal computer in that era, Apples and CP/M machines were dominant platforms reinforced by network effects. The idea of buying the same standard as other users was recommended in the trade press. For application developers, the idea of writing applications for the same standard as other developers was recommended as well. Noncompatible kinds of PCs declined in importance at the expense of the two leading platforms. While there was competition between these two platforms, network effects inertia was setting in around them. Users would be well served to choose one of these two platforms because many developers were making applications for them, and developers would be well served to choose them because they had

---

[8] While Apple sold both hardware and software, other firms sold widely distributed products such as spreadsheets and word processors.

[9] Thus the supply of complete CP/M systems was even more vertically disintegrated than that of Apples. In what follows, I shall often use vertically disintegrated supply as a shorthand for the vertically disintegrated supply of widely used components.

[10] Network effects is a large area of economics and very important to the PC industry. The most recent and complete survey is in Farrell and Klemperer (2001) in volume three of the *Handbook of Industrial Organization*. See also the online bibliography at Nicholas Economides' web site. A very accessible summary is in Shapiro and Varian's *Information Rules* (1998). The *Journal of Economic Perspectives* 1994 special issue had survey essays from three perspectives, notably Besen and Farrell (1994), Katz and Shapiro (1994), and Liebowitz and Margolis (1994). The idea goes back to Veblen's (1899) "bandwagon" theory of demand.

the most users. Demand for PCs at the beginning was quite different from later on. Early demanders were hobbyists, players of games, home users, and to some extent those in small business. The PC was a general-purpose technology. Its early entrepreneurial sellers had purposes in mind, including hobby use, games, multimedia authoring, personal liberation, and home use. For example, many technical features of the Apple II were designed with games in mind. The PC was also invented with the idea that unforeseen applications would be invented. It was designed to permit a wide range of applications not foreseen at the time. The openness and expandability of the Apple II come immediately to mind as an example.

4.2.2 Office applications

Two important innovations, the spreadsheet and the word-processing program, opened up a new and even larger market for PCs. We now think of this market, white collar workers, as the main market for PCs, but that was not obvious in the 1970s. Neither of these innovations came from Apple or the sellers of CP/M, nor were there close con-tractual links between the sellers of standard platforms and the inventors of the newest applications. Instead, the spreadsheet *VisiCalc* was invented by a student and commercialized by him and his partners. Dominant word-processing program *WordStar* also had entrepreneurial origins; the entrepreneur had been marketing director at a firm selling personal computers.

The impact of the invention of these office applications was to sub-stantially raise the demand for PCs. In particular, VisiCalc led to the sales of a large number of Apple II computers to white collar number crunchers, and WordStar led to the sales of many CP/M machines for the use of white collar typists.

Many people, both scholars and industry participants, have noted one of the general lessons about innovation and organization here. The open and modular design of PCs and the vertical disintegration of the PC industry were important to this innovation.

Innovation in the early PC industry arose from a wide number of different firms. Different complements were invented in different firms (and sometimes emerged from a competitive struggle among several firms).[11] Openness, modularity, and vertical disintegration worked to facilitate a positive feedback system. The invention and improvement of

---

[11] See Langlois (2002) and Langlois and Robertson (1992).

applications raised the demand for PCs, and the invention and improvement of PCs raised the demand for applications.

If we differentiate among classes of applications, distinguishing office applications from games, we can see a second, less familiar, lesson from this history. The PC was a general-purpose technology invented without any foresight about its most valuable applications like word-processing and spreadsheets. Indeed, many early inventors in the PC industry detested the idea of the PC serving as a tool for white collar workers doing ordinary bureaucratic work in corporations. Yet that was the most valuable use of the PC during much of the growth of the industry over the 1980s and early 1990s. The invention of the PC itself was recombined by applications inventors to make what we now know of as a familiar technology, the white collar office PC. It went beyond, and in many cases against, the goals of earlier inventors.

This is an important general point. The most economically important use of a general-purpose technology need not be determined by the inventors of the GPT, but rather by the inventors of complements, applications. Recombination of GPTs gives them new markets as well as new technical life.[12]

Recombination arose from the *ex post* flexibility and permissiveness of markets. Rather than following a path planned out by any firm or any group of firms, the PC industry followed a circuitous route to its most valuable growth market. Many inventors changed the direction of the industry by changing its relationship to markets.

This is illustrated in, Figure 4.2, which shows three flows of causation. Invention of general-purpose technologies enables the invention of useful applications. Applications invention and GPT invention have positive feedback. Applications can recombine GPTs, bringing them into new markets.

4.2.3   *A wave enabled*

The early sponsors of the Apple II and CP/M PC platforms benefited from the increased demand caused by applications innovation. They also suffered competitively after applications innovation changed the industry in a way that permitted new entry and competition against them.

---

[12] The idea that recombination is an important part of technical progress is in Schumpeter (1911). See also Weitzman (1998) for an economic theory of recombination and Fleming (2001) for a managerial view. Fleming also has cites to the historical literature. Varian (2003) makes the argument that recombination is important in computing.

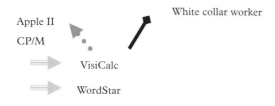

Apple II

CP/M

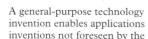

VisiCalc

WordStar

Enables

A general-purpose technology invention enables applications inventions not foreseen by the GPT inventor.

Distributed information.

Expands demand

A new application expands the demand for a GPT by making it useful to a new group of customers.

Sometime, a large new market.

Figure 4.2   Positive feedback

The successful commercialization of spreadsheets and word processors revealed a growth market; white collar workers in corporations. At the same time, rapid technical progress in microprocessors and other underlying technologies made improvements to PC possible. A new market plus new technological opportunity laid the foundation for creative destruction.

We sometimes think of the instigator of creative destruction as an entrepreneur or other new entity. In the first big wave of PC creative destruction, it was an outsider of another kind. IBM, long the dominant seller of corporate computing, saw the opportunity and entered. PCs would now be bought by IBM's traditional customers, corporations. A computer with "IBM" on it would sell well.

We sometimes think of the firms threatened by a wave of creative destruction as old and slow. The firms selling Apples and CP/M were, on average, five years old when the first wave of creative destruction arrived. It is wrong to think of them as uninteresting in or uninformed about the growth possibility afforded by new corporate customers, for they were pursuing it. Instead, the rapid change associated with an unanticipated shift in demand for their product, plus the entry of as formidable a market competitor as IBM, left them in a very difficult spot.

Early PCs were designed with hobbyists or home users in mind, not white collar workers. The first Apple II, for example, had forty columns of text, poor for word processing. More generally, the early PC design traded off ease-of-use and power in a different way than most corporate users would prefer. The gap between what PCs could do and what the newest demand segment wanted gave the young incumbent dominant firms a difficult technical and business problem.

Apple moved quickly to make a PC more suitable for white collar use, and introduced the Apple III before IBM entered. This was a sound strategy, but the Apple III was an implementation disaster, failing frequently in the field. The sellers of CP/M decided to innovate their way out of trouble, designing a radical improvement with a number of new features they thought would be valuable in corporations. This meant, however, that they were slow to market, leading at first to complaints from complementors and then to market disaster.

4.2.4   A wave

IBM entered with a vertically disintegrated structure and an incremental computer design. It was backward compatible with CP/M machines, meaning that applications and hardware add-ons for CP/M machines could easily be adapted for use with the IBM PC. Better microprocessors would have permitted a technical leap forward in the IBM PC. But IBM chose not to take that path, and the keys to IBM's creative destruction were IBM's reputation with corporate customers, quick "porting" of existing applications to the IBM PC and, later, new applications that ran only on the IBM PC.

Why was IBM able to enter the personal computer business so quickly? An important element was the vertically disintegrated structure of the existing PC industry. IBM invented a new computer, but contracted with existing PC industry firms for many of the key components. Indeed, the leading sellers of microprocessors, printers, disk drives, programming tools, and spreadsheet software worked with IBM. While the leading operating system vendor did not work with IBM, IBM was able to enter with a clone of their product, CP/M. (Formal intellectual property protection rights were weak, and the threatened lawsuit by the inventors of CP/M was ineffectual.) The leading word-processor vendor at first also refused to work with IBM, but switched after the IBM PC began to succeed.

There was another advantage to entrant IBM from the vertically disintegrated and open PC industry, which was backward compatibility. Users and developers could migrate to the IBM PC without losing their sunk investments. While the sellers of CP/M would have liked to prevent that, they were badly posed to do so. They could not prevent key

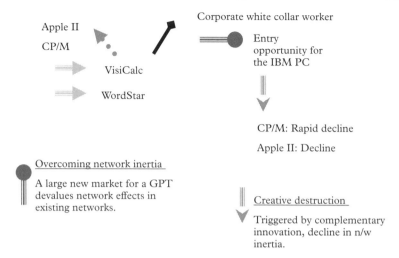

Figure 4.3   Triggers of creative destruction

complementors from working with IBM, since those key complementors were in other firms and linked, if at all, only by weak contracts.

The first great wave of creative destruction in the PC industry was unleashed. See Figure 4.3, which shows the devaluation of inertial barriers to entry caused by advances in a complement and the resulting possibilities for entry and creative destruction.

4.2.5   *Mixed incentives for openness and vertical disintegration*

The early PC industry's open and modular design and its vertical disintegration had dual consequences. They led to the positive feedback cycle of invention and improvement in the PC itself and in applications. They also were important in enabling creative destruction. This duality is central to understanding performance in the PC industry. There is a large difference in the private and social value of openness and vertical disintegration. Both parts of the duality, the positive feedback cycle and the creative destruction, benefited PC users. On the other hand, the incumbent firms selling CP/M and Apple IIs gained from the positive feedback cycle and lost from the creative destruction. (Entrants, on the other hand, gained from openness and vertical disintegration. Yet those structures were picked by the incumbents, not the entrants.) This gap between private and social incentives is essential to understanding the industry's history. It follows directly from the competitive nature of creative destruction, which looks better to society than to the existing firms whose rents are destroyed.

But let us be clear that the lesson here for Schumpeterian Economics is far more general than the narrow and specific point about "open architecture," which seems like a technical concept from computing. Instead, the point is about the role of a permissive, forward-looking system of innovation in which inventions can come from multiple sources. In short, it is about market innovation. Market innovation leads to increased social value. Uncontrolled market innovation means that existing firms sometimes get a large slice of that larger pie, and sometimes that they are the victims of creative destruction. Market innovation, like the competitive market system generally, is fabulous for consumers and a mixed bag for producers.

4.2.6   *Creative destruction in waves*

The sellers of WordStar and VisiCalc were instrumental in laying the groundwork for the wave of creative destruction, which quickly came to destroy their positions as well. The IBM PC led to a rapid increase in PC sales to corporations. That raised, substantially, the demand for spreadsheets and word processors. That was in the interest of spreadsheet and word-processor sellers generally. Yet by lowering entry barriers it permitted the success of new competitors, which was not in the interest of the incumbent sellers of WordStar and VisiCalc.

The mechanism by which entry barriers fell in spreadsheet and word-processor markets is specific to network effects markets. Each of VisiCalc and WordStar had enjoyed substantial entry barriers because of network effects inertia. A new spreadsheet or word-processor user would choose the same product as the large installed base of existing users, sharing knowledge and files with them. The network-effects advantages were substantial, so that entry even by a superior product would have difficulty in succeeding. That was changed by the rapid rise in the number of new spreadsheet and word processor demanders in the corporate world following introduction of the IBM PC. These new users were numerous, and worked in different kinds of firms than many of the early users. Rather than looking to the existing installed base of PC users for network effects, the new users could also look to one another. This change in the focus of network effects lowered entry barriers.

With lower entry barriers, superior products like Lotus 1-2-3 and WordPerfect entered the main applications markets and ultimately became the dominant products. The center of spreadsheet network effects moved to 1-2-3 and that of word-processing network effects to WordPerfect.

The series of events that led to the destruction of VisiCalc's and WordStar's position in a wave of new creation began with innovations in a complement, the PC, that shifted out demand rapidly. Because of vertical disintegration, the nature, timing, and size of that shift were outside the control of incumbents. Rapid increases in the demand for a product that arise by bringing in new kinds of customers favor all sellers, but they particularly favor entrants in network effects markets. Entry barriers fell, and VisiCalc and WordStar were subject to new competition.

This parallels the series of events in creative destruction of the Apple II and CP/M rents. In each case, innovation in a complement shifted out demand and changed its composition. In each case, the lower entry barriers permitted new competition. In each case, the new competition led ultimately to a change in market leadership, with destruction of existing market positions and creation of new market positions.

4.3   One wave after another

These general lessons about creative destruction apply not only to the first wave surrounding the introduction of the IBM PC, but also to later waves of creative destruction in the PC industry.

The vertically disintegrated supply of the IBM PC made IBM a more rapid entrant with partial and backward compatibility, as we have just seen. In the early stages of the IBM PC era, the openness and modularity also served to encourage complementors, such as the new spreadsheet and word-processor entrants Lotus and WordPerfect.

The greatest advantage of this structure to IBM came in getting established in the PC industry. Once the IBM PC was established as a standard, it was in IBM's interest to change to a less competitive structure. The vertically disintegrated structure and openness made the industry more competitive going forward. This was to the disadvantage of all incumbents, but particularly of IBM.[13] This is a familiar and general story. Competition for the market gives suppliers incentives close to those of demanders. *ex post*, established suppliers would like to prevent further competition.

---

[13] Many scholars and industry participants debate the wisdom of IBM's decision from a private-interests perspective. Little can be learned from this debate. IBM chose a risky strategy with large present benefits and large, uncertain future costs at the beginning of a wave of creative destruction. Bill Gates' view that the debate is merely backward-looking revisionist history may be found in Gates, Myhrvold and Rinearson (1995).

The disadvantage to IBM of the openness and vertical disintegration came through entry and competition. Because of the open and modular design of the IBM PC, other firms were able to imitate it. Because of the vertical disintegration, the same other firms were able to gain complementors despite IBM's wishes to the contrary. In this section, we first look at why the openness and vertical disintegration were self-reinforcing, and then at how they encouraged competition.

The *ex post* disadvantages to IBM were advantages to consumers. The advantages to consumers came, in the first instance, because a modular and open design in an industry with many sellers of widely distributed products permits creative destruction and competition for the market. Let me begin with a few examples drawn from PC hardware. I focus on PC hardware because it was IBM's primary market.

4.3.1   Divided technical leadership

IBM set standards for connecting the IBM PC to display monitors. These were quickly seen as inadequate by customers and complementors. A number of monitors were introduced. Entrepreneurial *Hercules* entered with a monitor card that could be plugged into the PC. Hercules' card did not use IBM standards for connecting PCs and monitors. Dominant spreadsheet firm Lotus quickly moved to support the Hercules standard, and made very popular versions of 1-2-3 that worked only with a card at least as functional as Hercules'. What had been part of the IBM standard for connecting monitors quickly became an industry standard.

Design standards inside the PC itself also moved outside the IBM's control. Again the mechanism involved complementors who sought rapid technical progress and the vertically disintegrated structure of the industry. PC applications grew too large for the original IBM PC's limited memory. Applications developers complained about the limitation and pressed IBM for technical progress. Because of the industry structure, IBM was not the only firm that could provide that technical progress. IBM's complementors selling widely used products could also do so. In the case of memory standards, it was not IBM, but complementors who introduced what was for a time the most successful design for adding large amounts of memory, the LIM standard – named after three firms, Lotus (L), Intel (I), and Microsoft (M).

The lesson of the Hercules and LIM standards is a general one. As complex a technology as a PC contains many interface standards. There is a natural tendency to think of a single firm, such as IBM, as "the" standard setter. With vertical disintegration of widely used products and openness, however, there are multiple potential standard setters. A dominant firm in a particular market, even IBM, faces rivalry in the setting of standards and in their improvement. This is a powerful force for continuing openness once it has been established. This argument is an important part of the reason Shane Greenstein and I called the structure of the PC industry "divided technical leadership."[14]

There are two directions of causation linking competition and openness in computing. We saw above that there is a causal flow from openness and vertical disintegration to competition. In the last two sections, we have seen flows of causation in the reverse direction, in which competition causes openness and vertical disintegration. Competition for the market gave IBM an incentive to adopt open standards and vertically disintegrated structure at the beginning of the IBM PC era. Vertical disintegration in the most widely used products and the resulting divided technical leadership favored open standards once the IBM PC was established.

This solves a puzzle about incentives. If consumers gain from openness and vertical disintegration, and firms sometimes lose from them, why then did firms choose them? A firm may be compelled to do so, for competitive reasons, *ex ante*. The perspective of the firm *ex post* will be to wish to undo the vertical disintegration and openness in order to avoid competition. IBM certainly decided that *ex post*, and made a series of efforts to decrease the openness of the IBM PC and to increase vertical integration into key components. In the competitive race around the introduction of the IBM PC, however, the firm had strong incentives for openness and vertical disintegration. These features were key to its entering quickly and with a PC that was partially backward compatible. And, as we shall see, *ex post* competitive forces made it difficult to go back to a closed architecture. The gap between IBM's incentives and social incentives for openness was real, but competitive forces pushed IBM toward proconsumer structures.

More generally, openness, vertical disintegration among widely used products, and dynamic competition can form a mutually reinforcing system. This is why the early PC industry did not quickly revert to a model with closed proprietary standards and a vertically integrated dominant firm.

---

[14] See Bresnahan and Greenstein (1999). A body of formal theory addresses related issues. See, for example, Farrell, Monroe, and Saloner (1998).

4.3.2   Clone competition

The emergence of "clones" to compete with IBM was encouraged by the openness and modularity and by the vertically disintegrated structure of the PC industry. Modularity and openness lowered the fixed costs of competing with IBM. A clone PC manufacturer needed to make only a PC. The many existing complements to an IBM PC would work with it.[15]

At first, customers viewed clones as inferior to IBM. How were users to be assured that a clone would be as reliable and well supported as an IBM, and how certain was it that the clone would run all IBM PC applications in the future as well as the present? This changed as, with the support of the widely distributed complementors, some sellers of clones began to eclipse IBM technically. The success of Compaq at shipping a PC that used a new Intel microprocessor, the 80386, before IBM did so is an important example. This competition offered benefits to customers. It also offered benefits to IBM's complementors. They benefited from an increase in the rate of technical progress in PCs and competitive falls in the price of PCs. Complementors gained directly by having a better PC to work with, and indirectly by having a larger market.

A few complementors (such as Intel, Lotus, and Microsoft) with widely distributed and influential products were in a position to encourage the development of clones. Since those complementors were vertically disintegrated from IBM, they had an incentive to encourage clones. These few complementors selling widely distributed products thus had both the opportunity and the incentive to encourage the emergence of clone competitors for IBM.

What had been an "IBM PC" standard became an "Industry Standard PC." Control of PC standard setting slipped away from IBM. After a while, IBM attempted to regain control of the PC platform. The firm introduced new and improved standards for the interface for add-in cards, for networking, for a new operating system, and so on. It was, however, too late, as the industry standard PC was established and successful. The PC market itself became highly competitive.

No individual firm replaced IBM in the PC market, not Compaq with its technical advance nor Dell with its new and successful model of assembly and distribution. Instead, market supply of PCs replaced IBM. This is a distinct form of creative destruction. Many of the PC-selling

---

[15] A system of compatibility testing grew up, with PC manufacturers and third parties offering assurances to consumers that buying a clone would get them something technically very similar to an IBM PC.

entrants were entrepreneurs, like Dell and Compaq, though others were established computer firms in the wider IT industry, like Hewlett Packard. They gained the support of IBM's complementors in the vertically disintegrated PC industry, and were able to offer customers a backward-compatible and improved version of the PC that worked with many, many other complementary inputs. Creative destruction here is as far removed from action by a single creator as is imaginable, and the rents destroyed at IBM were tiny compared with the aggregate possibilities for further growth.

The industry structure of the IBM PC industry had substantial advantages for PC customers. Much of its advantage to IBM was, however, *ex ante*. *Ex post* establishment of IBM PC network effects, IBM would have liked to, and attempted to, regain control of the standard so that it could block entry. IBM's inability to block entry contributed substantially to ongoing innovation and growth.

Many of the important firms involved in the early PC industry were consumed by creative destruction. The firms most important in triggering and carrying out the first wave of creative destruction, the sellers of WordStar, VisiCalc, and the IBM PC, were later consumed by creative destruction. The major applications vendors and IBM were both creators and among those whose positions were later destroyed. There were repeated changes in the industry's technical and market basis. When they were rapid, these changes led to occasions for creative destruction. While incumbent dominant firms lost, customers gained the opportunity to choose – at least during the wave of creative destruction – between the incumbents and new entrants.

One more round of rapid changes would lead to successful competition for the market against incumbent dominant firms in important markets.

4.3.3   Precedents to another wave

Over the 1980s, the PC industry anticipated the invention of easier-to-use PCs. The diffusion of PCs into ordinary white collar work meant that less computer-knowledgeable users were coming into the industry as customers. The key technical advance that would ultimately permit even wider use of the PC was the Graphical User Interface (GUI.) The mass-market, easy-to-use PC was long anticipated, but was not realized until the introduction of the GUI Windows version 3 in the early 1990s.

The successful Windows 3.0 was anticipated by many earlier but less-successful efforts. A number of efforts to make PCs much easier to use were introduced starting from the first half of the 1980s, including GUI software from IBM, from Microsoft (Windows version 1, 2), from the sellers of VisiCalc and from entrant entrepreneurs. Like Windows, many of these products were positioned as complements to the IBM PC. None had any real success.

Other efforts to improve ease of use went forward within applications themselves. On the IBM PC and its clones, Lotus, WordPerfect, Excel, and Word all took on many GUI features. (Excel and Word came with an early "runtime" version of Windows.) Since effective provision of ease of use by the most popular form of PC was lacking, applications provided it themselves.

The most successful GUI effort of the 1980s was Apple's Macintosh. Introduced in 1984, it was not compatible with either Apple IIs or IBM PCs. Though more successful than other GUI efforts, the Macintosh was always a distant second to IBM PCs (plus clones) in terms of demand.

The existence of Macintosh as a second-choice applications platform had several impacts. It gave users a distinct choice of products, with PCs offering lower prices and more choices of hardware and software but Macintosh offering greater ease of use. The existence of a reasonably successful second-choice PC standard, the Macintosh, meant that there were two platforms for business applications. On the Macintosh, the leading spreadsheet and word processor were Microsoft Excel and Microsoft Word, while on the IBM PC and clones, Lotus 1-2-3 and WordPerfect were the leading products.[16] We see once again that the important precedents for a wave of creative destruction arise in the marketplace. Their core is supply–demand mismatches. There was, in this case, little doubt about the long-run direction of the industry toward greater ease of use. All important firms attempted to move in that direction, and all had limited success. There was, however, considerable doubt about the precise form of that long-run change and about its timing. The ultimate transition came with Windows 3.0. The previous history makes clear that this product was a triumph of commercialization and of implementation rather than a brilliant leap of invention.

4.3.4   Consequences of a wave

The introduction of Windows 3.0 in the early 1990s marked the beginning of another wave of creative destruction in the PC industry.

---

[16] While at least Excel was clearly a better product than the market-leading Lotus 1-2-3, Excel and Word were not even close to being leading products on the more popular IBM PC platform.

The leading applications vendors, Lotus and WordPerfect, were ultimately replaced. Technical change in a complement to those applications brought in a large number of new users, many distinct in demand
characteristics from existing users. That lowered the entry barriers
previously held high by network effects.

The success of Windows 3.0 increased the number of customers for
PCs, and thus for the major applications categories. New users of PCs
entered the market in large numbers.

Their tastes for computing features were somewhat different than earlier
IBM PC users, putting more weight on ease of use. Complementary
advances in PC hardware, such as ongoing improvements in microprocessors, memory, and disk drive, meant that a cheap, GUI-based PC
was now available. Windows-based PCs were compatible with the earlier
IBM PC and its clones, and thus could run IBM PC applications. These
features meant that there was a large market for the new machines.

The implication of rapidly rising demand for the major applications
categories was lowered barriers to entry. The network effects leading to
inertia around WordPerfect and Lotus 1-2-3 were devalued. Capable
entrants were waiting in the wings. Word and Excel had been steadily
improving as applications running on a machine with a GUI, the
Macintosh. Once there was a popular GUI version of the PC, these
products entered and competed very successfully against WordPerfect
and Lotus.

There has been a loud debate in the PC industry about whether
Microsoft, seller of both Windows and the entrant applications, behaved
honorably in the wave of creative destruction in applications categories.[17] We should ignore that debate for purposes of understanding
the general analytical lessons of the applications creative destruction
wave of the early 1990s.

Some of the major lessons are one that we have seen before. Change
in a complement played a large role in creative destruction, changing
both long-run and short-run market conditions. Falls in entry barriers,
rather than merely an entrant product overtaking an incumbent one
technologically or in suitability to the market, determined the timing of
creative destruction.

Finally, the common identify of the seller of Word, Excel, and
Windows does have analytical meaning. The moralistic debate about
Microsoft's behavior in causing the transition from WordPerfect and
Lotus to Word and Excel is irrelevant to the analytical meaning. One

---

[17] Microsoft was a partner with IBM in a competitive effort to Windows for a while, and
both IBM and applications vendors said they were misled.

consequence of this wave of creative destruction was that vertical dis-integration among the widely distributed products in the PC industry was reduced. Among widely distributed and influential software products, it was reduced to zero. Microsoft had been the dominant operating system seller for a long time. Now Microsoft was also the seller of the most important widely distributed applications. This changed the structure of the PC industry in a critical way. There has not been another wave of creative destruction in the interim.

4.3.5   The origins of entrants

Where do these entrants come from? This question is a stepchild in Economics generally. Analysis at the firm level answers this question badly. We cannot answer it merely by talking about the creative genius of entrepreneurship. We cannot answer it by taking the perspective of the incumbent dominant firm, surprised by the new market or technical conditions which support entry. Entrants in industries that involve cumulative investment do not come out of nowhere; they come out of somewhere.

Word and Excel came out of the second-place PC of the earlier era, the Macintosh. Those products had had a profitable history, though in a far smaller market than WordPerfect and Lotus had on the IBM PC. Further, they were sold by a firm, Microsoft, who could see the advantages of entering and competing for the market. At the time of its successful entry into the business applications market in the early 1990s, Microsoft was a fifteen-year-old entrepreneurial firm; its efforts to enter the largest and most profitable applications markets had been failing for ten of those fifteen years.

Here, as in the earlier entry of IBM, we see the advantages of a large diverse information technology sector serving many different kinds of users. Firm's reputation, capital built up in related markets (IBM in corporate data centers) or product designs built up outside of the largest and most competitive markets (Word and Excel on Macintosh) are slowly growing assets for entrants. A diverse IT sector permits investments in these assets for one purpose; repositioned for another purpose in another market, they become entrants. The prior investment point is particularly important when competition for the market goes quickly compared with the rate at which firms can invest in new technologies and new modes of commercialization. While waves of creative destruction sometimes take a period of time in PC markets, that is usually because of supply constraints, and participant firms are almost always better off if they have made a subset of the appropriate prior investments.

Other important PC entrants were entrepreneurial firms. All the firms in the 1970s, as I pointed out above, were entrepreneurial. Many of the important entrants in the 1980s, such as Lotus, WordPerfect, Compaq, and Dell, were entrepreneurial as well. Similarly, many of the thousands of other important suppliers of PC hardware and software of the 1980s and 1990s not covered explicitly in this chapter were entrepreneurial start-ups.

When the nature of demand and technology are changing over time, creative destruction can replace existing firms and products with new ones. When it is difficult to understand exactly what kind of firm or product will work well in the future, there is a high social return to firm and product diversity.[18] Incumbent dominant firms contribute to that diversity; entrepreneurial start-ups contribute to that diversity; firms from elsewhere in the computing and communication industry contribute to that diversity; and finally, firms already in the PC industry in one market migrating to another market contribute to that diversity.

The combination of a diverse set of potential entrants and repeated waves of creative destruction meant that PC markets had effective competition for the market from the founding of the industry in 1975 through the early 1990s.

4.4   A wave rebuffed: sea change or seawall?

There has been one more occasion for a new wave of creative destruction, the widespread use of the Internet. A new and even larger market for the PC opened up as a result of new online technologies, notably the World Wide Web (WWW) and the browser. Entrepreneur *Netscape* commercialized the web browser, starting a wave of entrepreneurship and opportunity that surprised PC industry incumbents. No new competition came to the established PC categories, however. In this section I examine first the causes of this wave and then its (lack of) consequences.[19]

4.4.1   Precedents to a wave

In the mid-1990s, after the establishment of Windows, the leading platform for PC applications was "Wintel," that is, a PC of any brand

---

[18] This is the central point of Cohen and Malerba (2001). See also Evenson and Kislev (1976), Metcalfe (1998), Nelson (1993) and Nelson and Winter (1982).
[19] This section draws on materials made public in connection with the Microsoft antitrust trial in the United States. I worked in the Antitrust Division during that trial. It also draws on the research and journalistic literature about the antitrust trial and about firms in this era, notably Bank (2001), Ferguson (1999), and Cusumano and Yoffie (1998), and on the research and journalistic literature about Microsoft as a company, notably Cusumano and Shelby (1995), Stross (1997), and Manes and Andrews (1993).

running Microsoft Windows on Intel-architecture chips. To achieve backward compatibility with industry standard PCs, early Wintel machines (e.g. those running versions 3.0 and 3.1 of Windows) involved a number of design compromises. The next version, Windows 95, was a big step forward. Nonetheless, throughout the mid-and late-1990s, Wintel machines drew complaints from complementors and corporate customers. Two main complaints stood out; an acute problem of high "total cost of ownership," that is, high maintenance and update costs once a Wintel machine was installed. The second problem was chronic; it was then, as before, difficult to network Windows machines.

For Microsoft, there appeared to be time to solve these problems. No widely distributed complementary product was in a position to create a wave of creative destruction. Accordingly, the firm undertook to solve the problems of Windows by a series of incremental and largely backward-compatible steps that would lead customers from the industry standard PC of the late-1980s to the much more capable machines of the twenty-first century.

Intel faced a somewhat different problem with the future of Wintel machines. The transition to the more graphical Windows and to more graphical applications programs gave users a reason to upgrade their computer hardware, continuing a two-decade trend. Intel was seeking to enable new innovations in applications, possibly in the area of multimedia or entertainment, to continue the trend into the future. Intel also faced a somewhat different competitive environment than Microsoft, with cloner AMD posing a real competitive threat.

4.4.1.1 Anticipatory innovations   Firms in the PC industry, like those in the broader communications and computer industries, had long anticipated the applications we now associate with the Internet. For years before the widespread use of the Internet, they saw the benefits of connecting very large numbers of people to online resources. They saw the benefits of a universal commercial network connecting computers in markets. The diffusion of the PC to most white collar desks, and the creation of vast amounts of online data and information, increased the potential benefits of connectivity. The possibility of connecting computer networks to people at home for entertainment or marketing (electronic commerce) purposes had also been visible for some years.

Before the widespread use of the Internet, a wide number of distinct technologies were introduced in order to support those "online" applications. Some were top-down initiatives led by powerful central forces like a telephone company, a government, or a private–public consortium. These were mixed in success, with results ranging from nothing up to limited success – Minitel in France was probably the most successful in reaching a mass market. Others were proprietary, closed systems pushed by a single vendor. These varied, too, with limited success for some Electronic Data Interchange products for business-to-business electronic commerce and for AOL on the mass market side.

Of all the PC, computer, and communications firms, one had a strategy for online applications that is particularly well documented. That is, Microsoft, whose internal debates about online applications were made public as a result of a lawsuit. In the era before the commercialization of the browser, that firm confined its attention to proprietary, closed systems for linking the online world to the PC. Even as an "Internet mania" came to Microsoft's attention in 1994, it remained determinedly focused on a closed and proprietary online approach. Yet Microsoft is merely the best documented. Before the commercialization of the browser, many existing commercial computing and tele-communications firms attempted to supply general-purpose technologies for connecting to the online world that did not ignite mass market online applications. The last precursor for the surge of online applications we actually saw in the late-1990s was the Internet itself. That technology advanced in military, government, university, and related sites for twenty years without ever gaining mass use or its present commercial importance.

4.4.2   Beginnings of a wave

Two steps were very important in making the Internet a mass market technology, the WWW and the (web) browser. These are general-purpose technologies that were invented with specific purposes in mind. The WWW was first invented as a way for high-energy physicists to share data and results. It was designed to be open and general. The WWW led to a number of useful inventions within the low-value walls of academe. One of these was the browser, which put a simple graphical user interface on the WWW. Another general-purpose technology invented with a specific purpose in mind, the browser was also open and general.

Entrepreneur *Netscape* commercialized the browser. This was recombination on a grand scale. The mass-market commercial browser was a new complement to several existing assets. It was a complement to the WWW and the Internet. It was also a complement to the commercial PC and to the vast amounts of commercial data stored on large computers.

The browser was a modular component. It worked with the existing WWW and Internet. Netscape's commercialization strategy was to write browsers that worked with all kinds of PCs. This offered browser users backward compatibility. They could continue to use their existing PC and have access to a great deal of online information. To get access to a wide variety of information from a wide variety of sources stored on a wide variety of different computers, the user had only to get one piece of software, the browser. Invention and commercialization of the browser dramatically raised the demand for PCs. PC users had easy access to new online information if they had a browser. An "Internet mania" grew up, in which there was a great deal more information on the WWW because there were more WWW users, and more users because there was more information. These network effects led to rapid growth, and the browser became a very successful PC application.

Once they were online, users began to demand communications applications such as email, online entertainment, wider opportunities to shop, be entertained and be informed, sharing photos, and instant messaging. Here the second advantage of the modular and open WWW and browser came into play. A provider of information, entertainment, or services to consumers did not need to set up an online network to connect to them. Nor did they need to form a contractual relationship with some kind of proprietary service. Instead, they needed only to connect their computer to the Internet. Major technical and organizational barriers to the supply of networked content and applications had been removed. More complex network effects between online applications and users began to form.

These new applications made PCs far more attractive to some classes of potential users, such as communications-oriented home users, than they had been earlier. The demand for PCs grew rapidly. The new users of PCs were different from the white collar workers who had been the mainstay of PC demand for a decade and a half. Ironically, they were closer to the users originally forecast for the PC back at the beginning – home users, multimedia users, communications users, and so on.

These are exactly the kinds of developments that had triggered waves of creative destruction earlier in the history of the PC industry. Existing PC industry-dominant products, such as the applications Word and Excel (by now combined into Office) or the Windows operating system, would have their network effects devalued by the rapid arrival of a large number of new users whose demand was distinct from existing users. From the perspective of the Windows operating system, the arrival of a large class of new applications – online ones – threatened to devalue its network effects even further. Just as we saw in earlier waves of creative destruction, the browser both raised the demand for the PC by recombining it and opened new competitive threats to existing PC dominant firms and products by lowering entry barriers.

4.4.2.1 Reactions of existing firms Many existing computer and tele-communications firms embraced the new opportunity. For example, *Sun*, manufacturer of computers used as "servers" in networks, introduced a new applications development platform called *Java*.

The PC industry firms with widely distributed products were reduced to two, Intel (microprocessors) and Microsoft (operating system and widely distributed applications). They responded to the new opportunity very differently.

Intel embraced the Internet as an opportunity to work with a new class of complementors. They formed an alliance with Sun, sellers of Java, for example, whose goal from an Intel perspective was to encourage development of a number of new, microprocessor-intensive, applications for PCs. (Java's virtues did not include economizing on microprocessor activity.) They sought to add multimedia features to their microprocessors to support new, consumer-oriented applications. They benefited from a remarkable increase in the demand for PCs as the new home and communications-oriented market segments took off. Microsoft, by contrast, decided that the potential wave of creative destruction was a threat to its dominant position in the operating system and in major applications. The widespread use of the Internet would raise demand for Microsoft's products. Yet the rapid change brought about by innovation in a complement, the browser, was also a problem for Microsoft, along the lines we have seen throughout this chapter. Microsoft diagnosed two problems. First, the open and modular structure of the WWW and the browser posed a competitive threat to Microsoft. The loss of control of standards to connect PCs to the online world was the first thing troubling Microsoft; rapid innovation by a large number of different firms, they reasoned, could lead to the setting of open standards. With a successful browser sold by an independent firm would come recreation of a vertically disintegrated structure for the PC industry. Microsoft feared the restoration of the competitive situation that had long prevailed in the PC business. The second thing that troubled Microsoft was the potential fall in entry barriers if products and technologies like the browser and Java were to succeed. There was an imminent threat of falling entry barriers as new and diverse users came into the PC market. Entrants were waiting in the wings; Linux, for example, had advanced considerably serving a very different kind of customer than PCs.

To make matters more difficult for the incumbent dominant firm, new applications invention was opening a gap between the existing PC and the PC that new customers would really like – just as the invention of the word processor and the spreadsheet had, earlier. Once again, the gap between what existing PCs could do and the desires of the newest demand segment gave the incumbent dominant firm a difficult technical and business problem.

The possible reversion to open standards and vertical disintegration were discussed in extensive detail inside Microsoft. So, too, was the threat of new competition from lowered entry barriers, and the gap between Microsoft's existing products and what the growing demand segment wanted.[20]

4.4.3   End of a wave

The potential wave of creative destruction in the PC business set off by the widespread use of the Internet did not occur. Rather than failing, it was blocked. While it had been entirely surprised by the success of the browser, once Microsoft saw the threat it responded with alacrity, energy, and focus. The threat of creative destruction gives the established firm powerful incentives.

Microsoft reacted to two powerful incentives. First, as Schumpeter emphasized, the threat of creative destruction gives powerful innovation incentives. Microsoft recognized a second incentive, which is that preventing the competitive threat of creative destruction can preserve a dominant position. Microsoft went down both paths, with very different results.

The first path illustrates the differences in resources and capabilities between an established dominant firm and an entrepreneurial start-up. Microsoft moved thousands of people into a new division to compete with Netscape. Working rapidly, that division eventually succeeded in catching up to Netscape in product quality. The catchup was too little and too late, however, for Microsoft to eclipse Netscape. Marketing officials in both firms observed at the time that Microsoft's browser quality improvements were insufficient in and of themselves.[21]

---

[20] I have quoted some of the internal discussion along these lines in Bresnahan (2001) and in Bresnahan (2002) I am the second person to analyze the PC industry along the lines of this chapter. The first, Bill Gates, who is quoted in the cited papers, has made several tens of billions of dollars more than I from this analysis.

[21] The new division was shut down and the browser effort moved into the division that sells Windows after this became clear.

This was a remarkable stretch of business history. There is little surprise in the observation that an entrepreneurial entrant, Netscape, opened up a new market ignored by an incumbent dominant firm, Microsoft. What is surprising is that, even after Microsoft counterattacked in full force, it was unable to win the browser-standard-setting race against the newly founded entrepreneurial firm. Even against the superior resources and capabilities of an established and successful firm, and even after the key activities shifted to commercialization and to incremental improvements, the innovativeness of the entrepreneurial firm won out.

Microsoft reached the conclusion that it could not succeed in the effort to set browser standards merely by innovating in its own product. It needed to do more to avoid creative destruction.

Microsoft's second path was to prevent independent outside creation in order to avoid destruction. Microsoft first made contractual offers to Netscape and Sun, the sellers of Java, to avoid the threat of competition (as basic cartel theory predicts it should have). Sun accepted on the condition that Microsoft work with standard (non-Windows-only) Java, only later to sue claiming that Microsoft had violated the contract.[22] Netscape declined, anticipating that a contractual collaboration would weaken their position. Having failed with its own product and with the offer of contract, Microsoft prevented the widespread distribution of the Netscape browser (and Sun's Java[23]) and prevented third-party complementors from working with Netscape (or with Sun's Java). Crucial distributors, such as manufacturers of PCs and Internet service providers, were blocked from distributing the threatening technologies. Complementors such as applications developers and even Intel were blocked from technical collaboration with sellers of the threatening technologies.

One reason Microsoft could compel other firms not to work with entrants was that the openness and vertical disintegration of the PC industry had declined. Windows was less open than earlier PC operating systems had been, for Microsoft kept the information about interaction with Windows under tight proprietary control. While it gave that information out to many complementors, it could withhold the information from firms who cooperated with the Internet entrepreneurs. Second, the only firm selling a very widely distributed software product in any of the markets listed in Figure 4.1 was Microsoft. Microsoft's

---

[22] The suit was recently settled with a large payment to Sun.
[23] Sun's lawsuit asserted that Microsoft distributed its own version of Java in violation of the contract.

effort on this second path succeeded. Without widespread effective distribution, the independent browser and Java failed. Without widespread distribution and many complementors, mass-market network effects could not take root. Microsoft's actions on the second path were illegal under the antitrust laws, but that is not important for our current inquiry. The important lesson for our present inquiry is not *how* the opportunity for creative destruction was blocked but merely *that* it was blocked.

The blocked distribution of the Internet innovations meant that the threatened wave of creative destruction never came to pass. The failure of the Internet revolution to re-establish vertical disintegration in the PC industry also means that the conditions for another wave of creative destruction are unpromising today.[24]

4.4.3.1 Lessons of blocked creative destruction   Microsoft's successful effort to block creative destruction competition against its dominant positions carries the same analytical lessons as do earlier waves of creative destruction that succeeded. Creative destruction is not in the interests of established dominant firms. Microsoft blocked the widespread distribution of products subject to network effects to stifle one wave of creative destruction. It also prevented re-establishment of the vertical disintegration of widely used PC products and technologies that could have led to new waves of creative destruction.

Would Microsoft's Windows or Office have been swept away by creative destruction if the distribution of threatening new technologies had not been blocked? This question cannot reasonably be answered with yes or no. The disruptive changes associated with the browser and Java would have lowered entry barriers into the operating system and business applications markets. Whether an entrant or entrants would have succeeded is unsure, one too many steps away from the historical record.

At a minimum, as Schumpeter emphasized, an entrant or entrants would have put valuable competitive pressure on Microsoft to innovate in a proconsumer direction, pressure that has been lacking since the end of the browser war in 1998.

4.5 Conclusion

Any study of creative destruction in the PC industry should answer two basic positive economics questions. Why were there a series of waves of

---

[24] While the European Union's decree in the Microsoft case is proconsumer, it falls far short of reestablishing competitive conditions in the PC industry. The earlier US decree is ineffectual.

creative destruction in this industry over two decades? Why did creative destruction cease?

The first question is answered by the interaction between entry barriers in PC industry markets and the vertical disintegration of supply of widely used products and technologies. Network effects lead to entry barriers around established products. Under vertical disintegration, technical progress in complements can lower entry barriers. That technical progress goes forward whether a particular established firm thinks of it or not and whether it is in their interest or not. That explains the inventive power of the industry, with innovation in widely distributed components spread over a large and diverse body of firms. It also explains the creative destruction. Vertical disintegration over the first twenty years of the industry enhanced the rate of innovation; it enhanced it partly through a Schumpeterian process, repeatedly destroying the rents of established firms.

The waves of creative destruction ended with the decline, in the 1990s, in vertical disintegration of widely used PC products and technologies. The widespread use of the Internet threatened to restore the industry to its more innovative and competitive form by adding new markets and vertical disintegration among the widely used products, but that was not to be.

In making this argument, I am implicitly discarding several ideas. One story of declines in creative destruction might be maturation. That would mean that creative destruction ceased because opportunities for it disappeared. A related maturation story would be that, after a series of trials, the market had selected the best firms for all products. Both of these maturation stories are belied by the events surrounding the widespread use of the Internet, where outsiders invented and commercialized important new technologies. Outsider supply was still far too important for this to be a time of maturation.

A second idea I am discarding – more a habit of mind than an idea, really – is that the boundaries of the firm are determined entirely by the efficient organization of supply. Vertical integration of widely used products in the PC industry matters for more than just the efficiency, either static or dynamic, of the products and technologies that might be supplied together or separately. Instead, vertical disintegration lowers entry barriers, permitting creative destruction.

There is a very general point here. Whatever the conditions supporting creative destruction in any industry, suppliers and demanders have a gap in their incentives. Both suppliers and demanders benefit from the value-increasing part of creative destruction. There is a gap between supplier and demander incentives insofar as conditions change to permit competition that destroys producer rents – and creative destruction is a great destroyer of producer rents. In the PC industry, this means that the value of vertical disintegration within the widely distributed products is higher for consumers than for producers.[25]

A third idea I am discarding – really another habit of mind – is that the timing of waves of creative destruction can be entirely explained by comparing the incumbent dominant firms with outsiders. In the PC industry, superior technology or market organization by outsiders is necessary but not sufficient for creative destruction. Instead, entry barriers must fall as well.

Caution is advisable in drawing general lessons about creative destruction from any industry study. Caution is all the more advisable here since the PC industry is extreme in several dimensions, including the pace of technical change. An advantage of looking at the PC industry is that the intervals between opportunities for creative destruction have been short in calendar time because of the rapid pace of change.

Why has vertical disintegration been so important for creative destruction? There are general lessons in that. Many of the critical transitions in the PC industry followed a circuitous route in which decentralized invention of complements moved the industry from one role to another in incremental steps. That is, decentralization of invention led to recombination.

The importance of decentralization provides a positive explanation of recombination. Many scholars, from Adam Smith to Schumpeter to our own colleagues, have made the normative argument that recombination of existing technologies is a valuable form of technical progress. From a positive economics perspective, recombination is linked to decentralization among innovators. To be sure, recombination economizes on the past stock of invention by re-using it (the normative theory). Recombination also economizes on knowledge about the future direction of technical progress; decentralization and *ex post* flexibility rather than on plan or contract create the circuitous path to recombinant growth.

In the PC industry, attempts at creative destruction rarely come solely from the inventiveness of the entrant. Instead, cumulated change in complementary products and technologies leads, through recombination, to a long-run opportunity for an improvement. Rapid change in complementary products and technologies lowers entry barriers, creating immediate opportunities for entry and competition for the market.

---

[25] Vertical disintegration of niche products, such as software applications for narrow groups of users, is in the interest of both sellers and buyers in this industry.

That helps explain the high rate of creative destruction in the PC industry over its first twenty years. A PC is a system made up of many different technologies and products. From the perspective of any single technology or product, there are a number of external changes accumulating, and a number of potential loci for rapid innovation that create new entry opportunities. The PC industry is obviously extreme both along that dimension and in the frequency of creative destruction in its early days. Yet it is clearly a general point that economic organization that permits decentralization in invention can lead toward creative destruction.

Why do established firms miss waves of creative destruction? Part of the answer in the PC industry is that some waves are very difficult to foresee. To the extent that a wave of creative destruction involves a circuitous route through a number of inventive steps, it is not foreseen by *any* firm. The decentralized invention process open to all firms in the economy dominates a single firms' efforts. Central planning, whether by governments or by a single established firm, is dominated by the market system.

When that has not been the explanation, the problem facing incumbent dominant firms has sometimes been that they have the wrong knowledge, and sometimes that they have the wrong incentives. Yet the sense of wrong knowledge and the sense of wrong incentives are subtle. For example, the applications-dominant firms of the late-1980s knew that a transition to an easier-to-use PC was coming; it was the time, form, and suppliers of that transition they could not foresee (to their peril). Similarly, Microsoft had forecast the transition to a market in which PCs were connected online; it was the value of the Internet specifically, and of an open and modular approach to online connection, that they missed. In the PC industry, many waves have been difficult to foresee in adequate detail to guide firm strategy.

These examples, and the others we have seen, push us away from a simple answer to whether we need creative destruction because established firms are inadequately innovative or have limited incentives? Competitive supply has two long-established advantages in economics. It gives suppliers better incentives. It also takes advantage of the capabilities of multiple firms. Creative destruction competition is like other competition in this regard.

Creative destruction in the PC industry is preceded by fundamental changes in the supply–demand match. The change often arises from technical progress in a complement. We have seen a number of occasions on which technical progress in applications led to a mismatch between supply and demand in general-purpose components of the PC.

We have also seen occasions on which progress in the GPT components realized applications mismatches, and occasions on which progress in some GPT components realized mismatches for other GPT components. The essential feature of all of them is that technical progress in a complement can change the market situation of a particular product. Thus, to understand creative destruction in one PC market, one must look at other PC markets. One implication is that the entrants themselves are not the only creators behind creative destruction in the PC industry. The innovations by complementors that create new opportunities and lower entry barriers are central. The generality of this particular point will vary with the degree to which markets are linked. Another implication is that vertical disintegration and open systems support creative destruction, while vertical integration and proprietary system are an impediment to it. Creative destruction has conditions. We can analyze when it is likely to happen. It is not merely the limits on human brilliance that matter (though of course they do). Instead, there are long-run supply and demand issues. There are short-run entry barrier issues. There are, of course, issues of the relative capabilities of incumbents and entrants. All of these are amenable to analysis. Some of the elements of that analysis are specific to industries (like the PC here) while others are general.

In the PC industry, we should understand the timing of creative destruction as being driven by two forces. One is the forces for stasis associated with existing positions. The other is technical progress that enables new positions. At that level of abstraction, the point is perfectly general. What the PC example brings to the fore is that the force for stasis is not only efficient assets accumulated by existing firms, but also entry barriers associated with their status as incumbents.

Creative destruction occurs in markets. That is not a statement of the answer; it is a statement of the question. We need to undertake market analysis to understand creative destruction. But we can. Perhaps the most general and important general point from the PC industry is that the analysis of market competition and the analysis of creating value by introducing new goods into markets takes us so far in understanding creative destruction.

References

Bank, D. 2001. *Breaking Windows: How Bill Gates Fumbled the Future of Microsoft*. New York, N.Y.: Free Press.
Besen, S.M. and Farrell, J. 1994. "Choosing How to Compete: Strategies and Tactics in Standardization," *Journal of Economic Perspectives* 8: 117–31

Bresnahan, T. 2001. "Network Effects and Microsoft" available at www.stanford.edu/~tbres

Bresnahan, T. 2002. "The Economics of the Microsoft Case" available at www.stanford.edu/~tbres

Bresnahan, T. and Greenstein, S. 1999. "Technological Competition and the Structure of the Computer Industry," *Journal of Industrial Economics* 47: 1–47

Campbell-Kelly, M. and Aspray, W. 1996. *Computer: A History of the Information Machine*. New York, N.Y.: Basic Books

Chandler Jr., A.D. 2001. *Inventing the Electronic Century: The Epic Story of the Consumer Electronics & Computer Science Industries*. New York, N.Y: Free Press

Christensen, C.M. 1997. *The Innovator's Dilemma*. Boston, MA: Harvard Business School Press

Cohen, W.M. and Malerba, F. 2001. "Is the Tendency to Variation a Chief Cause of Progress?," *Industrial & Corporate Change* 10 (3): 587–608

Cusumano, M.A. and Selby, R.W. 1995. *Microsoft Secrets: How the World's Most Powerful Software Company Creates Technology, Shapes Markets, and Manages People*. New York, N.Y.: Free Press

Cusumano, M.A. and Yoffie, D.B. 1998. *Competing on Internet Time: Lessons from Netscape and its battle with Microsoft*. New York, N.Y.: Free Press

Evenson, R. and Kislev, Y. 1976. "A Stochastic Model of Applied Research," Journal of Political Economy 84: 265–81

Farrell, J. and Klemperer, P. 2001. "Coordination and Lock-In: Competition With Switching Costs and Network Effects," *Handbook of Industrial Organization* Forthcoming 3: 1–90

Farrell, J., Monroe, H. and Saloner, G. 1998. "The Vertical Organization of Industry: Systems Competition versus Component Competition," *Journal of Economics & Management Strategy* 7 (2) 143–82

Ferguson, C. 1999. "*No Prisoners, a winner's tale of greed and glory in the Internet wars.*" Times Business New York

Fleming, L. 2001. "Recombinant Uncertainty in Technological Search," *Management Science* 47 (1) 117–32

Freiberger, P. and Swaine, M. 2000. *Fire in the Valley: the Making of the Personal Computer*. 2nd edn. New York, N.Y.: McGraw-Hill

Gans, J., Hsu, D., and Stern, S. 2002. "When Does Start-Up Innovation Spur the Gale of Creative Destruction?" *RAND Journal of Economics* 33 (4): 571–86

Gates, B., Myrhvold, N., and Rinearson, P. 1995. *The Road Ahead*. New York: Viking.

Henderson, R. 1993. "Underinvestment and Incompetence as Responses to Radical Innovation: Evidence from the Photolithographic Industry," *RAND Journal of Economics* 24 (2): 248–70

Henderson, R. and Kim, B.C. 1990. "Architectural Innovation: The Reconfiguration of Existing Product Technologies and the Failure of Established Firms," *Administrative Science Quarterly* 35: 9–30

Katz, M. and Shapiro, C. 1994. "Systems Competition and Network Effects," *Journal of Economic Perspectives* 8: 93–115

Langlois, R . 2002. "Modularity in Technology and Organization," *Journal of Economic Behavior and Organization* 49 (1): 19–37

Langlois, R.N. and Robertson, P.L. 1992. "Networks and Innovation in a Modular System: Lessons From the Microcomputer and Stereo Component Industries," *Research Policy* 21: 297–313

Liebowitz, S.J. and. Margolis, S.E. 1994. "Network Externality: An Uncommon Tragedy," *Journal of Economic Perspectives* 8 (Spring): 133–50

Liebowitz, S.J. and Margolis, S.E. 1999. *Winners, Losers and Microsoft: Competition and Antitrust in High Technology*. Oakland: Independent Institute

Manes, S. and Andrews, P. 1993. *Gates: How Microsoft's Mogul Reinvented an Industry – and Made Himself the Richest Man in America*. 1st edn., New York, N.Y.: Doubleday

Metcalfe, J.S. 1998. *Evolutionary economics and creative destruction*. Vol. 1. London, New York: Routledge

Nelson, R.R. 1993. "Uncertainty, Learning, and the Economics of Parallel Research and Development Efforts" in E. Mansfield and E. Mansfield (eds.), *International Library of Critical Writings in Economics*, vol. 31: 444–57

Nelson, R.R. and Winter, S.G. 1982. *An Evolutionary Theory of Economic Change*. Cambridge, MA.: Belknap Press of Harvard University Press

Reinganum, J.F. 1989. "The Timing of Innovation: Research, Development, and Diffusion," in *Handbook of Industrial Organization*, Schmalensee and Willig (eds.), New York: Elsevier Science Publishing Co., Inc.

Schmalensee, R.L. 2000. "Antitrust Issues in Schumpeterian Industries," *American Economic Review* 90 (May): 192–96

Schumpeter, J. 1942. *Capitalism, Socialism, and Democracy*, pp.82–85 (page cite to 1975 Harper edition.)

Schumpeter, J. 1911. *The Theory of Economic Development*, New York: Oxford University Press

Shapiro, C. and Varian, H.R. 1998. *Information Rules: A Strategic Guide to the Network Economy*. Boston, MA: Harvard Business School Press

Stross, R.E. 1997. *The Microsoft Way: The Real Story of How the Company Outsmarts its Competition*. Reading, MA: Addison-Wesley

Varian, H.R. 2001 (Revised March 30, 2003). *Economics of Information Technology*, Mottiote Lecture Bocconi University Milan, Italy

Veblen, T. 1899. *The Theory of the Leisure Class: An Economic Study of Institutions*. New York, NY: C. Scribner's Sons

Weitzman, M.L. 1998. "Recombinant Growth," *Quarterly Journal of Economics* 113 (2): 331–60

Comments to Chapters 3 and 4:

Stemming the tide of creative destruction?

Ashish Arora

The two contributions in this section connect to Schumpeter's *Capitalism Socialism and Democracy* very differently. Cohen and Sauerman try to explain what went wrong with Schumpeter's prophecy of the demise of capitalism, and use that to discuss the importance of nonpecuniary motives of researchers. Bresnahan's focus is on the process of creative destruction in the computer industry. Implicitly, he too departs from Schumpeter by pointing out how monopoly can stem the tide of creative destruction.

Both of the chapters in this section, though dissimilar in their focus, are united in one respect – both seek to understand some fundamental aspects of what makes capitalist systems superior to alternatives. Cohen and Sauerman focus on the individual inventor and his incentives. Bresnahan focuses on the working of markets and the role of related markets in the value chain. In so doing, both dissent from a strict reading of Schumpeter, but in the end, neither ends in heresy.

Cohen and Sauerman suggest that Schumpeter predicted that the capitalism demise would come because large business organizations would be unable to innovate – the replacement of the entrepreneur by the modern corporation would rob capitalism of its engine, and perhaps (reading between the lines), stifle innovation and economic growth. But the bigger point they make is that economists (though not Schumpeter himself, as evidenced below) have tended to construe incentives very narrowly as pecuniary incentives. Instead, men are motivated by a variety of incentives, including the intrinsic pleasure of invention and discovery and the gratification from the acknowledgment of peers and society at large.

One possible interpretation of their argument is that Schumpeter believed that capitalism's decline would arise from its economic failure.

I am grateful to Tim Bresnahan, Wes Cohen, and Steven Klepper for helpful comments but remain responsible for all remaining errors.

As Cohen and Sauerman note in footnote 2, "On one theme he is consistent. He retains the belief that the entrepreneurial incentive is key to technological progress and economic growth in the long run, and that that incentive is best preserved at a much smaller scale of enterprise than provided by the large industrial firm." Since virtually everything can be (and apparently, has been) read into Schumpeter, a definitive answer is not possible. Instead, I will content myself with noting that the best known parts of *Capitalism Socialism and Democracy* would not support this interpretation. Instead, the very success of capitalism, which involves the creation of large enterprises, leads to a situation where "Progress itself may be mechanized as well as the management of a stationary economy, and this mechanization of progress may affect entrepreneurship and capitalist society nearly as much as the cessation of economic progress would" (CSD: 131). In short, entrepreneurs would not be needed, in the ultimate act of deskilling. Just as Taylorism and Fordism were replacing craftsmen with repeated (and eventually, in some cases, automated) detail operation, so Schumpeter also feared for the entrepreneurs (Braverman, 1975). For, as Schumpeter reminds us, "This function (the entrepreneurial function) does not essentially consist in either inventing anything or otherwise creating the conditions that the enterprise exploits. It consists in getting things done" (CSD: 132). Further, if the failure of large corporations to provide "high-powered incentives" to entrepreneurship was how Schumpeter thought modern capitalism would die, it seems implausible that victory would go to Socialism; Socialism surely would be incapable of doing better on that front. Indeed, Schumpeter thought that the prime appeal of Socialism would lie elsewhere. "Socialist bread may well taste sweeter to them (*convinced socialist*) than capitalist bread simply because it is socialist bread, *and it would do so even if they found mice in it*." (CSD. 191. Emphases mine)

Exegetical disputes notwithstanding, in pointing out that the economists' focus on pecuniary incentives may be excessively narrow, Cohen and Sauerman are firmly part of Schumpeter's legacy. Entrepreneurs are motivated not merely by money but the prestige and distinction that money will make possible. But as the citation below shows, if it is true that money can buy anything, then the narrow focus, though not descriptively accurate, may nonetheless be analytically right.

In capitalist society, social recognition of performance or social prestige carries a strong economic connotation both because pecuniary gain is the typical index of success, according to capitalist standards, and because most of the paraphernalia of social prestige ... have to be bought. ... This prestige or distinction value of private wealth has of course always been recognized by economists. ... *And it is clear that among the incentives to supernormal performance this is one of the most important.* " (CSD: 208. Emphases mine.)

Cohen and Sauerman make a compelling case that the neglect of intrinsic motives and of non pecuniary extrinsic ones is a blind spot for economics models, particularly for models of R&D management. The latter motivation is interesting (and revisionist) for Schumpeter was concerned about the incentives of the entrepreneur, not the salaried (or stock-optioned) inventor. But it is the latter that is the focus of the chapter.

The second question is, therefore, what would change if economists were to admit that man does not live by bread alone (and that the other things he does live by cannot be readily bought)? Can some known facts perhaps be viewed in a new light or some apparent anomalies explained? There are some obvious candidates, such as skunkworks, where engineers and scientists tinker in secrecy on pet projects in apparent violation of the corporate will. Similarly, such incentives may explain the willingness to contribute to open-source software projects.[1] More intriguingly, is the superior performance of firms that allow their researchers more freedom to publish due to spillovers from academia (Gambardella, 1995; Cockburn and Henderson, 1998) or the ability to attract better scientists (Stern, 2004), who are presumed to have higher intrinsic value for autonomy and peer recognition and thus motivated to work harder?

But one must be careful, for the logical thread can tangle quickly. Consider the example of Shockley, who wanted to keep his researchers focused on fundamental research (presumably appealing to the intrinsic motives for discovery) only to find that many key researchers cared more about impact and about making money – they wanted " ... to solve concrete problems, see those solutions implemented in practice, and make some money" (Cohen and Sauerman, Chapter 3, this volume p. 85). We have apparently a case of a manager who placed too much faith in the nonpecuniary motive, or, at the very least, a group of technologists who, in a reversal of the apparent motto of the dope peddler in the Tom Lehrer song, wanted to do good by doing well.[2]

---

[1] There is a delicious irony in tenured economics professors speculating on why scientists and engineers will do research even if they do not have to. The mystery is perhaps artificial. As Paul David once remarked, one only had to look at how economics departments were run to recognize that the "rational" economic model was not a good description of how even its most devout proselytizers behaved.

[2] I am relying upon Cohen and Saurman's characterisation of the issue here. It is possible that this was a simple disagreement about strategy, with Schockley betting on fundamental research as a more profitable route to profits.

Even more interesting is the example of information trading by engineers (von Hippel, 1987). Though most economists have focused on the apparent anomaly of engineers passing information to employees of rival firms, the real anomaly may be in the "don't ask don't tell" policy followed by managers regarding their engineers trading tips with peers at competitor firms. Managers clearly think that overall they (and the firm) benefit from these informal exchanges. Why, then, not formalize it? To go one step further, if monetized exchange dominates barter elsewhere, why is an extreme form of barter (information for information) the norm here? Though we do not get the answer, there is a hint that incentives may be the answer: monetization may conflict with norms of collegiality and may reduce cooperative incentives.

If Cohen and Sauerman address why the advent and dominance of large corporations did not halt innovation, Bresnahan's piece analyzes when the gales of creative destruction in modern capitalist economies turn into a gentle breeze, scarcely capable of ruffling a floppy disk, let alone an established IT firm. In keeping with Schumpeter's legacy, monopoly power in markets concerns Bresnahan not because monopolies price above marginal cost. Indeed, there are many reasons to believe that sustained horizontal competition – competition between firms producing the same type of good – is difficult to sustain in a world with large sunk costs and strong network effects. It may also be socially undesirable in the same way that a single local telephone company is better, in that the fixed costs are amortized over a larger output. Instead, Bresnahan's real concern is about entry and entry barriers in the IT industry, and how monopoly power in one market may be used to raise entry barriers in other related markets.

Bresnahan's argument is simple. Cast in Schumpeterian language it goes something like this. Schumpeterian competition – from the new technology or the new type of organization – which strikes at the very foundations of the existing firms, is unlikely to threaten an established incumbent in the IT industry if the incumbent also dominates related markets. Put differently, potential entry is more likely to actually materialize, if the entrants have a foothold in a related market. The related market may be an existing market or a new one opened up by entrepreneurship. If so (and more on why this should be so), it follows that the proper role for antitrust policy in Schumpeterian industries is to ensure that no firm gets to dominate too many related markets. Not only do firms with experience in related markets make more formidable rivals for incumbents elsewhere in other markets, the competition among these incumbents also lowers entry barriers for start-ups and other disrupters.

Why should this be so? In the computer industry, with its large sunk costs and network effects, challenges to established products are difficult,[3] but not impossible, as Microsoft defenders, most notably Liebowitz and Margolis (1999), have argued. Visicalc was replaced as market leader by Lotus, which was replaced by Quatro, which was replaced by Excel in the spreadsheet market for PCs. Similarly, Wordstar was replaced by WordPerfect, which was replaced by Word in the word-processor market for PCs. Neither Excel nor Word have faced a serious threat in the last decade or so. Could this simply reflect the innate superiority of these products? And is it a mere coincidence that these are sold by Microsoft, which also dominates the operating system for PCs? Is Microsoft's bundling of these products into the Office Suite entirely a benign happenstance for consumers? Bresnahan's answers, at least as I infer them from his analysis, are No, No, and No.

The first key element of the analysis is that in a dynamic industry, opportunities for entry arise time and again as technology progresses. Improvements in the price performance ratios bring new consumers into the market. For instance, as improvements in PCs and associated software made PCs useful for office workers, new users for spreadsheets and word processors arose. These new users were not locked into the established product, and in some cases, network effects also weighed less heavily on their choices since the new users were distinct from the existing hobbyists.[4]

Which firms succeeded in capturing these new users? Here, the second element of the analysis is the ability of *de novo* start-ups to benefit from the relative modularity of the computer technology. Lotus and WordPerfect could succeed as stand-alone products. Open standards meant that users could "mix-and-match"; in the ugly jargon favored by the IT industry, a "best-of-breed" strategy was viable. Open standards are vital; else a potential entrant would have to develop not only a new

---

[3] Of course, where these effects are small, as in the PC hardware market, even IBM succumbed to price competition from imitators. Schumpeterian competition, from new organizational forms, in the form of Compaq and then Dell, still continued apace.

[4] The study of standards and the implications of their openness or otherwise has been a cottage industry in economics for some time. As is the case with virtually all game theoretic models, there are no robust results regarding the implications for entry and competition. However, as Farrell, Monroe, and Saloner (1998) have shown, a model where firms can choose whether to open their standards (and potentially cede one of the markets to a rival) openness tends to produce higher social welfare and lower profits. Arora and Bokhari (2006), using a model with myopic entry and price-taking firms where firms also choose whether to specialize or enter as an integrated firm, show that markets where incumbents tend to be vertically disintegrated see more rapid entry and lower costs and prices.

spreadsheet but also a compatible cooperating system and a compatible word processor.

The third key piece of the analysis has to do with successful entrants. As Steven Klepper has emphasized in other work, successful entrants are more likely to arise either as spin-offs from existing (successful) incumbents in the same or related markets (cf. Klepper and Simon, 2000; Klepper, 2001). For instance, spin-offs from successful automobile firms were likely to survive longer than spin-offs from less successful incumbents or *de novo* entrants. More importantly for this discussion, successful firms were more likely to arise from firms that had success in related industries. For instance, TV producers spawned by successful radio producers were more likely to survive than other TV producers. So also in computers. Word and Excel were both successful products on the Apple platform, and when Microsoft introduced a user-friendly interface for the PC in the revolutionary Windows 3.1 operating system, were well positioned to take on the established incumbent products, which were slow in adapting to the new operating system. In other words, the success of Word and Excel (though not necessarily their continued survival) is the quintessence of Schumpeterian competition.

The failure to take advantage of established standards and modularity had grave consequences. Some will remember the technologically advanced NeXT computers (a start-up led by Steve Jobs), which debuted in the early 1990s and which embodied advanced graphical and communication technology that would become available on PCs only several years later, but was a commercial failure. The problem, in part, was that NeXT had a new display technology, a new operating system based on Unix, and a hardware platform different from the standard Intel platform. It also required large and expensive hard drives. That, and other seemingly minor problems, doomed the company.[5] Nor was large size and experience always enough to overcome a smaller but better-established incumbent, as IBM found out in its bid to dislodge Windows with the OS2. Understanding customer needs and savvy marketing remained a key capability.

Now to the denouement. The exception that tested the rule was the rise of the Internet and, with it, the browser. The incumbent, Microsoft, was able to ride out the gale because it was able to use its technical control, buttressed by intellectual property rights, over key related markets including the operating system, and the office suite to close

---

[5] Eventually, Apple purchased the company in 1996 and used the NeXT operating system as the basis for future Mac operating systems.

technical standards and use its market power to browbeat suppliers in other markets into withholding cooperation with potential rivals. All this bought time, which Microsoft used effectively to catch up and develop competing products.

To paraphrase Bresnahan, the waves of creative destruction (to switch from an aerial to an aquatic metaphor) are ineffective if the seawalls of entry barriers are tall enough. Divided technical leadership – where different firms dominate different segments of the value-and-technology chain – is what is needed to pull down the walls. This logically implies open standards and modular knowledge.

Leaders in related markets are also a ready source of potential entrants to exploit new opportunities, as Netscape apparently once was and as Google now threatens to be. A key contribution of Bresnahan's analysis is to focus our attention on the entire value chain in an industry, rather than merely individual product markets. The chemical industry provides additional supporting evidence. As detailed in Arora and Gambardella (1998), after World War II, oil refining and the production of synthetic fibers and plastics came to share a common technical base in the United States, giving it an initial advantage. European organic chemical firms, which had pioneered organic chemicals technologies using coal-based feedstocks and acetylene chemistry, were initially at a disadvantage. Europe lacked oil and, more importantly, European firms lacked experience with using liquid oil feedstocks and the technology for processing ethylene. However, they were able to overcome the initial disadvantage, in large measure because they could obtain the process technologies quite readily. The key was vertical separation between oil and chemical firms, and more importantly, between firms responsible for production and those responsible for engineering and constructing plants. Specialized engineering firms, which arose to provide plant construction and design services to chemical firms, led the way in licensing petrochemical technologies to Europe, and later to Japan and other countries (Freeman, 1968). Prodded by this competition, even large and well-established chemical firms began to license, led by Union Carbide, a leading licensor of polyethylene and polypropylene technology (Arora, Fosfuri, and Gambardella, 2001).

The rapidity with which the European chemical industry made the transition to petrochemicals is remarkable. In the United Kingdom, for instance, only nine percent of the total organic chemical production was based on oil and natural gas, and the proportion rose to sixty-three percent by 1960. The market for technology dramatically changed the competitive position of firms; accumulated production experience of incumbents could not deter entrants from successfully competing with incumbents (Lieberman, 1989). The chemical industry thus provides a clear example of the benefits of vertically disintegrated industry structures in promoting entry and competition, and in technology diffusion.

The other important issue, and a more central part of the Schumpeterian legacy, is the role of "economic experiments." That is, after all, what entrepreneurs do. In Bresnahan's schema, these experiments involve technical innovation in hardware and software; they also involve new organizations and marketing strategies targeted at new sets of buyers. Some succeed and many do not, but the successes redound to larger societal gains.

A prerequisite for such experiments is variety, or put negatively, the absence of decisive central control. Under socialism, the problem is not merely in the inefficient allocation of resources, but in the absence of the conditions for genuine economic experiments. For, as Nathan Rosenberg pointed out, a planning bureau would inevitably skimp on spending money on what might appear crazy hare-brained schemes (Rosenberg, 1992). The importance of decentralization of leadership is no less vital in politics. Rosenberg and Birdzell (1986) favorably contrast the political fragmentation of Western Europe in promoting technical and economic (and cultural) progress in contrast to the political centralization in China. Political fragmentation has its costs, to be sure, as the Balkans remind us every day. But the costs of political centralization are no less severe, and perhaps even more so.

Bresnahan forcefully argues that the same might happen under capitalism when a single entity dominates critical standards and controls the likely sources of such experimentation. The way to avoid such concentration is not to try to break up monopolies in individual markets but to ensure that such monopolies do not raise entry barriers too far. In particular, the interaction between market power and intellectual property rights can lead to perverse consequences. In the computer industry, there is a real danger that control over *de facto* standards, bolstered by copyright (and now, patent) protection, can dangerously raise entry barriers. Even though the U.S. antitrust authorities did not take the opportunity presented to them, recent events indicate that all may not be lost. Perhaps, like Schumpeterian competition, antitrust action disciplines even when it is merely a threat.

References

Arora, Ashish and Bokhari, Farasat 2006. Open Versus Closed Firms and the Dynamics of Industry Evolution. Forthcoming, *Journal of Industrial Economics*

Arora, Ashish and Gambardella, Alfonso 1998. "Evolution of Industry Structure in the Chemical Industry" in Arora, Landau, and Rosenberg (eds.), *Chemicals and Long Term Economic Growth*. New York: John Wiley and Sons

Arora, Ashish, Fosfuri, Andrea, and Gambardella, Alfonso 2001. *Market for Technology: The Economics of Innovation and Corporate Strategy*. Cambridge, MA: MIT Press

Braverman, Harry. 1975. *Labor And Monopoly Capital; The Degradation of Work in the Twentieth Century*. New York: Monthly Review Press

Cockburn, Iain, and Henderson, Rebecca 1998. "Absorptive Capacity, Coauthoring Behavior, and the Organization of Research in Drug Discovery," *Journal of Industrial Economics* 46 (2): 157–82

Farrell, Joseph, Monroe, Hunter, and Saloner, Garth 1998. "The Vertical Organization of Industry: Systems Competition versus Component Competition," *Journal of Economics & Management Strategy* 7 (2): 143–82

Freeman, C. 1968. "Chemical Process Plant: Innovation and the World Market," *National Institute Economic Review* 74: 931–41

Gambardella, Alfonso 1995. *Science and Innovation: The US Pharmaceutical Industry During the 1980s*. Cambridge: Cambridge University Press

Klepper, Steven 2001. *"Firm Capabilities and Industry Evolution: The Case of the US Automobile Industry," Mimeo*. Pittsburgh: Carnegie Mellon University

Klepper, Steven and Simon, Kenneth 2000. "Dominance by Birthright: Entry of Prior Radio Producers and Competitive Ramifications in the US Television Receiver Industry," *Strategic Management Journal* 21: 997–1016

Lieberman, M. 1989. "The Learning Curve, Technology Barriers to Entry, and Competitive Survival in the Chemical Processing Industries," *Strategic Management Journal* 10 (5): 431–47

Lieborwitz, Stan and Margolis, Steven 1999. *'Winners, Losers & Microsoft: Competition and Antitrust in High Technology'*, Oakland: Independent Institute

Rosenberg, Nathan 1992. "Economic Experiments," *Industrial and Corporate Change* 1: 181–203

Rosenberg, Nathan and Birdzell, Lawrence 1986. *How the West Grew Rich*. New York: Basic Books

Stern, Scott 2004. "Do Scientists Pay to Be Scientists?," *Management Science* 50 (6): 835–53

von Hippel, Eric 1987. "Cooperation Between Rivals: Informal Know-How Trading," *Research Policy* 16: 291–302

Part 3

Innovation and industrial dynamics

5 Statistical regularities in the evolution of industries: a guide through some evidence and challenges for the theory

Giovanni Dosi

Introduction

Fundamental drivers of the evolution of contemporary economies are the activities of search, discovery and economic exploitation of new products, new production processes and new organizational arrangements within and amongst business firms. Such processes ultimately entail the emergence and development of novel bodies of technological knowledge, novel 'ways of doing things' and novel organizational set-ups. Indeed the identification of the sources of change and the 'political economy' of their economic selection continues to be a major challenge for all analysts of socio-economic change. Knitted together, however, comes also the understanding of the statistical properties that such processes might possibly display. This work focuses on the latter, concerning specifically the patterns of industrial evolution. Three basic questions in particular are addressed here:

(1) Are there distinct characteristics of the micro entities (*in primis*, business firms) and their distributions which systematically persist over time?
(2) How do such characteristics within the population of competing firms affect their relative evolutionary success over time? And, in

Support to the research by the Italian Ministry of Education, University and Research (MIUR, Project 2002132413_001) is gratefully acknowledged. The work builds on ongoing exciting research collaboration with Giulio Bottazzi, Elena Cefis and Angelo Secchi: the reader will indeed notice the widespread influence of Bottazzi's analyses on this work. It benefited from insightful discussions with Bronwyn Hall, Mariana Mazzucato, Sid Winter and from several comments of the participants to the Schumpeter Conference, Milan – in particular Steven Klepper and John Sutton. A skillful research assistance has been provided by Marco Grazzi. This research would have not been possible without the precious help of the Italian Statistical Office (ISTAT) and in particular of Andrea Mancini and Roberto Monducci.

particular, what are the ultimate outcomes in terms of growth and profitability performances?

(3) Amongst the foregoing statistical properties and relations between them, which ones are invariant across industries, and, conversely, which ones depend on the technological and market characteristics of particular sectors?

Note that the answer to these questions has also major implications with respect to the empirical validation of evolutionary theories of industrial change. After all, such theories focus on the twin processes of technological and organizational learning, on the one hand, and market selection, on the other, as the central drivers of industrial change.

If this is so, one ought to be able to robustly detect also in the empirical data the marks of variables and processes which are so crucial for the theory – including, for example, the footprints of firm-specific knowledge accumulation, competition-based selection, and industry-specific *regimes* of learning. Hence the discussion of the evidence which follows can also be read as an assessment of the elements of empirical corroboration of evolutionary interpretations of economic change, together with a series of challenges which the theory still faces.

At the same time, the increasing availability of longitudinal panels of firm-level data is likely to shed new light also on old questions raised in the old 'structuralist' and 'structure–conduct–performance' perspectives in industrial economics concerning for example the relationships between firm size, industrial concentration and the ability to exercise 'monopoly power' and thus extract 'super-normal' profits.

In order to address these questions we proceed in a sort of 'inductive' manner. We start by examining some basic features of the distributions of firms sizes, growth rates and profitability (Section 5.1). Next, Section 5.2 considers some evidence on the underlying inter-firm heterogeneity – particularly with regard to technological innovativeness and productivity – and their relationships with corporate performances.

Finally, Section 5.3 recalls the basic elements of an evolutionary interpretation of the evidence. Together with important points of corroboration of such a view – including those regarding a profound heterogeneity of firms at all levels of observation – one also faces standing challenges – *in primis*, concerning the purported role of markets as effective selection devices.

Some *caveats*. Concerning the sources of evidence, while this work draws on multiple secondary sources, it heavily relies upon the data banks analysed by the research groups of which I am or have recently been part. These data regard (1) longitudinal micro-evidence on

Italian manufacturing (the MICRO.1 data from the Italian Statistical Office, ISTAT), (2) U.S. manufacturing (COMPUSTAT data), and (3) the world pharmaceutical industry (the PHID data bank organized by Fabio Pammolli at EPRIS, Florence).

Moreover, the discussion which follows largely neglects most phenomena concerning 'life cycle' properties of industries, which would require a much greater disaggregation and much longer time spans (for a through discussion on the subject, see Klepper, 1997). Neither do I address explicitly the 'stylized facts' on entry and exit dynamics (cf. the recent survey by Bartelsman, Scarpetta and Schivardi, 2005). Rather, this work is restricted to the distributions of sizes and performances of incumbents, their dynamics and their relations with their underlying technological characteristics.

5.1   Firm sizes, growth rates and profitabilities

Let me begin by considering the old and new evidence concerning industrial structures together with two common performance variables, namely corporate growth and profitabilities.

5.1.1   Size distributions

A first, extremely robust, 'stylized fact' regards the quite wide variability in firm sizes. More precisely, one observes – throughout industrial history and across all countries – right-skewed distributions of firm sizes[1]: within a large literature, see Steindl (1965), Hart and Prais (1956), Ijiri and Simon (1977), all the way to Stanley et al. (1996), Bottazzi et al. (2006), and Bottazzi and Secchi (2003b).

Here Figure 5.1 presents the distribution of Italian firms with more than twenty employees. Here, *size* is measured in terms of value added but alternative proxies such as sales and number of employees yield a very similar picture. Irrespective of the precise form of the density function, the intuitive message is the coexistence of many relatively small firms with quite a few large and very large ones – indeed in a number much higher than one would predict on the ground of any Gaussian shape. In turn, all this militates against any naive notion of some 'optimal size' around which empirical distributions should be expected to fluctuate. Notice that, as a consequence, also any theory of production centred around invariant U-shaped cost curves, familiar in

---

[1] This property as well as few other ones that we shall discuss below apply also to *plant* distributions. However, in this essay we shall mostly focus on firms which as such may well be composed of several plants.

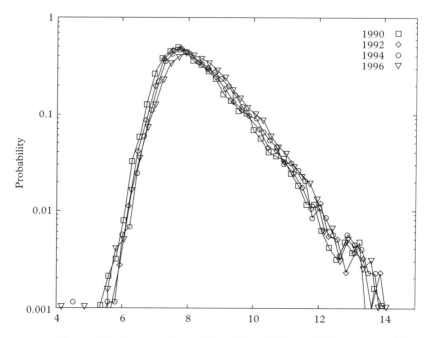

Figure 5.1   Empirical densities of log (VA_i) in different years (size measured in terms of value added). *Source* Bottazzi G., Cefis E., Dosi G., and Secchi A. (2006)

microeconomic theory, lose a lot of plausibility. Were they the rule, one ought to reasonably expect also a tendency to converge to such technologically optimal equilibrium sizes.[2]

Plausible candidates to the representation of the empirical size distributions are the log-normal, Pareto and Yule ones. Certainly, the full account of the distributions suffers from serious problems in offering also an exhaustive coverage for the smallest firms. Recent attempts to do that, such as Axtell (2001) on the population of U.S. firms, lend support to a 'power-law' distribution linking firm sizes probability densities with the size ranking of firms themselves (cf. Figure 5.2).[3]

---

[2] The literature does present interpretations which try to reconcile standard production theory with such an evidence. My personal view is that they tend to range between the implausible and the incredible – the latter including Lucas (1978), suggesting that the observed distributions are the outcome of an optimal allocation of managerial skills-.

[3] The (cumulative) probability density function of a Pareto distribution of discrete random variables is

$$(S)\ \Pr[S \geq S_i] = \left(\frac{S_0}{S_i}\right)^{\alpha} \quad S_i \geq S_0 \tag{5.1}$$

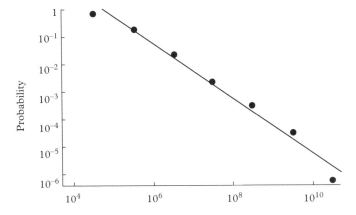

Figure 5.2   Cumulative distribution of US firms by receipts (logs, 1997 $). *Source* Axtell (2001)

The evidence discussed so far concerns *aggregate manufacturing* firm size distributions. Are these properties *robust to disaggregation*? An increasing body of finer sectoral data suggest that in fact *they are not*.

Corroborating a conjecture put forward in Dosi et al. (1995) and further explored in Marsili (2001), aggregate 'well-behaved' Pareto-type distributions may well be a *puzzling outcome of sheer aggregation* among diverse manufacturing sectors, characterized by diverse regimes of technological learning and market interactions which do *not* display Paretian size distributions. While some sectors present distributions rather similar to the aggregate ones, others are unimodal symmetric and almost log-normal and yet others are bi-modal or even multi-modal. Figures 5.3 and 5.4, taken from Bottazzi et al. (2006) on three Italian manufacturing sectors, vividly illustrate such inter-sectoral diversity.

The more recent evidence (e.g. on Italy, see Bottazzi et al. (2006)) based on extensive micro panels does robustly confirm an older "stylized fact" regarding the remarkable inter-sectoral differences in concentration ratios (cf. the thorough overview in Schmalensee (1989), and also the inter-country comparison in Pryor (1972)).

Together, however, the same evidence appear to go against the conventional wisdom according to which sectoral concentration should go together with (sectoral) average firm sizes: in fact the data analysed by Bottazzi et al. (2006) suggest the lack of any correlation whatsoever.

---

where S_0 is the smallest firm size and S_i is the size of the *i*-firm, as increasingly ranked. Under the restriction that $\alpha \cong 1$, this is known as *Zipf Law*. Note that, generally, the Pareto description is generally restricted to the upper tail of the distribution (for which one also finds more reliable data).

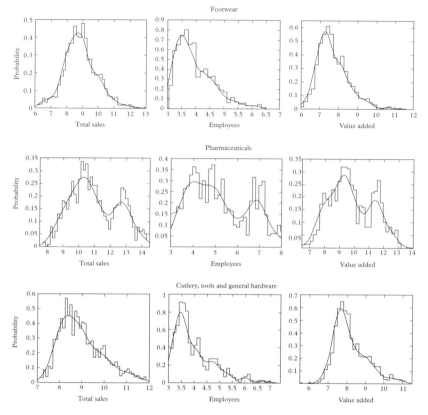

Figure 5.3   Densities of log (S_i), log (L_i) and log (Va_i) in different Italian manufacturing sectors. *Source* Bottazzi G., Cefis E., Dosi G., and Secchi A. (2006)

Finally, admittedly circumstantial evidence hints at a plausible oligopolistic core versus fringe firms separation in several sectors – indirectly supported by the mentioned bimodality of size distributions.[4]

Come as it may, *industrial structures* – in this case proxied by size distributions – are the outcomes of the growth dynamics undergone by every entity in the industrial population (jointly, of course, with birth and death processes).

What about such growth processes?

---

[4] Indeed, an important research task ahead concerns the transition probabilities between "core" and "fringe".

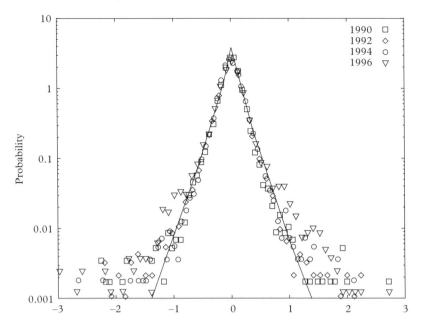

Figure 5.4   Growth-rate distributions in different years. Size measured in terms of Value added. Italian aggregate manufacturing. *Source* Bottazzi G., Cefis E., Dosi G., and Secchi A. (2006)

5.1.2   *Corporate growth rates*

It is handy to start the analysis of firm growth processes by setting a sort of 'straw man' which also happens to be a classic in the literature, namely the so called *Gibrat Law* (cf. Gibrat (1931), Simon and Bonini (1958), Kalecki (1945), Steindl (1965), Ijiri and Simon (1977) and Sutton (1997)).

Let

$$s_i(t+1) = \alpha + \theta_i s_i(t) + \varepsilon_i(t) \tag{5.2}$$

where $s_i(\cdot)$ are the log sizes of firm i at times t, $t+1$ and α captures the sector-wide (both nominal and real) components of growth.

Gibrat law in its strong form suggests that

(a) $\theta_i = 1$ for every i,
    and
(b) $\varepsilon_i(t)$ is an independent identically and normally distributed random variable with zero mean

Table 5.1 *Selected emprirical studies on Gibrat's Law*

Study	Methodology	Controls	Data	Results
Mansfield, 1962	Logarithmic specification	None	About 1,000 US firms in steel, petro leum and tires over 1916–57.	Gibrat's law fails to hold in about 50% of cases; smaller firms grow faster.
Brusco–Giovannetti–Malagoli, 1979	Logarithmic specification	None	1,250 Italian firms in ceramics, mechanical and textiles over 1966–77.	Gibrat's law fails to hold in most cases when only surrived firms are included.
Kumar, 1985	Logarithmic specification	Persistence	1,747 UK quoted firms in manufact. and services over 1960–76.	Smaller firms grow faster. Smaller firms grow faster.
Hall, 1987	Growth-rate regression	Sample selection, heteroskedasticity	1,778 US manufact. firms over 1972–79 and 1976–83 (only incumbents)	Smaller firms grow faster.
Evans, 1987a and 1987b	Growth-rate regression	Sample selection, heteroskedasticity	42,339 US manufacturing firms, subdivided in 100 sectors.	Smaller firms grow faster in 89 industries out of 100.
Contint-Revellt, 1989	Growth-rate regression	Persistence	1,170 Italian firms over 1980–86 (only incumbents).	Moderate evidence that smaller firms grow faster.
Dunne–Roberts–Samuelson, 1989	Growth-rate regression with grouping procedure	None	219,754 US manufacturing plants over 1967–82 (only entrants).	Smaller firms grow faster.
Wagner, 1992	Logarithmic specification	Persistence	About 7,000 West German manufact. plants over 1978–89; (only incumbents).	Gibrat's law fails to hold, but no evidence that smaller firms grow faster.

Study	Specification	Issues	Sample	Result
Dunne–Hughes, 1994	Logarithmic specification	Sample selection, heteroskedasticity, persistence	2,149 UK companies over 1980–85 (only incumbents).	Smaller firms grow faster.
Mata, 1994	Growth-rate regression	Sample selection, heteroskedasticity	3,308 Portuguese manufacturing firms over 1983–87 (only entrants).	Smaller firms grow faster.
Solinas, 1995	Logarithmic specification	None	5,128 Italian firms over 1983–88 (only entrants).	Once the sample is limited to companies with at least one employee, smaller firms grow faster.
Hart–Oulton, 1996	Logarithmic specification	Heteroskedasticity, persistence	87,109 UK companies over 1989–93 (only incumbents).	Smaller firms grow faster.
Tschoegl, 1996	Logarithmic specification, growth-rate regression	Heteroskedasticity, persistence	66 Japanese regional banks over 1954–93 (only incumbents).	Moderate evidence that smaller firms grow faster.
Weiss, 1998	Logarithmic specification	Sample selection, heteroskedasticity, persistence	43,685 Austrian farms over 1986–90 (only incumbents).	Smaller firms grow faster.
Harhoff–Stahl–Woywode, 1998,	Growth-rate regression	Sample selection, heteroskedasticity	10,902 West German firms over 1989–94 (only incumbents).	Smaller firms grow faster.
Almus–Nerlinger, 1999	Logarithmic specification	Persistence	39,355 West German manufacturing firms over 1989–96 (only incumbents).	Smaller firms grow faster.

Source: Lotti F., Santarelli E., Vivarelli M. (2003), to which the reader is referred also for the full references to the mentioned works

Hypothesis (a) states the 'law of proportionate effects': growth is a *multiplicative* process independent of initial conditions. In other words, there are no systematic scale effects.

Note that were one to find $\theta_i > 1$ one ought to observe a persistent tendency toward monopoly. Conversely, $\theta_i < 1$ would be evidence corroborating regression-to-the-mean, and, indirectly, witness for some underlying 'optimal size' attractor.[5]

A good deal of evidence is summarized in Table 5.1, borrowed from Lotti, Santarelli and Vivarelli (2003).

Overall, hypothesis (a) which is indeed the object of most inquiries gets a mixed support:

(1) Most often, smaller firms – on average – grow faster (under the *caveat* that one generally considers small *surviving* firms);
(2) Otherwise, no strikingly robust relationship appears between size and average rates of growth (cf. Mansfiled (1962), Hall (1987), Kumar (1985), Bottazzi et al. (2006) and Bottazzi and Secchi (2003b), among others).
(3) The relationship between size and growth is modulated by the age of firms themselves – broadly speaking, with age exerting *negative* effects of growth rates, but *positive* effects on survival probabilities, at least after some post-infancy threshold (cf. Evans (1987a and b))[6].

Note that such pieces of evidence are easily consistent with evolutionary theories of industrial change. Indeed an evolutionary interpretation would be rather at odds with a notion of convergence to some invariant 'optimal' size, with decreasing returns above it. Conversely, it is rather agnostic on the precise specification of *non-decreasing* returns. In particular, it does not have any difficulty in accepting a world characterized by *nearly constant returns to scale*, (i.e. by values of θ_i in eq. 5.2 on average not too far from one) jointly with drivers of firm growth on average uncorrelated with size itself.

Conversely, precious clues on the basic characteristics of the processes of market competition and corporate growth are offered by the statistical properties of the 'error term' ($\varepsilon_i(t)$ in eq. 5.2). Note in this

---

[5] More rigorously, with $\theta < 1$ there exist a limit distribution with finite variance (if ϵ has a finite variance). In turn, any properly instructed economist would conjecture that such a distribution should display a good part of its mass around the 'optimal size' value. That is, intuitively even under the persistent arrival of 'disturbances' of several origins and several magnitudes, with $\theta < 1$ one may still easily conjecture some 'fundamental' driving tendency toward some underlying 'optimal structure' – whatever that means.

[6] Moreover, the relationship between size and growth appears to be influenced by the stage of development of particular industries along their life cycles: cf. Geroski and Mazzucato (2002).

Table 5.2 *Growth variability / firm-size relations: 'Scaling Law': σ $(g|s) \approx S^{\beta}$*

$\rightarrow$ *Aggregate manufacturing, U.S. data*

- Amaral et al. (1997): $\beta \approx -.2 \pm 0.03$
- Bottazzi and Secchi (2003b): $\beta \approx -0.19 \pm 0.01$

$\rightarrow$ *International pharmaceutical industry*

- Bottazzi et al. (2001): $\beta \approx -0.2 \pm 0.02$
- De Fabritiis Pammolli and Riccaboni (2003): $\beta \approx -0.17 \pm 0.05$

$\rightarrow$ *Aggregate and sectoral manufacturing, Italian data*

- Bottazzi et al. (2002): $\beta \approx .0$

respect that the absence of any structure in the growth process (as in fact argued by Geroski (2000)) would be very damaging indeed to evolutionary theories of industrial change. In fact, if one were to find corroboration to hypothesis (b) according to which – to recall – growth would be driven by a multiple, small 'atom-less' uncorrelated shocks, this would come as bad news to evolutionary interpretations whose basic building blocks comprise the twin notions of (1) persistent heterogeneity among agents, and (2) systematic processes of competitive selection among them.

What properties in fact the statistics on firm growth display?

Growth variability Since the early insights from Hymer and Pashigian (1962), a quite robust (albeit not unanimous) evidence suggests that the *variance* of firms growth rates *falls* as firms sizes increase (cf. Table 5.2 for a concise summary). Interestingly, however, it falls less than proportionally.

Why is that?

An interpretation is that the variance-scale relation depends on the *diversification-size* relation. In fact, firms grow by both expanding within their incumbent lines of business and by diversifying into new ones. In turn, if market dynamics across activities are not perfectly correlated and if size goes together with an increasing number of lines of business in which a firm operates, then one should indeed expect a lower variance for bigger firm sizes.[7]

---

[7] The relationship between diversification and growth variance might also explain the absence of such a scaling in the Italian evidence (Bottazzi et al. (2002)), probably due to

In the absence of any correlation in market dynamics across lines of business and with a number of lines of business proportional to size, one should expect to see the variance fall with the square root of size (that is, to observe a coefficient β in table 5.2 of around −0.5). However, most of the evidence suggests a coefficient of around −0.2, as such suggesting either non-proportionality in the relation diversification size or correlation between markets or a mixture of both. In fact, in Bottazzi et al. (2001) and Bottazzi and Secchi (2003b), one begins to disentangle the issue on the grounds of disaggregate data on the pharmaceutical industry, showing, at least in this case, that the scaling coefficient is entirely due to a less than proportional increase in the number of markets in which firms are active as a function of their size. Moreover, Bottazzi and Secchi (2006b) offers an explanation of such diversification patterns in terms of a *branching process* which is intuitively consistent with *capability-driven patterns of diversification*. As capability-based theories of the firm would predict, the expansion into new activities builds incrementally upon the knowledge and the complementary assets accumulated within existing ones (see also the conjectures in Teece et al. (1994) on the ensuing 'coherence' in the diversification profiles).

Growth rates distributions   One of the most important pieces of evidence able to throw some light on the underlying drivers of corporate growth regards the distribution of growth rates themselves.

For convenience consider again the normalized (log) size

$$s_i = \log S_i \, (t) \; - <\log S(t) >$$

where $<\log S(t)>$ ($\equiv 1/N \sum_i \log S_i(t)$) is the mean log size. The variable of interest is thus the normalized growth $g_i(t) = s_i(t+1) - s_i(t)$

The evidence suggests an extremely robust stylized fact: growth rates display distributions which are *at least exponential (Laplace) or even fatter in their tails* (see Stanley et al. (1996a) and Bottazzi and Secchi (2003b) on U.S. data; Bottazzi et al. (2001) on the international pharmaceutical industry; Bottazzi, Cefis and Dosi (2002) and Bottazzi et al. (2006) on the Italian industry).

Figures 5.4 and 5.5 present some examples from Italian data.

This property holds across (1) levels of aggregation; (2) countries; (3) different measures of size (e.g. sales, employees, value added, assets),

---

low degrees of diversification of Italian firms, *as they appear in the statistics*. Anecdotal evidence suggests in fact that diversification events often entail the formation of a new legal entity (also due to fiscal reasons) rather than the development of new lines of business within the original company.

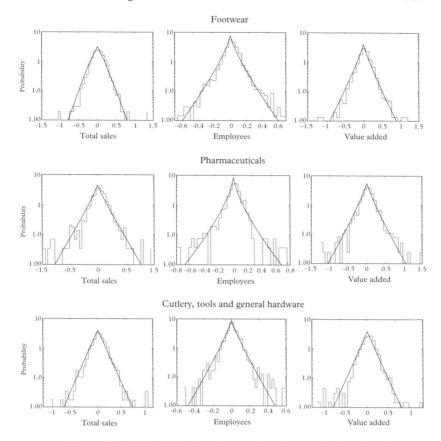

Figure 5.5   Probability densities and maximum likelihood estimation of firm growth rates *g* in three different Italian sectors. *Source* Bottazzi G., Cefis E., Dosi G., and Secchi A. (2006)

even if (4) one observes some (moderate) variations across sectors with respect to the distribution parameters.

Note that such statistical properties of growth rates are indeed good news for an evolutionary analyst. The generalized presence of fat tails in the distribution implies much more structure in the growth dynamics than generally assumed. More specifically, ubiquitous fat tails are a sign of some underlying correlating mechanism which one would rule out if growth events were normally distributed, small and independent. In Bottazzi et al. (2006) we conjecture that such mechanisms are likely to be of two types. First, the very process of competition induces correlation. Market shares must obviously add up to one: someone's gain is someone else's loss. Second, in an evolutionary world one should indeed expect 'lumpy' growth events (of both positive and negative sign) such as the introduction of new products, the construction/closure of plants and entry to and exit from particular markets.[8]

Autocorrelation in growth rates   Another piece of evidence on the structure of growth processes concerns the possible autocorrelations over time. Here the variable under study is in the first difference $g_i(t + \tau) - g_i(t)$, where, as above, the g_i (.) are the (normalized) growth rates of each firm i. Begin by noting that ideally one would like to have time series long enough to describe the properties of the sample path of each firm on the grounds of the conjecture that the evolutionary pattern of each firm ought to be specific to each entity in its interactions with the population of other firms which happen to compete in that particular market in those particular times – all bearing distinctly different technological, organizational and strategic features.

Well short of that, one generally has to be content with *sectoral averages* in the differences $<g_i(t+\tau) - g_i(t)>$, under different autoregressive lags.

Interestingly, in an industry for which one has reasonable longitudinal panel data at different levels of disaggregation – namely the international drugs industry – one does find a robust autocorrelation structure. For example, firm-level growth rates exhibit a long-lasting positive autocorrelation, statistically significant up to the seventh lag (cf. Bottazzi et al. (2001)).

Broader, inevitably coarser, evidence typically on three-digit sectors (as such already aggregates of a quite large number of lines of business) like that in Bottazzi and Secchi (2003b) on U.S. manufacturing displays (1) only a relatively short autoregressive structure (typically with one-lag-only significance); and (2) a good deal of inter-sectoral variability.

At similar levels of aggregation, the Italian panel of manufacturing firms often displays average autocorrelations which are quite small (around $|.1|$), and significant if at all only at the first lag (Cf. Bottazzi et al. (2006)). Even in this case, however, the data suggest *highly heterogeneous firm-specific autocorrelation profiles* within each sector. This is confirmed by 'bootstrapping' exercises involving the comparison between the distribution of *actual* firm-specific coefficients with any

---

[8] Suggestive attempts to model increasing-return dynamics yielding the observed fat-tailed distribution are in Bottazzi and Secchi (2003a) and (2006b).

'virtual' one obtained by randomly scrambling actual growth rates over the same (but randomly drawn) firms. The two distributions turn out to be significantly different, meaning that there are *systematic but idiosyncratic differences* in autocorrelation structures, which are not captured by sectoral average autocorrelation coefficients (cf. Bottazzi, Cefis, and Dosi (2002)).[9]

5.1.3   *Profitabilities and their dynamics*

Together with corporate growth, profitability is another crucial measure of revealed corporate performances. There are three major intertwined issues here, namely (1) the revealed *inter-firm differences* in profitability proxies; (2) their persistence over time; and (3) the properties of their patterns of change.

Some due premises. I strongly believe that simpler measures are better measures because they reduce theory-driven biases. So, for example, derivations of profitability measures from purported technological relations which nobody has actually seen such as Cobb-Douglas and alike are likely to lead to blind alleys (those on this point in Dosi and Grazzi (2006)).

Given that, let me stick to the simplest possible measure of profitability, aiming at the same time at the highest possible degree of sectoral disaggregation.

Consider the variables

$$\text{gom}_i(t) = \log(\text{GOM}_i(t)) - log(GOM(t))$$
$$\text{GOM}_i\,(t) = \text{VA}_i(t) - W_i(t)$$

where

$$\text{GOM}_i = \textit{gross operating margins}$$
$$\text{VA}_i = \textit{value added}$$
$$\text{W}_i = \textit{total wage costs}$$

and, as above, $< \ldots >$ stands for the sectoral averages.

If capital/output ratios are not too different across firms – as they should not be – the more one refines the sectoral disaggregation, then the simple *MOL* measure should not be too biased a proxy for 'true'

---

[9] Revealing complementary evidence to the same effect suggests that even growth paths, *conditioned on size*, tend to be significantly *firm-specific*: cf. Cefis, Ciccarelli, and Orsenigo (2002).

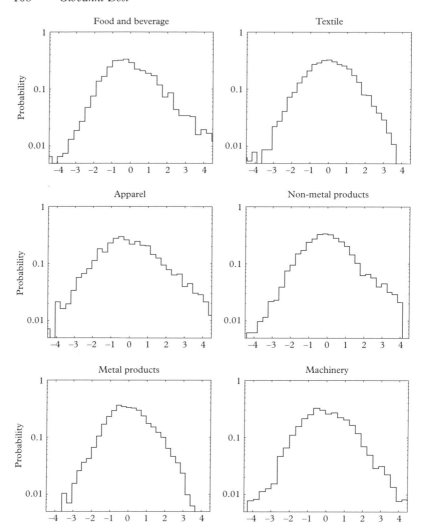

Figure 5.6   Distribution of (normalized) gross margins by sectors.
Source Our elaboration of Italian (ISTAT MICRO.1) data: cf. text on
the data description

profitabilities. Figure 5.6 offers some impressive evidence drawn from
the Italian sample on inter-firm profitability asymmetries: the reader is
indeed invited to appreciate the width of the support of the density
distributions going well beyond, say, ten to one ratios in profitability
margins between the best and the worse performers.

Given that, a crucial property regards the persistence of such differentials. After all, evidence on low persistence could simply suggest that capitalism involves daring and heroic efforts by multitudes of firms which happen to make many mistakes as well as reap huge rewards, but markets are there to help and quickly redress individual mistakes and wash away abnormal rents. It turns out that this view does not quite match the evidence.

There is indeed a wide literature on the *persistence of profitability differences across firms*: see, among others, Müller (1986) and (1990), Cubbin and Geroski (1987), Geroski and Jacquemin (1988), Goddard and Wilson (1999), Cable and Jackson (2003), Cefis (2003b), Gschwandter (2004). The Italian evidence strongly supports the persistence view.

As shown in Table 5.3, the autocorrelation over time in profit margins is extremely high in all manufacturing sectors, with just a relatively mild tendency of mean-reversion, revealed by both the negative coefficient on the first differences and the value of the autoregressive coefficient on the *levels* slightly lower than unity.

Finally note that, interestingly, the rates of change in profit margins display distributions which are again fat-tailed, at least exponential, or even fatter-tailed: see Figure 5.7, displaying the growth rates of the normalized margins, $g_{\mathrm{gom}_i}(t) = \mathrm{gom}_i(t + 1) - \mathrm{gom}_i(t)$. The sectors shown in the figure are chosen simply to illustrate the point that the property holds across activities that are very different in terms of technologies and forms of corporate organizations.

For the interpretation of such an evidence let me refer the reader back to the discussion of a similar evidence in the case of growth rates of companies as such. Again we find here the mark of powerful underlying correlation mechanisms which tend to induce 'coarse grained' shocks upon profitabilities.

Recalling our previous argument, consider – as a term of comparison – a process of variation in profitabilities of individual firms driven by little idiosyncratic shocks occurring all the time, independent from each other. A caricatural way of illustrating it is by depicting a multitude of producers which all survive near equilibrium (i.e. in the conventional definitions, near a zero-profit steady state), while being nonetheless continuously hit by small and uncorrelated profit opportunities (e.g. one or few unexpected or uniformed customers; some small advances on products characteristics, etc.) If such shocks are uncorrelated, again for the law of large numbers, summing up over, say, years, one should expect normally distributed changes. Not getting it as such is a revealing evidence on 'drivers of change' which are more 'lumpy' and more powerfully correlated with each other.

Table 5.3 *Autocorrelation of gross margin levels and growth rates*

| | ISIC CODES | LEVELS | | DIFFERENCES | |
		AR(1) coefficient	Standard deviation	AR(1) coefficient	Standard deviation
Food products and beverages	15	0,9516	0,0060	−0,3424	0,0190
Textiles	17	0,9117	0,0067	−0,2953	0,0163
Wearing apparel, dressing and dyeing of fur	18	0,9328	0,0087	−0,3992	0,0254
Tanning and dressing of leather; luggage, footwear	19	0,8781	0,0144	−0,2984	0,0307
Manufacturing of woods and related products	20	0,9430	0,0115	−0,3658	0,0332
Paper and allied products	21	0,9432	0,0104	−0,2935	0,0298
Printing and publishing	22	0,9471	0,0098	−0,3917	0,0293
Chemicals and allied products	24	0,9370	0,0081	−0,3315	0,0216
Rubber and miscellaneous plastics products	25	0,9203	0,0085	−0,2917	0,0208
Other non-metallic mineral products	26	0,9430	0,0067	0,3337	0,0190
Basic metals	27	0,9149	0,0101	−0,3245	0,0248
Fabricated metal products, except machinery	28	0,9213	0,0064	−0,2957	0,0155
Industrial machinery and equipment	29	0,9207	0,0055	−0,3215	0,0137
Electrical machinery and apparatus	31	0,9649	0,0086	−0,2382	0,0265
Forniture and other N.E.C. manu- facturing industries	36	0,9210	0,0080	−0,3349	0,0185

Source: Our elaborations on Italian (ISTAT MICRO .1) data. The selected sectors are those which include more than 200 firms

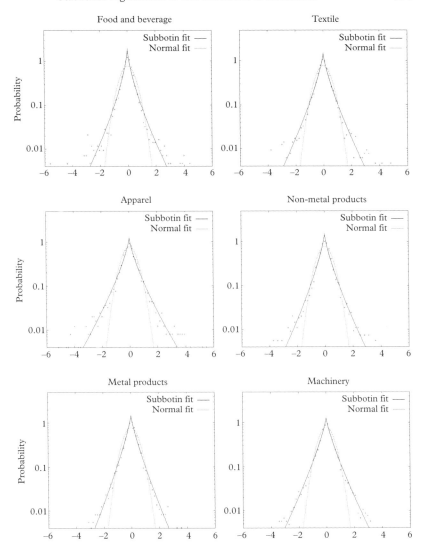

Figure 5.7 Distribution of gross margins growth rates by sectors, Italian data. Each figure displays the maximum likelihood estimates of Subbotin distributions (cf. Bottazzi and Secchi (2006a)) (The fit of a normal distribution is added just to highlight how much were fatter than the observed data are)

*5.1.4   The statistical structure of industrial evolution:
some concluding remarks*

There are possibly two major messages that come from the whole discussion so far.

The *first*, more methodological one, is that there is a rich statistical structure in the dynamics of industries which has remained largely neglected until recently, as long as most analyses simply focused on *average* relations between corporate performances and corporate characteristics, or just between firm sizes and firm rates of growth. Indeed, the revealed structure of the stochastic processes describing industrial evolution bear the familiar signs of all *complex system dynamics*, including the fat-tailed distributions in the rates of changes of all variables of interest. That, in turn, is likely to witness for the existence of some underlying correlation mechanism, which makes the system (in our case, each industry) 'self-organizing' in its growth process. In most respects, the statistical evidence on industrial change corroborates the exciting conjecture that evolutionary phenomena tend to generically undergo 'non-Gaussian' lives – influenced by persistent (positive or negative) interactions amongst agents within and across relevant populations.

Second, but relatedly, the core indicators of corporate performances discussed in this section – that is growth and profitability – reveal a widespread and profound *heterogeneity* across firms that persist over time notwithstanding the competition process. Given all that, a natural question concerns the sources of such heterogeneities themselves.

5.2   Behind heterogeneous performances: innovation and production efficiency

Straightforward candidates for the explanation of the differences in corporate performances are in fact (1) differences in the ability to innovate and/or adopt innovation developed elsewhere regarding product characteristics and production processes; (2) different organizational arrangements; (3) different production efficiencies.

Needless to say, the three sets of variables are profoundly related. Technological innovations typically involve also changes in the organization of production. Different ways of searching for innovations imply distinct organizational arrangements regarding the relationships amongst different corporate tasks (e.g. R&D, production, sales). And, most obviously, technological and organizational innovations ultimately shape the degrees of efficiency in which inputs happen to generate outputs.

With that in mind, let me offer some telegraphic overview of the evidence concerning the patterns of technological innovation, on the one hand, and production efficiencies on the other. (I am forced to neglect here the role of organizational variables. In fact, *organizational capabilities* are intimately linked with the very process of technological innovation and with production efficiencies: cf. among others the discussions to which I have contributed in Dosi, Nelson and Winter (2000) and Dosi, and Marengo Faillo (2005)).

5.2.1   Technological innovativeness

A rich and wide literature in the field of economics of innovation does indeed suggest that firms deeply differ also in their ability to innovate: for detailed surveys and discussions, see Freeman (1994), Freeman and Soete (1997), Nelson (1981) and (1991), Pavitt (1999), Dosi, Orsenigo and Sylos Labini (2005), Dosi (1988).

(1) Innovative capabilities appear to be highly asymmetric, with a rather small number of firms in each sector responsible for a good deal of innovation output.

(2) Somewhat similar considerations apply to the *adoption* of innovations (in the form of new production inputs, machinery, etc.) revealing asymmetric capabilities of learning and 'creative adaptation' and entailing long-lasting logistic-shape profiles of diffusion.

(3) Differential degrees of innovativeness are generally persistent over time and often reveal a small 'core' of systematic innovators (together with the foregoing broad critical surveys, see more specifically Cefis (2003a and c)).

(4) Relatedly, while the arrivals of major innovations are rare events, they are not independently distributed across firms. Rather, recent evidence suggests that they tend to arrive in firm-specific "packets" of different sizes.[10]

In terms of intuitive comparisons of such evidence with the predictions of evolutionary theorizing, heterogeneity in innovative/initiative abilities is indeed a robust piece of corroborating evidence. And so is the evidence on micro-correlation of innovative events, well in tune with an evolutionary notion of few, high-capability, persistent innovators.

---

[10] On the statistical properties of the discrete innovations, *in general*, cf. Silverberg (2003) arguing for a secular Poisson-type process. However, at a much finer level of observation the firm-specific patterns of innovation are not likely to be Poisson-distributed. Rather, as one shows in Bottazzi et al. (2001) in the case of the pharmaceutical industry, few firms 'draw' relatively large 'packets' of innovations well described by Bose–Einstein (rather than Poisson) statistics.

On a much larger scale, the persistent asymmetries across countries, even within the same lines of business, cry out in favour of profound heterogeneities in learning and searching capabilities.

5.2.2   *Production efficiencies*

As well known, there are two straightforward measures of production efficiency, namely labour and total factor productivity (TFP).

It should come as no surprise at this point of the discussion that, despite its obvious limitations, I tend to prefer a measure based on the net output (that is the 'real' value added) per employee or, even better, per worked hours. The reason for this preference lies in the dubious elements which make up conventional production functions, in turn the instrument necessary to yield the TFP measure. This is not the place to discuss the issue. Suffice to mention, first, that technologies as we know them essentially involve *complementarities* among inputs – so that it makes little sense to separate the 'contribution' of each 'factor' to the final output. To paraphrase on a suggestive metaphor suggested by Dick Nelson, it makes as much sense as trying to disentangle the separate contributions of butter, sugar, eggs, and so on to the taste of a cake. (Again I am forced to refer to Dosi and Grazzi (2006) for more details.)

Second, but related, one typically lives in a technological world characterized by micro-coefficients which are fixed in the short term (i.e. each firm basically masters the technique actually in use), while in the longer term techniques change essentially because of learning and technical progress. Conversely, if this is the case, it does not make much sense to distinguish changes *along* any purported production function versus changes of the function itself.

Come as it may, an overwhelming evidence *concerning both labour productivity and TFP* and at all levels of disaggregation suggests widespread differences in production efficiency *across firms and across plants* which tend to be persistent *over time*: see, among others, Nelson (1981), Baily, Hulten and Compbell (1992), Baldwin (1995), Bartelsman and Doms (2000), Foster, Haltiwanger and Krizan (2001), Jensen and McGuckin (1997), Power (1998).

Our Italian data are well in tune with such stylised facts. Figure 5.8 presents the distribution of (normalized) value added per employee, that is

$$\pi_i(t) = \log \Pi_i(t) - <\log \Pi_i(t)>$$

whereby

$$\Pi_i(t) = VA_i/N_i;$$

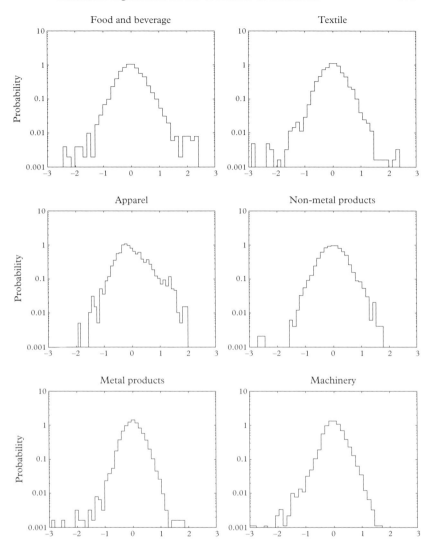

Figure 5.8   Distribution of labour productivity by sectors. *Source*: our elaboration on Italian (ISTAT MICRO.1) data

and

$<\log \Pi_i(t)> \equiv$ mean (log) value added (VA) per employee (N) averaged over all firms in any particular sector.

Moreover, as shown in Table 5.4, productivity differentials are quite stable over time with some mild regression-to-the-mean tendency.

Table 5.4 *Autocorrelation of labour productivity levels and growth rates*

(Labour productivity=VA/#employees)	ISIC CODES	LEVELS		DIFFERENCES	
		AR(1) coefficient	Standard deviation	AR(1) coefficient	Standard deviation
Food products and beverages	15	0,8619	0,0092	-0,2641	0,0208
Textiles	17	0,8699	0,0076	-0,2770	0,0171
Wearing apparel, dressing and dyeing of fur	18	0,9285	0,0087	-0,3428	0,0250
Tanning and dressing of leather; luggage, footwear	19	0,8932	0,0158	-0,3123	0,0398
Manufacturing of woods and related products	20	0,8357	0,0154	-0,3254	0,0312
Paper and allied products	21	0,8772	0,0140	-0,2348	0,0030
Printing and publishing	22	0,8391	0,0127	-0,1596	0,0319
Chemicals and allied products	24	0,7947	0,0132	-0,1883	0,0234
Rubber and miscellaneous plastics products	25	0,8920	0,0108	-0,2831	0,0244
Other non-metallic mineral products	26	0,9057	0,0077	0,3065	0,0195
Basic metals	27	0,8583	0,0135	-0,1645	0,0270
Fabricated metal products, except machinery	28	0,8572	0,0079	-0,3299	0,1580
Industrial machinery and equipment	29	0,8098	0,0079	-0,3177	0,0143
Electrical machinery and apparatus	31	0,8534	0,0119	-0,1072	0,0236
Manufacturing of other transport equipment	35	0,7518	0,0299	-0,3490	0,0481
Forniture and other N.E.C. manufacturing industries	36	0,8609	0,0093	-0,3512	0,0187

Source: Our elaborations on Italian (ISTAT MICRO .1) data

Also at the level of input efficiencies the broad picture is characterized by general and profound heterogeneity across firms.

As Griliches and Mairesse (1997) vividly put it

"we ... thought that one could reduce heterogeneity by going down from general mixtures as 'total manufacturing' to something more coherent, such as 'petroleum refining' or 'the manufacture of cement'. But something like Mandelbrot's fractal phenomenon seem to be at work here also: the observed variability – heterogeneity does not really decline as we cut our data finer and finer. There is a sense in which different bakeries are just as much different from each others as the steel industry is from the machinery industry."

For evolutionary scholars, heterogeneity in the degrees of innovativeness and production efficiencies should not come as a surprise. Indeed, this is what one ought to expect to be the outcome of idiosyncratic capabilities (or lack of them), mistake-ridden learning and forms of path-dependent adaptation. Differences in innovative abilities and efficiencies (together with differences in organizational set-ups and behaviours) ought to make up the distinct corporate 'identities' which in turn should somehow influence those corporate performances discussed in the previous section.

But do they? How? And on what time scales?

5.2.3   Corporate capabilities, competition and performances

Let us distinguish between profitability and growth indicators of performances.

The positive impact of innovativeness upon corporate profitabilities appears to be well documented: see Geroski, Machin and van Reenen (1993), Cefis (2004), Cefis and Ciccarelli (2005), Roberts (1999), among others; see also Kremp and Mairesse (2004) on the relationship between innovation and productivity.

Together our Italian data highlight a positive relationship between profit margins and relative labour productivities (that is, normalized with the respective sectoral means): see Figure 5.9.

At the same time, the impact of both innovativeness and production efficiency upon growth performances appears to be somewhat more controversial. Mainly North-American evidence, mostly at *plant* level, does suggest that increasing output shares in high-productivity plants and decreasing shares of output in low-productivity ones are very important drivers in the growth of average productivities, even if the process of displacement of lower efficiency plants is rather slow (cf. the evidence discussed in Baily et al. (1992) and Baldwin (1995)).

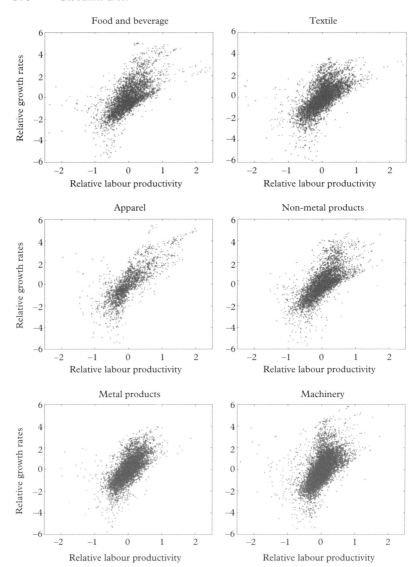

Figure 5.9 Gross margins and (normalized) labour productivity, 1989–1997. *Source*: Our elaboration on Italian (ISTAT MICRO.1) data

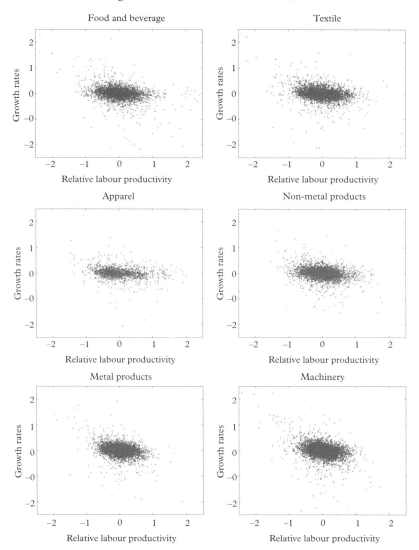

Figure 5.10 Labour productivity and growth rates (measured in terms of sales), 1989–1997. *Source*: Our elaboration on Italian (ISTAT MICRO.1) data (Growth of Value Added)

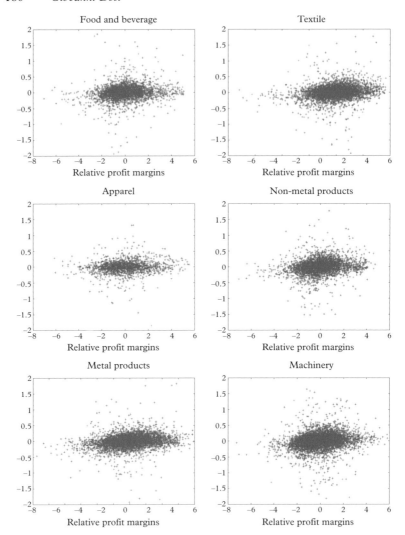

Figure 5.11   Growth rates and profit margins in different manufacturing sectors. *Source* Our elaboration on Italian (ISTAT MICRO.1) data. (Growth of Value Added)

Firm-level data are less straightforward. For example, our Italian data show

(1) A weak or non-existent relationship between growth however measured (e.g. in terms of Value Added, Employment or Sales)

and relative productivities (see Figure 5.10): more efficient firms do *not* grow more;

(2) Even when some positive relation between efficiency and growth appears, this is almost exclusively due to the impact of few *outliers* (the very best and the very worst);

(3) Similarly, no systematic relation appears between (relative) profit margins and (relative) growth rates (cf. Figure 5.11).

(4) Moreover, the evidence from other data sets such as the international pharmaceutical industry shows that more innovative firms do *not* grow more (Bottazzi et al. (2001)). Rather the industry constantly displays the coexistence of heterogeneous types of firms (e.g. innovators versus imitators).

The implications of all these empirical regularities, if confirmed by the observation of other countries and other industries are far-reaching. Let us consider them from an evolutionary perspective.

5.3   Evolutionary interpretations: corroborations and challenges by way of a conclusion

How well does the whole statistical story reviewed in this essay fit with evolutionary interpretations?[*]

Certainly, the recurrent evidence at all levels of observation of *inter-firm heterogeneity* and its persistence over time is well in tune with an evolutionary notion of idiosyncratic learning, innovation (or lack of it) and adaptation.

Heterogeneous firms compete with each other and, given the prevailing input and output prices obtain different returns. Putting it in a different language, they obtain different 'quasi-rent' on conversely losses below the notional 'pure competition' profitability. At the same time, even leaving aside any entry or mortality phenomenon, surviving incumbents undergo changes in their market shares and therefore in their relative (and, of course, absolute) sizes.

In all that, the evidence increasingly reveals a rich structure in the processes of learning, competition and growth.

Various mechanisms of correlation – together with the 'sunkness' and indivisibilities of many technological events and investment

---

[*] Here, with evolutionary interpretation we mean that body of literature focusing on economic change as an evolutionary process driven by technological and organizational change which finds one of its central roots in Nelson and Winter (1982). See also Winter (1984) and for a discussion of its main building, Dosi and Winter (2002).

decisions – yield a rather structured process of change in most variable of interest – for example size, productivity, and profitability – also revealed by the 'fat-tailedness' of the respective growth rates.

At the same time, market selection – the other central tenet, together with learning, of evolutionary interpretations of economic change – do not seem to work particularly well, at least on the yearly timescale at which statistics are reported (while the available time series are not generally long enough to precisely assess what happens in the long run). Conversely, diverse degrees of efficiencies and innovativeness seem to yield primarily relatively persistent profitability differentials.

That is, contemporary markets do not appear to be too effective selectors delivering rewards and punishments according to differential efficiencies. Moreover, the absence of any strong relationship between profitability and growth militates against the 'naively Schumpeterian' (or for that matter 'classic') notion that profits feed growth (by plausibly feeding investments).

Finally, the same evidence appears to run against the conjecture, put forward in the 1960s and 1970s by the 'managerial' theories of the firm on a trade-off between profitability and growth, with 'managerialized' firms trying to maximize growth subject to a minimum profit constraint.[11]

In turn, the very fact that market selection might play less of a role than that assumed in many models of evolutionary inspiration, if confirmed, is as such an important advance in the understanding of how markets work (or do not).

More generally, the increasing availability of longitudinal panel data with an array of variables describing both the 'inner features' and the performances of individual firms begins to unveil the rich statistical structure of the processes of industrial evolution. In that, one can go a long way, I have tried to show, with little or no use of (typically unobservable) strategic variables. One has just begun. Ahead lie, first, exercises of 'evolutionary accounting' trying to disentangle the relative role of entry, market selection and incumbent learning as drivers of industrial change. Together, second, it is of paramount importance to try to condition the observed performance profiles of individual firms upon their underlying technological and organizational 'identities'.

There is indeed a whole world to be discovered resting somewhere in between the 'pure stochasticity' of a Gibrat-type framework, on the one extreme, and, the *ex post* rationalization of whatever observation in terms of sophisticated hyper-rational behaviours, on the other.

---

[11] In fact the absence of such a trade-off had been already noted by Barna (1962). Note also that this proposition is orthogonal to the finding that current growth appears to be correlated with *future* long-term profitability (cf. Geroski, Machin and Walters (1997)).

References

Amaral, L.A.N., Buldyrev S.V., Havlin, S., Salinger, M.A., Stanley, H.E., and Stanley, M.H.R. 1997. 'Scaling behavior in economics: the problem of quantifying company growth', *Physica A*, 244: 1–24

Axtell, R.L. 2001. 'Zipf distribution of U.S. firm sizes', *Science* 293: 1818–20

Baily, M.N., Hulten, C., and Campbell, D. 1992. 'Productivity dynamics in manufacturing plants', *Brookings Papers on Economic Activity. Microeconomics*, 187–267 (with comments by Bresnehan, T. and Caves, R.E.)

Baldwin, J.R. 1995. *The Dynamics of Industrial Competition*. Cambridge: Cambridge University Press

Barna, T. 1962. *Investment and Growth Policies in British Industrial Firm*. Cambridge: Cambridge University Press

Bartelsman, E. and Doms, M. 2000. 'Understanding productivity: lessons from longitudinal microdata', *Journal of Economic Literature* 38: 569–94

Bartelsman, E., Scarpetta, S., and Schivardi, F. 2005. 'Comparative analysis of firm – demographics and survival: micro level evidence for the OECD countries', *Industrial and Corporate Change* 14: 365–91

Bottazzi, G. and Secchi, A. 2003a. 'Why are distributions of firm growth rates tent-shaped?', *Economics Letters* 80: 415–20

Bottazzi, G. and Secchi, A. 2003b. 'Common properties and sectoral specificities in the dynamics of U.S. manufacturing companies', *Review of Industrial Organization* 23: 217–32

Bottazzi, G. and Secchi, A. 2005. 'Growth and diversification patters of the worldwide pharmaceutical industry', *Review of Industrial Organization* 26: 195–216

Bottazzi, G. and Secchi, A. 2006a. 'Explaining the distribution of firms growth rates', *Rand Journal of Economics*, forthcoming

Bottazzi, G. and Secchi, A. 2006b. 'Firm diversification and the Law of Proportional Effect', *Industrial and Corporate Change, forthcoming*

Bottazzi, G., Cefis, E., and Dosi, G. 2002. 'Corporate growth and industrial structure. Some evidence from the Italian manufacturing industries', *Industrial and Corporate Change* 11: 705–23

Bottazzi, G., Cefis, E., Dosi, G., and Secchi, A. 2006. 'Invariances and Diversities in the Evolution of Manufacturing Industries'. *Small Business Economics*, forthcoming

Bottazzi, G., Dosi, G., Lippi, M., Pammolli, F., and Riccaboni, M. 2001. 'Innovation and corporate growth in the evolution of the drug industry', *International Journal of Industrial Organization* 19: 1161–87

Cable, J.R. and Jackson, R.G. 2003. 'The persistence of profits in the long-run: a new approach', *American Economic Review* 93: 1075–90

Cefis, E. 2003a. 'Is there persistence in innovative activities?', *International Journal of Industrial Organization* 21: 482–515

Cefis, E. 2003b. 'Persistence in innovation and profitability', *Rivista Internazionale di Scienze Sociali* 110: 19–37

Cefis, E. 2003c. 'Is there persistence in innovative activities?', *International Journal of Industrial Organization* 21: 489–515

Cefis, E. 2004. *Persistent asymmetries in firm performances*. Bergamo: Department. of Economics, mimeo

Cefis, E. and Ciccarelli, M. 2005. 'Profit differentials and innovation', *Economics of innovation and new technologies* 14(1–2): 43–61

Cefis, E., Ciccarelli, M., and Orsenigo, L. 2002. 'From Gibrat's Legacy to Gibrat's Fallacy. A Bayesian Approach to Study the Growth of Firms', WP–AD 2002–19, IVIE, University of Alicante, Alicante

Cubbin, J. and Geroski, P. 1987. 'The convergence in profits in the long-run: inter-firm and inter-industry comparisons', *Journal of Industrial Economics* 35: 427–42

De Fabritiis, G., Pammolli, F., and Riccaboni, M. 2003. 'On the size and growth of business firms', *Physica A* 324: 38–44

Dosi, G. 1988. 'Sources, procedures and Microeconomic Effects of Innovation', *Journal of Economic Literature* 26: 1120–71

Dosi, G., Nelson, R., and Winter, S. (eds.) 2000. *The Nature and Dynamics of Organizational Capabilities*. Oxford/New York: Oxford University Press

Dosi, G., Marsili, O., Orsenigo, L., and Salvatore, R. 1995. 'Learning market selection and the evolution of industrial structures', *Small Business Economics* 7: 411–36

Dosi, G., Faillo, M. and Marengo, L. 2005. 'Organizational capabilities, patterns of knowledge accumulation and governance structures in business firms: an introduction', forthcoming in Touffut, J-P. (ed.), *Organizational Innovation within Firms*, Cheltenham, UK and Brookfield, U.S.: Edward Elgar, (currently available as Sant'Anna School of Advanced Studies, LEM Working Paper 2003/11)

Dosi, G. and Grazzi, M. 2006. 'Technologies at problem-solving procedures and Technologies at input-output relations: some perspectives on the theory of production', *Industrial and Corporate Change* 15: 173–202.

Dosi, G., Orsenigo, L., and Sylos Labini, M. 2005. 'Technology and the economy', in Smelser and Swedberg (eds.) *The Handbook of Economic Sociology*, 2nd edn, Princeton, NJ: Princeton University Press, Russell Sage Foundation

Dosi, G. and Winter, S. 2002. 'Interpreting economic change: evolution, structures and games', in Augier, M. and March, J. (eds.), *Choice, Change and Organizations: Essays in Memory of Richard M. Cyert*, Cheltenham: Edward Elgar.

Evans, D.S. 1987a. 'The relationship between firm growth, size and age: estimates for 100 manufacturing industries', *Journal of Industrial Economics* 35: 567–81

Evans, D.S. 1987b. 'Tests of alternative theories of firm growth', *Journal of Political Economy* 95: 657–74

Foster, L., Haltiwanger, J.C. and Krizan, C.J. 2001. 'Aggregate productivity growth: lessons from microeconomic evidence', in Dean, E., Harper, M., and Hulten, C. (eds.), *New Developments in productivity Analysis*. Chicago: Chicago University Press.

Freeman, C. 1994. 'The economics of technical change', *Cambridge Journal of Economics* 18: 463–514

Freeman, C. and Soete, L. 1997. *The Economics of Industrial Innovation*. London: Pinter, 2nd edn.

Geroski, P.A. 2000. 'The growth of firms in theory and in practice' in Foss and Mahnke (eds.), *New Directions in Economics Strategy Research*, Oxford: Oxford University Press

Geroski, P. and Jacquemin, A. 1988. 'The persistence of profits: a European comparison', *Economic Journal* 98: 357–89

Geroski, P. and Mazzucato, M. 2002. 'Learning and the sources of corporate growth', *Industrial and Corporate Change* 11: 623–44

Geroski, P., Machin, S.J., and van Reenen, J. 1993. 'The profitability of innovating firms', *Rand Journal of Economics* 24: 198–211

Geroski, P., Machin, S.J. and Walter, C.F. 1997. 'Corporate growth and profitability', *Journal of Industrial Economics* 45: 171–89

Gibrat, R. 1931. *Les inégalités économiques*. Paris: Librairie du Recueil Sirey

Goddard, J.A. and Wilson, J.O.S. 1999. 'The persistence of profit: as new empirical interpretation', *International Journal of Industrial Organization* 17: 663–87

Griliches, Z. and Mairesse, J. 1997. 'Production function: the search for identification', in Steiner Strøm (ed.), *Econometrics and Economic Theory in the Twentieth Century: the Ragner Frisch Centennial Symposium*, Cambridge: Cambridge University Press

Gschwandtner, A. 2004. *Profit persistence in the 'Very' Long Run: Evidence from Survivors and Exiters*. Vienna: University of Vienna, Department of Economics, WP 0401

Hall, B.H. 1987. 'The relationship between firm size and firm growth in the U.S. manufacturing sector', *Journal of Industrial Economics* 35: 583–606

Hart, P.E. and Prais, J.S. 1956. 'The analysis of business concentration', *Journal of the Royal Statistical Society* 119: 150–91

Hart, P.E. and Dulton, N. 1996. 'Firm size and rate of growth', *Economic Journal* 106: 1242–52

Hymer, S. and Pashigian, P. 1962. 'Firm size and rate of growth', *Journal of Political Economy* 70: 556–69

Ijiri, Y. and Simon, H.A. 1977. *Skew Distributions and the Sizes of Business Firms*. Amsterdam: North Holland

Kalecki, M. 1945. 'On the Gibrat distribution', *Econometrica* 13: 161–70

Klepper, S. 1997. 'Industry life cycles', *Industrial and Corporate Change* 6: 145–81

Kremp, E. and Mairesse, J. 2004. '*Knowledge Management, Innovation and Productivity: A Firm Level Exploration Based on French Manufacturing Data*'. Cambridge MA: NBER, WP 10237

Kumar, M.S. 1985. 'Growth, acquisition activity and firm size: evidence from the United Kingdom', *Journal of Industrial Economics* 33: 171–96

Jensen, B. and McGuckin, H. 1997. 'Firm performance and evolution: empirical regularities in the US micro data', *Industrial and Corporate Change* 6: 25–47

Lotti, F., Santarelli, E. and Vivarelli, M. 2003. 'Does Gibrat's law hold in the case of small, young, firms?', *Industrial and Corporate Change* 13: 213–35

Lucas, R.E. 1978. 'On the size distribution of business firms', *Bell Journal of Economics* 9: 508–23

Mansfield, E. 1962. 'Entry, Gibrat's law, innovation and the growth of firms', *American Economic Review* 52: 1023–51

Marsili, O. 2001. *The Anatomy and Evolution of Industries*. Cheltenham: Edward Elgar

Mueller, D. 1986. *Profits in the Long-Run*. Cambridge: Cambridge University Press

Mueller, D. (ed.) 1990. *The Dynamic of Company Profits. An International Comparison*. Cambridge: Cambridge University Press

Nelson, R. 1981. 'Research on productivity differences: dead ends and new departures', *Journal of Economic Literature* 19: 1029–64

Nelson, R. 1991. 'Why do firm differ and how does it matter?', *Strategic Management Journal* 12: 61–74

Nelson, R. and Winter, S. 1982. *An Evolutionary Theory of Economic Change*. Cambridge MA: The Belknap Press of Harvard University Press

Pavitt, K. 1999. *Technology, Management and Systems of Innovation*. Cheltenham: Edward Elgar

Power, L. 1998. 'The missing link: technology, investment and productivity', *Review of Economics and Statistics* 80: 300–13

Pryor, F.L. 1972. 'An international comparison of concentration ratios', *The Review of Economics and Statistics* 54: 130–40

Robert, P.W. 1999. 'Product innovation, product-market competition and persistent profitability in the U.S: pharmaceutical industry', *Strategic Management Journal* 20: 655–70

Schmalensee, R. 1989. 'Inter-industry studies of structure and performance' in Schmalensee and Willig (eds.), *Handbook of Industrial Organization*. Amsterdam: North-Holland

Silverberg, G. 2003. 'Breaking the waves: a Poisson regression approach to Schumpeterian clustering of basic innovations', *Cambridge Journal of Economics* 27(5): 671

Simon, H.A. and Bonini, C.P. 1958. 'The size distribution of American firms', *American Economic Review* 48: 607–17

Stanley, M.H.R., Amaral, L.A.N., Buldyrev S.V., Havlin, S., Leschhorn, H., Maass, P., Salinger, M.A, and Stanley, H.E. 1996a. 'Scaling behavior in the growth of companies', *Nature* 379: 804–06

Stanley, M.H.R., Buldyrev, S.V., Havlin, S., Mantegna, R., Salinger, M.A. and Stanley, H.E. 1996b. 'Zipf plots and the size distribution of firms', *Economic Letters* 49: 453–57

Steindl, J. 1965. *Random Processes and the Growth of Firms*. London: Griffin

Sutton, J. 1997. 'Gibrat's legacy', *Journal of Economic Literature* 35: 40–59

Teece, D., Rumelt, R., Dosi, G., and Winter, S. 1994. 'Understanding corporate coherence: theory and evidence', *Journal of Economic Behavior and Organization* 23: 1–30

Winter, S. 1984. 'Schumpeterian competition in alternative technological regimes', *Journal of Economic Behavior and Organization* 5: 287–320

6 Spin-off entry in high-tech industries: motives and consequences

Steven Klepper and Peter Thompson

Introduction

In recent years interest has grown in the phenomenon of entrepreneurship. One does not have to look beyond Silicon Valley to see the importance of new enterprises, which seemingly have played a key role in the region's vitality. But where do new enterprises come from? Surprisingly, little is known about the origin of entrants, especially new enterprises. This is perhaps a legacy of the way entry is typically modeled in theories of competition. It has always been assumed that if entry is profitable, it will occur. It is not at all clear, though, whether such confidence is justified [Geroski (1995)].

Recent work suggests that entrants are quite diverse at birth, and their pre-entry experience persistently affects their performance [Carroll et al. (1996), Geroski, Mata, and Portugal (2002), Klepper (2002a, 2002b), Klepper and Simons (2000), Thompson (2005)]. One class of entrants that perform distinctly well in some industries are firms founded by employees of incumbent firms in the same industry [Klepper (2002b), Agarwal et al. (2004), Walsh, Kirchhoff, and Boylan (1996)]. We shall call these firms spin-offs. While in some instances spin-offs are sponsored or linked to their "parent" firm, generally the founders of spin-offs do not maintain any link to their prior employers.

In some industries, spin-offs are legion. Indeed, in the semiconductor industry so many spin-offs can be traced back to one firm alone, Fairchild Semiconductor, that they have been dubbed Fairchildren. Opinions differ greatly about the contribution of spin-offs to innovation and economic growth. Some perceive spin-offs as parasites, feeding off the

Klepper gratefully acknowledges support from the Economics Program of the National Science Foundation, Grant No. SES-0111429.

innovative efforts of their unwitting "parents."[1] Scholars who interpret spin-offs as parasites fear that spin-offs can undermine the ability of their parents to appropriate the returns of their innovative efforts, thereby undermining the incentives of incumbents to innovate. Others see spin-offs as the font of innovation, compensating for the inertia that plagues many incumbents. To them, the Fairchildren jumped a sinking ship and led the semiconductor industry to new glory, fueling the juggernaut known as Silicon Valley.

Where does the truth about spin-offs lie? The answer presumably lies in a better understanding of the motives of spin-offs in innovative, high-tech industries and the process governing their formation. Why, in fact, do employees of high-tech firms leave to found firms in the same industry? Is it mainly to exploit innovations they worked on for their employers? Is it mainly because of the inability of their employers to perceive and/or act upon promising technological developments in their industry? Alternatively, might spin-offs be a natural outcome of a world in which people have different perceptions about the best paths for organizations to follow? The main purpose of this chapter is to explore the motivations behind spin-offs and the role they play in economic growth.

In Section 6.1, we review empirical studies of spin-offs, many of which are quite recent, and extract a set of common patterns from the studies. One particularly prominent theme of the empirical literature is that spin-offs arise out of disagreements within existing firms that lead frustrated employees to pursue their ideas in their own firms. The existing theoretical literature, reviewed in Section 6.2, has ignored the role of disagreements. In Section 6.3, we therefore develop the foundations of a model of spin-off formation driven by disagreements. Doing so proves to be rather challenging, because Aumann (1976) all but precludes disagreements among rational actors that talk to each other. We introduce a minimal degree of nonrationality, based on the concept of *solipsism*, and ask whether such a concept is capable of generating predictions consistent with the empirical literature. We show that the concept is indeed able to account for a number of distinctive empirical regularities concerning spin-offs. However, some predictions of the model are at odds with the data. In Section 6.4, we therefore conclude with a discussion of new directions for development of our model.

---

[1] Ironically, this is the view of Intel, Fairchild's most famous spin-off. Intel goes to great lengths to harass employees that leave to start their own firms [Jackson (1998, pp. 211–338)].

6.1   Empirical regularities concerning spin-offs

New studies of high-tech spin-offs in the automobile [Klepper (2003, 2004a,b), laser (Klepper and Sleeper (2005)], and disk drive (Franco and Filson (2000), Agarwal et al. (2004)) industries have added greatly to our knowledge about spin-offs. Using unique sources to identify all industry entrants and their characteristics, including their pre-entry backgrounds, these studies analyze the factors influencing the rate at which firms spawn spin-offs and the performance of the spin-offs. Another high-tech industry where spin-offs were prominent is semiconductors, and Brittain and Freeman (1986) study the factors influencing the rate of spin-offs from semiconductor firms in Silicon Valley. The only other high-tech industry where spin-offs have been considered is biotech. Stuart and Sorenson (2003) exploit data on the location of all biotech start-ups and on biotech firms that were acquired or engaged in IPOs to make inferences about the impetus for spin-offs without having to trace the heritage of the biotech entrants. Mitton (1990) also studies the origin of biotech start-ups in San Diego. Outside of the high-tech sector, Phillips (2002) studies spin-offs from Silicon Valley law firms. Cutting across industries, Gompers et al. (2003) use data on venture capital (VC) financed start-ups to analyze the rate at which publicly traded firms spawned VC-financed spin-offs. We review the main findings from these studies.

We consider first the automobile industry, which began in 1895. Through 1966 there were 725 entrants into the industry, nearly all of which entered before 1926. Spin-offs accounted for twenty percent of the entrants, with the percentage of spin-off entrants rising over time. Spin-offs performed comparably with entrants with pre-entry experience in industries related to autos and substantially better than the majority of entrants without any relevant pre-entry experience, and by the 1910s spin-offs produced a majority of the leading makes of automobiles. The top four firms in terms of the number of spin-offs spawned were the four early leaders of the industry, Olds Motor Works, Cadillac, Ford, and Buick (which was the cornerstone of General Motors when it was formed). Among all firms, the spin-off rate was greater in firms that produced leading makes of automobiles and that survived longer. The firm spin-off rate also increased with age through age 14 and then declined, was greater in firms that were acquired by either auto or nonauto firms (in a short window around the time of the acquisition), and was greater in firms located in the Detroit area, where the industry was heavily agglomerated [Klepper (2004b)]. The performance of the spin-offs in terms of their longevity was positively related to the performance of their parents, with eleven of the thirteen spin-offs that produced leading makes of automobiles descending from Olds, Cadillac, Ford, and Buick/GM [Klepper (2004b)]. In a detailed study of these thirteen spin-offs, Klepper (2003) found that nine were formed by top-level engineers and managers as a result of disagreements within the parent firm about the kinds of cars to produce or about the management of the firm, with the spin-offs sometimes continuing efforts their parent initiated but then abandoned. The thirteen spin-offs played an important role in the technological advance of the industry, accounting for a majority of the fifty major innovations in the industry from 1902 to 1925 that were not introduced by the two leading firms, Ford and Buick/ General Motors.

The laser industry began in 1961, and through 1994 spin-offs accounted for sixty-nine or seventeen percent of the 465 entrants whose backgrounds could be traced. The spin-offs survived much longer than other start-ups and comparably with diversifying entrants with prior experience in industrial electronics, who were the longest-lived diversifying entrants. Dividing lasers into eight main categories (and a residual), Klepper and Sleeper (2005) note that spin-offs typically specialized initially in a type of laser produced by their parent firm. Firms tended to remain specialized and produce a narrow range of laser types, with spin-offs accounting for many of the leading producers of each type of laser as well as the top two firms in the industry, Spectra Physics and Coherent. Similar to autos, for each type of laser the rate at which firms spawned spin-offs rose to age 14 and then declined, was greater in firms acquired by laser or no laser firms (in a short window around the time of the acquisition), and was greater in Silicon Valley firms, where the industry was modestly agglomerated. Using reports from a monthly trade journal and supplemented by interviews with founders, Klepper and Sleeper (2005) discuss the impetus for eight spin-offs that were illustrative of the factors underlying spin-offs in each of the eight main laser types. Each had a founding team with at least one high-level technical manager, and some also had founders with high-level managerial backgrounds in marketing and operations. In five of the eight spin-offs, the founders left to develop a technology they worked on in their parent firm but the parent chose not to develop, with three of the spin-offs licensing technology from their parent. Two of the other three spin-offs were formed after the parent was acquired, in one instance to service customers the parent abandoned after being moved and in the other to compete directly with the parent.

In the disk-drive industry, of the 153 entrants from 1977 to 1997, twenty-six percent were spin-offs. Five major "architectural" innovations that reduced the size of disk drives and opened up new markets servicing smaller computers were introduced in the period 1977–1997. All five were pioneered by spin-offs, who displaced the industry leaders and survived longer than entrants with other backgrounds [Agarwal et al. (2004)]. All spin-offs had at least one founder with a high-level technical background and sometimes other founders with a high-level marketing or production background, similar to autos and lasers. The rate of spin-offs was greater in firms with better disk drives and that were quicker to produce the new drives. These firms in turn had spin-offs with better disk drives and that were quicker to enter subsequent new markets, consistent with better firms having better spin-offs. Older firms that entered by 1976 had a lower rate of spin-offs, but otherwise age did not affect the firm spin-off rate. Christensen's (1993) analysis of the slowness of incumbents to introduce the smaller disk drives is revealing about the impetus for the leading spin-offs. On the basis of over sixty interviews with executives, Christensen (1993, pp. 562–63) found that leading incumbent firms conceived and developed prototypes of the smaller disk drives but then abandoned them when their customers showed little interest in them. Engineers that worked on the smaller drives then left in frustration to start their own firms, which ended up pioneering the drives. Judging from King and Tucci's (2002) analysis of entry into new disk-drive markets, though, this was not a general tendency. They found that more experienced firms were actually *more* likely to enter new markets at every point in time, consistent with their finding that the sales of entrants in the new markets increased with their prior experience.

In Silicon Valley semiconductor firms studied by Brittain and Freeman (1986), firms that produced a wider array of semiconductor devices had higher spin-off rates, similar to the findings for lasers. Firms that were earlier entrants into new product groups also had higher spin-off rates, similar to disk drives. Both findings are consistent with better firms having higher spin-off rates. The spin-off rate was greater in firms whose growth had slowed and in firms that were acquired by non-semiconductor firms or that hired a new CEO from outside the semiconductor industry. Stuart and Sorenson (2003) analyze the effect of acquisitions and IPOs on the rate of formation of biotech firms. They found higher start-up rates near regions where biotech firms were acquired or engaged in IPOs, which they presumed was due to spin-offs. These effects were present only in states with greater restrictions on the enforcement of noncompete covenants, and the effect of acquisitions on start-ups was restricted to acquisitions where the acquirer came from outside of the biotech industry. Mitton (1990) documents how in

San Diego biotech firms, control changes resulting from acquisitions by nonbiotech firms led to "cultural" differences that spurred top-level managers to leave to start their own biotech spin-offs. Mitton also found that most of the San Diego biotech spin-offs were formed to develop technologies their parents declined to pursue.[2] The oldest parents of spin-offs in Mitton's study were ten years old, and through age ten the rate at which they spawned spin-offs increased with age, consistent with the findings for autos and lasers.

Phillips' (2002) analysis of spin-offs from Silicon Valley law firms focused on how spin-offs affected the performance of their parents, but he also analyzed factors influencing the performance of spin-offs and briefly the factors influencing the rate at which firms spawned spin-offs. Firm spin-off rates increased through the age bracket nine to fifteen after which they declined. The length of survival was greater for spin-offs whose founders had greater status in their parent firm and less for spin-offs from failing firms. He also found that firms had higher hazards after spin-offs than comparable firms without spin-offs.

Gompers et al. (2003) found that publicly traded firms in Silicon Valley and Massachusetts, both hotbeds of entrepreneurial start-ups, and firms that were themselves VC-financed had higher rates of VC-financed spin-offs. The former result is consistent with the higher rate of auto spin-offs in Detroit and laser spin-offs in Silicon Valley and may simply reflect the greater ease of forming a founding team and securing advice and financial support in regions with a larger number of related start-ups. The spin-offs of Silicon Valley and Massachusetts firms and VC-backed firms were less likely to be engaged in technologies related to their parents than the spin-offs of other firms. It was also found that less diversified firms had a higher spin-off rate and that slowed growth heightened firm spin-off rates, similar to the Silicon Valley semiconductor firms. Firm spin-off rates also declined with age, which is consistent with the findings for autos and lasers if publicly traded firms were generally over fourteen years old.

Certain patterns consistently emerge from the various industries studied. Around twenty percent of all entrants were spin-offs, and the spin-offs were distinctly good performers. They generally had at least one founder who was a high-level technical manager and sometimes also had founders with high-level marketing and operational experience. Better firms had a higher spin-off rate, and their spin-offs were better performers. Disagreements over what technologies to develop and

---

[2] In contrast to autos, lasers, and disk drives, however, forty percent of the start-ups were financed in part by their parents.

sometimes about management practices were the principal impetus for the leading spin-offs. Spin-offs were less likely to occur in older firms, but it appears that initially the spin-off rate increased with firm age. Acquisitions induced an increase in the likelihood of spin-offs during a short window around the time of the acquisition, especially when the acquirer was from another industry. A new CEO from another industry, an IPO, and slowed growth also appear to have increased the rate of firm spin-offs.

6.2   Existing theories of spin-off formation

How well do existing theories explain the common patterns in the various studies of high-tech spin-offs? With the growing interest in spin-offs, various models have been proposed to explain them.[3] These models tend to fall into three camps. In the first, an employee makes a serendipitous discovery of some economic value. This discovery is in principle more valuable to the incumbent firm than it would be to a start-up, but information asymmetries of one form or another frequently persuade the employee to implement the discovery through his own start-up rather than reveal it to his employer.[4] In the second type of model, the discovery is common knowledge within the firm but it is less valuable to the incumbent than it would be to a start-up, because its implementation would cannibalize existing rents or because the firm has limited competency to evaluate the idea, particularly when the idea is tangential to the firm's main activities.[5] In the third type of model, employees learn from their employers about how to profitably compete in their industry, especially when their employer is successful. They exploit this knowledge by setting up their own firm in the same industry.[6]

---

[3] The traditional explanation for who becomes an entrepreneur is based on ability [Lucas (1978), Holmes and Schmitz (1990)]. While it is obvious that ability may be enhanced by working for an incumbent in the industry [Irigoyen (2002)], this literature does not explain why only a small fraction of the many employees with the requisite experience and entrepreneurial ability leave their employers to found a new firm.

[4] See Anton and Yao (1995), Wiggins (1995), Bankman and Gilson (1999), Gromb and Scharfstein (2002), Hellman (2002), and Amador and Landier (2003). Common themes are (1) firms cannot commit to a contingent contract that adequately rewards the employee for a discovery and the subsequent employee effort needed to implement it, and (2) noncontingent contracts that are *ex ante* acceptable to the firm will not always be sufficient to prevent a departure by the employee.

[5] See Pakes and Nitzan (1983), Tushman and Anderson (1986), Henderson and Clark (1990), Christensen (1993), Klepper and Sleeper (2005), and Cassiman and Ueda (2002).

[6] See Franco and Filson (2000) and Agarwal et al. (2004).

None of these theories seems to capture the process underlying most spin-offs. The first type of theory predicts that employees that found spin-offs will not reveal their ideas to their employers. However, the evidence suggests that at least among the leading spin-offs, the employer often knows precisely what the employee's idea is but does not want to pursue it. Indeed, a common theme of studies that focus on the motives of spin-offs is that they arise from frustration by employees over rejection of their ideas by their employers [Garvin (1983, p. 6), Lindholm (1994, p. 163)].[7] Moreover, it is not clear how the first type of theory can explain why acquisitions, IPOs, and slowed growth should heighten the chance of spin-offs or why age should have a nonmonotonic effect on the firm spin-off rate.

The second type of model is consistent with firms rejecting the ideas of their employees. This could happen if established firms are unable to evaluate certain types of ideas that are not in their core areas. With the success of firms no doubt dependent on their ability to evaluate ideas of their employees, it might be expected that such spin-offs would be more likely in less successful firms. Yet the evidence strongly points in the opposite direction, with more successful firms having higher spin-off rates. Alternatively, in the second type of model the ideas might be rejected because they would cannibalize the firm's sales and hence profits. But many of the ideas that spin-offs from the leading auto and laser firms pursued were actually initiated and first worked on by their parents, which is consistent with cannibalization fears only if the firms could not anticipate where the ideas would lead. This is hard to rule out from the limited evidence on spin-offs. But if this were important, one would imagine that most employees would be able to understand why their ideas were rejected and would not be so frustrated with their employers. It is also not clear why fears of cannibalization would be heightened when a firm was acquired or would be related to the age of the firm, thus providing no explanation for these findings.

The third type of model featuring learning is consistent with one aspect of the findings about spin-offs, namely the tendency for better firms to have higher spin-off rates and better-performing spin-offs. But these theories imply that spin-offs should do similar things to their parents based on their common knowledge. Yet the evidence indicates that the leading spin-offs commonly pursue ideas their parents rejected. So what exactly are the spin-offs learning from their parents, and why

---

[7] It is possible that employees get their ideas rejected because they do not want to fully reveal them in order to protect them from being copied by their employers. However, the frustration expressed by so many employees when their ideas are rejected suggests that partial revelation is not the problem.

should disagreements be the impetus for so many spin-offs? Moreover, if learning is the motive for spin-offs, why would spin-offs be more likely when firms are acquired? Acquisitions conceivably could promote learning. However, acquisitions by firms in the same industry might be expected to promote the most learning, yet spin-offs seem especially likely when firms are acquired by firms in other industries. Moreover, the increase in spin-off formation associated with acquisitions seems to be concentrated in too short a window around the time of acquisitions to be consistent with learning theories.

Thus, all three types of explanations for spin-offs come up short in important ways. The fact that many leading spin-offs arise out of disagreements within their parent firms is difficult for existing theories to accommodate. Existing theories also do not address why acquisitions, particularly by firms from another industry, increase spin-off rates or why age should affect spin-off rates. In the next section we propose a new model of spin-offs based on disagreements, and we explore its ability to accommodate the empirical evidence.

6.3   Disagreements and spin-offs: a new model

To model disagreements, we need to confront the fact that firms are not unitary actors but are composed of decision-makers with potentially different views about what the firm should do. Accordingly, we assume that a firm is composed of multiple individuals, each of which has some influence on the firm's decisions. At any given moment, optimal choices are not known, but over time they are slowly learned as employees receive signals about the true environment facing the firm. The signals are noisy and differ across members of the firm, leading to disagreements about what the firm should do. The firm's choice is a weighted average of the choices favored by each of its decision-makers, with weights based on the positions of individuals and perhaps their ownership share of the firm. Employees leave to start their own firms when their view of what the firm should do differs sufficiently from the firm's choice about how to proceed.

The essence of the model is that employees want to be involved with a firm that does what they believe to be the "right thing." In our view, sufficient attention has already been paid to contracting difficulties caused by employees being all too willing to screw their employers by not revealing discoveries they were paid to make, and by employers being unable to commit to state-contingent rewards that would persuade employees to be honest. We assume, in contrast, that individuals at higher echelons of firms are concerned with the value of the firm's activities, rather than with how they can manipulate their own payoff through deception or omission. On the other side of the same coin, we assume that firms have no interest in ripping off their senior employees.[8] We assume that there is considerable uncertainty about what the "right thing" is, and this generates genuine disagreements about the choice of strategy for a firm. These disagreements need not be predicated upon any particular discovery, whether private or common knowledge.

There is, however, an intellectual challenge in modeling disagreements. Presumably each individual must, in the language of Bayesian learning, reveal his posterior mean to his colleagues if he wants to participate in the decision-making process. But this revelation also allows each individual to infer precisely his colleagues' private signals. Efficiently incorporating these signals into his own beliefs, the revised posterior mean will then be the same for everyone. Aumann (1976) has shown under very general conditions that if the posteriors of two Bayesians with common priors are common knowledge, their posteriors must be the same; Geanakoplos and Polemarcharkis (1982) show further that if two agents with common priors exchange their efficient posteriors back and forth they will arrive at the common knowledge posterior; and McKelvey and Page (1986) have extended these results to n individuals.

These results leave two ways in which disagreements can persist. First, one can drop the common prior assumption. Although it has a substantial tradition in behavioral finance, beginning with Harrison and Kreps (1978), there has been some debate about whether doing so is reasonable [Aumann (1987, 1988), Gul (1988), Morris (1995)].[9] The consensus seems to be against dropping the common prior. Moreover, in our setting it seems reasonable to assume that firms are formed by individuals that want to work together because they hold similar beliefs. Second, one can drop the efficiency of individuals' updating algorithms. Some authors have assumed that individuals are overconfident in the sense that the posterior mean is a biased estimate of the true mean.[10] Others have assumed that decision-makers

---

[8] This is not a radical approach. It simply applies to the higher echelons of the firm Akerlof's (1982) widely admired but widely ignored sociological characterization of the workplace, where "the average worker works harder than is necessary according to the firm's work rules, and in return for this donation of goodwill and effort, he expects a fair wage from the firm."

[9] See Van den Steen (2001) for an extensive review.

[10] This is the central assumption in Amador and Landier (2003). However, the assumption is subsequently justified by appeal to issues of asymmetric information and moral hazard. See Malmendier and Tate (2002, 2003) for applications of the assumption to corporate investment and acquisitions.

overweight the information content of their private signals relative to publicly available information. This second approach has also been dubbed a form of overconfidence.

Our approach is of the second type. However, we propose a somewhat different nomenclature. We reserve the term "overconfident" to refer to individuals who underestimate the noise of any signals, whether their own or those inferred from their colleagues. We term asymmetric weighting of private and non-private signals (in favor of the former) as "solipsism." The distinction between overconfidence and solipsism has substance: solipsism is a necessary condition for disagreement; over-confidence simply magnifies the size of disagreement.

There is a large empirical literature supporting the assumption of overconfidence and, to a lesser extent, what we have called solipsism. De Bondt and Thaler (1995) have gone so far as to claim that "perhaps the most robust finding in the psychology of judgment is that people are overconfident." Evidence of overconfidence has been reported among diverse professions, including entrepreneurs [Cooper, Woo, and Dunkelberg (1988)] and managers [Russo and Schoemaker (1992)], although entrepreneurs exhibit much more overconfidence than managers [Busenitz and Barney (1997)]. Odean (1998) and Daniel, Hirshleifer, and Subrahmanyam (1998) cite many other examples, and different forms of overconfidence. Our assumption that individuals overweight private information relative to public information has found support in the laboratory [Anderson and Holt (1996)] and among financial analysts [Chen and Jiang (2003)]. Their findings are consistent with the broader notion that people expect good things (e.g. receiving accurate signals) to happen to them more often than they do to others [Weinstein (1980), Kunda (1987)].

6.3.1   The dynamics of disagreement

Our formal model is an extension of the Jovanovic and Nyarko (1995) Bayesian learning model to teams of decision-makers. Suppose a firm is operated by n individuals, each of which has some degree of decision-making authority or influence. Each individual is concerned with maximizing the expected value of the firm, rather than his own private returns. All individuals know that firm value is given by $v = -(\theta - x)^2$, where x is the activity undertaken by the firm and θ is a target. No one knows the target, but at time t individual $i = 1, 2, \ldots, n$, believes it is a draw from a normal distribution with mean θ_{it} and variance σ_{it}^2. Given his beliefs, i calculates that the optimal strategy is $x = \theta_{it}$, yielding an expected payoff of $v_{it} = -\sigma_{it}^2$. The activity actually chosen by the firm is a compromise, $x = \overline{\theta}_t = \sum_{i=1}^{n} \phi_i \theta_{it}$, of everyone's beliefs. The parameters ϕ_i are time-invariant weights attached to individual expectations, with $\sum_{i=1}^{n} \phi_i = 1$. The weight ϕ_i can be interpreted as i's decision-making influence.

Individual i's expected value of the compromise decision is

$$
\begin{aligned}
E_{it}[v] &= -E_{it}\left[(\theta - \overline{\theta}_t)^2\right] \\
&= -E_{it}\left[((\theta - \theta_{it}) + (\theta_{it} - \overline{\theta}_t))^2\right] \\
&= -E_{it}\left[(\theta - \theta_{it})^2\right] - (\theta_{it} - \overline{\theta}_t)^2 \\
&= -\sigma_{it}^2 - (\theta_{it} - \overline{\theta}_t)^2,
\end{aligned}
\tag{6.1}
$$

where in the third line the fact that any Bayesian posterior is unbiased implies $E_{it}(\theta - \theta_{it}) = 0$. From i's perspective the firm will do worse the more he disagrees with the firm's decision, and he may want to do something about it. We focus attention on i's possible departure to form his own start-up. Doing so will be attractive if the cost is not too high and if by so doing he can operate a firm using a strategy closer to his own beliefs. Let the cost be k (we shall call this the entry cost), and assume that i can form a new team consisting of individuals holding exactly the same subjective beliefs as his.[11] The expected value of the spin-off is then $E_{it}[w] = -k - \sigma_{it}^2$, so i prefers to strike out on his own whenever

$$
z_{it}^2 \equiv (\theta_{it} - \overline{\theta}_t)^2 \geq k.
\tag{6.2}
$$

For any set of weights, $\{\phi_i\}_{i=1}^{n}$, the decision to leave depends only on the subjective means, and not at all on the precision of those beliefs. Understanding the spin-off process is therefore reduced to the task of understanding how expected values come to differ sufficiently to induce individuals to strike out on their own. This difference will, however, be related to the precision of beliefs.

To see how subjective means come to differ over time, assume that the firm is founded at time 0 by a group of n individuals, all of whom share the same prior that θ is a random draw from $N(0, \sigma_\theta^2)$. Once each period, these individuals receive private and noisy signals, $S_{it}\theta + \varepsilon_{it}$, where the ε_{it} are random draws from a normal distribution with zero mean and variance σ_ε^2. Although all signals have variance σ_ε^2, each individual believes his own signals to have variance $\gamma\sigma_\varepsilon^2$ and his colleagues' signals to have variance $\gamma\beta\sigma_\varepsilon^2$. Individuals are labeled overconfident if $\gamma < 1$, and

---

[11] The assumption reflects the notion that firms are formed by groups of individuals with a common prior.

solipsistic if $\beta > 1$. In the limit as $\beta \to \infty$ individuals only respond to their own private signals.

Individual i's posterior after receiving t private signals is normal with mean

$$\tilde{\theta}_{it} = \frac{t\sigma_\theta^2 \overline{S}_{it}}{t\sigma_\theta^2 + \gamma\sigma_\varepsilon^2} \tag{6.3}$$

and variance

$$\tilde{\sigma}_{it}^2 = \frac{\gamma\sigma_\theta^2\sigma_\varepsilon^2}{t\sigma_\theta^2 + \gamma\sigma_\varepsilon^2} \tag{6.4}$$

where $\overline{S}_{it} = t^{-1}\sum_{\tau=1}^{t} S_{i\tau}$ is the mean of i's private signals to date t.

As β and γ are common to all decision-makers, the common-knowledge beliefs arrived at after repeatedly exchanging posteriors are the same as would be obtained if each individual's private signals were directly observable to his colleagues. In period t, therefore, individual i forms beliefs as though he has observed t private signals and $(n-1)t$ signals from his colleagues. Standard Bayesian formulae for normal conjugates then imply that i's expectation of the target is

$$\theta_{it} = \frac{t\sigma_\theta^2}{(\beta + n - 1)t\sigma_\theta^2 + \beta\gamma\sigma_\varepsilon^2}\left(\beta\overline{S}_{it} + \sum_{j\neq i}\overline{S}_{jt}\right) \tag{6.5}$$

with posterior variance

$$\sigma_{it}^2 = \frac{\beta\gamma\sigma_\theta^2\sigma_\varepsilon^2}{(\beta + n - 1)t\sigma_\theta^2 + \beta\gamma\sigma_\varepsilon^2} \tag{6.6}$$

The firm's decision is a weighted average of each team member's subjective mean:

$$\begin{aligned}
\overline{\theta}_t &= \frac{t\sigma_\theta^2}{(\beta + n - 1)t\sigma_\theta^2 + \beta\gamma\sigma_\varepsilon^2}\left(\beta\sum_{i=1}^{n}\phi_i\overline{S}_{it} + \sum_{i=1}^{n}\phi_i\sum_{j\neq i}\overline{S}_{jt}\right) \\
&= \frac{t\sigma_\theta^2}{(\beta + n - 1)t\sigma_\theta^2 + \beta\gamma\sigma_\varepsilon^2}\sum_{i=1}^{n}(1 + (\beta - 1)\phi_i)\overline{S}_{it}.
\end{aligned} \tag{6.7}$$

Hence,

$$z_{it} = \frac{(\beta - 1)t\sigma_\theta^2}{(\beta + n - 1)t\sigma_\theta^2 + \beta\gamma\sigma_\varepsilon^2}\left(\overline{S}_{it} - \sum_{i=1}^{n}\phi_i\overline{S}_{it}\right). \tag{6.8}$$

If $\beta = 1$, then $z_{it} \equiv 0$. That is, without solipsism, disagreement is not possible.

6.3.2   *The hazard of spin-off formation*

The empirical evidence shows that the likelihood that a firm spawns a spin-off initially increases with age, but declines with age for older firms. In this subsection, we show that this age profile for spin-off formation is predicted by the model.

The mean signals $\bar{S}_{it}$, $i = 1, 2, \ldots, n$, are normally distributed and independent across individuals, each with (unknown) mean θ and (true) variance σ_ε^2 / t. It then follows that z_{it} is normal with zero mean and variance

$$\text{var}(z_{it}) = \frac{(\beta - 1)^2 \, t\sigma_\theta^4 \sigma_\varepsilon^2 (1 - 2\phi_i + H)}{\left((\beta + n - 1)t\sigma_\theta^2 + \beta\gamma\sigma_\varepsilon^2 \right)^2}, \tag{6.9}$$

where $H = \Sigma_{i=1}^n \phi_i^2$ is the Herfindahl index of decision-making authorities.[12] Hence, i prefers to strike out on his own if

$$\frac{\chi_{it}^2 (\beta - 1)^2 \, t\sigma_\theta^4 \sigma_\varepsilon^2 (1 - 2\phi_i + H)}{\left((\beta + n - 1)t\sigma_\theta^2 + \beta\gamma\sigma_\varepsilon^2 \right)^2} \geq k,$$

where χ_{it}^2 is a chi-squared random variable with one degree of freedom. Rearranging, exit is preferred by i if

$$\chi_{it}^2 \geq \frac{k\left((\beta + n - 1)t\sigma_\theta^2 + \beta\gamma\sigma_\varepsilon^2 \right)^2}{(\beta - 1)^2 \, t\sigma_\theta^4 \sigma_\varepsilon^2 (1 - 2\phi_i + H)}. \tag{6.10}$$

Figure 6.1 illustrates inequality (6.10). The curve **AA** traces out the right-hand side as a function of t. It is u-shaped, with a minimum at $t = \beta\gamma\sigma_\varepsilon^2 / ((\beta + n - 1)\sigma_\theta^2)$, and limits of $+\infty$ at $t = 0$ and as $t \to \infty$. The unbounded limits reflect the facts that all individuals begin with the same prior and that they will eventually learn the common true parameter value if they receive a sufficient number of signals. Thus, for t small enough and large enough, it is unlikely that a draw from the χ_{it}^2 distribution will be large enough to induce a spin-off. Note also that for $\beta = 1$, the right-hand side of (6.10) is infinite. Figure 6.1 also plots sequences of the left-hand side of (6.10) for two individuals. The sequence for individual i first exceeds **AA** at point **a**, which is when he

---

[12] As $(1 - 2\phi_i + H) \equiv (1 - \phi_i)^2 + \Sigma_{j \neq 1} \phi_j^2$, it is clearly positive.

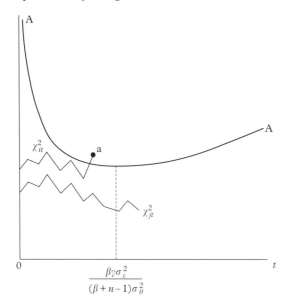

Figure 6.1   The dynamics of spin-off formation

departs to form his own firm. The sequence for individual j never crosses **AA**, so this individual does not depart the firm. It is clear from Figure 6.1 that the hazard of i forming a spin-off must start out at zero, then rise to some strictly positive value (because a x^2 random variable has unbounded support), and eventually decline back to zero as the firm ages.

PROPOSITION 1. *The hazard of spin-off formation initially rises from zero, but eventually asymptotically declines to zero.*

Assuming $\beta > 1$, most parameter changes induce an unambiguously signed change in the probability that at any given t inequality (6.10) is satisfied. The right-hand side of (6.10) is strictly increasing in γ, k, and ϕ_i for all positive t, and strictly decreasing in H and σ_θ^2 for all positive t. In contrast, changes in σ_ε^2 have an ambiguous effect on the right hand side of (6.10), which is increasing [decreasing] in for $t < [>]\sigma_\varepsilon^2/\sigma_\theta^2$. Let p_t denote the unconditional probability that at any given t inequality (6.10) is satisfied. Then the hazard of spin-off formation in period t is $h_t = p_t/(1 - \sum_{i=1}^{t} p_{t-i})$, and it has the following properties:

PROPOSITION 2. *For any $t = 1, 2, 3, \ldots$, and $\beta > 1$, the spin-off hazard is strictly decreasing in γ, k, and ϕ_i, and strictly increasing in H and σ_θ^2. The* *hazard is strictly decreasing in for $\sigma_\varepsilon^2/\sigma_\theta^2$, and is ambiguously related to σ_ε^2 thereafter.*

Intuitively, spin-offs are more likely in industries with low entry costs and high uncertainty about the appropriate activity. Low entry costs just make it easier to leave. The greater the uncertainty about θ, the more attention individuals pay to their private signals and the more likely they are to disagree. Overconfidence matters only if $\beta > 1$, in which case spin-offs are decreasing in γ. Conditional on the concentration of decision-making authority, individuals with less authority are more likely to leave. Finally, conditional on ϕ_i, greater concentration of decision-making authority induces higher spin-off formation rates. Concentration of decision-making authority matters even if individual i has no authority, because increased concentration makes the "compromise decision" more erratic. Changes in σ_ε^2 have an ambiguous effect on spin-off formation.[13]

6.3.3 The effect of acquisitions

A robust result from the empirical studies is that spin-offs are more likely to occur around the time that a firm is acquired, and this is particularly the case if the acquirer comes from outside the industry. This phenomenon is a natural consequence of our theory. Acquisitions change the distribution of decision-making authority. New individuals are brought from outside the acquired firm and reorganizations take place inside it. Consider a situation in which an individual i is currently not planning to leave a firm that is then acquired. If the reorientation of the firm moves the firm's decision toward i's beliefs, he still will not leave and there is no consequence for the spin-off hazard. But if i's decision-making influence is reduced, the firm's decision moves away from i's beliefs, possibly enough to induce immediate exit.

PROPOSITION 3. *Acquisitions that change the distribution of decision-making authority induce a short-term spike in the hazard of spin-off formation.*

---

[13] The ambiguity arises from two countervailing effects of an increase in the precision of the signals. On the one hand, more precise signals induce greater sensitivity of the posterior mean to signals and increase the likelihood that posteriors diverge. On the other hand, more precise signals increase the rate of learning, so that the posteriors of all individuals converge on the true parameter value more quickly. The former effect dominates for t small, while the latter dominates for larger values of t. Thus, when learning is more difficult, in the sense that σ_ε^2 is larger, spin-offs are likely to occur later in a parent firm's life.

In some instance an acquisition may eliminate a maverick CEO and prevent impending departures. But acquisitions that accomplish this must be undertaken *before* frustrated employees have left, requiring some remarkable prescience on the part of the acquiring firm. Such cases are likely to be rare.

6.3.4   Spin-off quality

Assume that at time t individual i forms a spin-off, which implies that there is a sufficiently large distance between θ_{it} and $\bar{\theta}_t$ to justify the entry cost. The value of the spin-off is then $w_{it} = -(\theta-\theta_{it})^2 - k$. But even ignoring the entry cost, it turns out that spin-offs will, on average, be bad ideas, in the sense that the expected initial quality of i's firm is lower than the current quality of its parent. The intuition is simply that the mean of $(\tilde{n}1)$ posteriors is likely to be closer to the true value than a single posterior that diverges from this mean.

More formally, recall that $\sqrt{k}$ is the smallest distance between i's posterior mean and the compromise decision necessary to induce i to form a spin-off. Then the expected quality of the spin-off, $E[w_{it}] + k$, is less than the expected quality of the parent, $E[v_t]$, if

$$E\left[-(\theta - \theta_{it})^2 + (\theta - \bar{\theta}_t)^2 \Big| \theta_{it} - \bar{\theta}_t| \ge \sqrt{k}\right] < 0. \tag{6.11}$$

It is easy to show that inequality (6.11) holds for $k=0$, because in this special case we have

$$\begin{aligned}
&E\left[-(\theta - \theta_{it})^2 + (\theta - \bar{\theta}_t)^2 \Big| |\theta_{it} - \bar{\theta}_t| \ge \sqrt{k}\right]\Big|_{k=0} \\
&= E[\bar{\theta}_t^2] - E[\theta_{it}^2] \\
&= \frac{t\sigma_\theta^4 \sigma_\varepsilon^2}{\left((\beta + n - 1)t\sigma_\theta^2 + \beta\gamma\sigma_\varepsilon^2\right)^2} \\
&\quad \times \left(\sum_{i=1}^{n}(1 + 2(\beta - 1)\phi_i)^2 - (\beta^2 + (n - 1))\right) \\
&= \frac{(\beta - 1)^2(H - 1)t\sigma_\theta^4\sigma_\varepsilon^2}{\left((\beta + n - 1)t\sigma_\theta^2 + \beta\gamma\sigma_\varepsilon^2\right)^2} \\
&< 0,
\end{aligned} \tag{6.12}$$

where the second line exploits the fact that $E(\theta) = 0$. Proving the same for the general case appears to be infeasible, because we need to further condition on the difference between θ_{it} and $\bar{\theta}_t$.[14] In Appendix A we therefore evaluate the properties of (6.11) numerically. The results are summarized in Figure 6.2. In all cases, the expected value of the spin-off is less than the expected value of the parent, although for some parameter values the difference is small.

PROPOSITION 4. *For $\beta > 1$,* (a) *The initial quality of a spin-off is on average lower than the quality of its parent.* (b) *The average gap between the quality of a parent and its spin-off at the time the spin-off is formed is increasing in β and σ_θ^2, decreasing in γ and k, and exhibits a u-shape as ϕ_i, n, t, and σ_ε^2 increase.*

As one might anticipate, greater degrees of solipsism and over-confidence lead to spin-offs of relatively poor quality. Spin-offs also have lower quality relative to their parents in industries where the variance of the prior, σ_θ^2, is large. As Jovanovic and Nyarko (1995) have shown, σ_θ^2 is positively related to the amount there is to learn over the life of a firm or technology, the amount that is actually learned over any interval of time, as well as to the amount of inequality in efficiency among a cohort of firms. Hence, *ceteris paribus*, in industries with greater inequality and more learning, spin-offs are also likely to have lower relative quality.

Recall that high entry barriers require large disagreements to induce spin-offs, and large disagreements also imply large differences in the strategies chosen by parent and spin-off. One might therefore expect the relative quality of spin-offs to be declining in the size of entry barriers. Somewhat surprisingly, the opposite is true. It turns out that the effect of high entry barriers on the minimum difference in beliefs necessary to induce a spin-off is more than offset by the requirement that departing individuals must also be more confident about their beliefs when entry barriers are high. Thus, the possibility of highly misleading signals is lower the higher are entry barriers.

Greater solipsism, greater overconfidence, greater prior uncertainty about the "right thing to do," and lower entry costs all raise the hazard of spin-off formation and reduce quality, suggesting that we would expect to observe a negative association between rates of spin-off formation and spin-off quality. However, this association must be tempered by the presence of several non-monotonic comparative statics results. Increases in i's decision-making authority, in the size of the managerial team, in the signal noise, and in the age of the parent all have non-monotonic effects on spin-off quality, causing it first to fall and then

---

[14] The term $\bar{\theta}_t^2 - \theta_{it}^2$ consists of sums of products of positively correlated normal variables. Some analytical results and approximations are known for the distribution of such products only in special cases [Craig (1936), Aroian (1947)], and modern work has resorted to numerical and Monte Carlo methods [e.g. Ware and Lad (2003)].

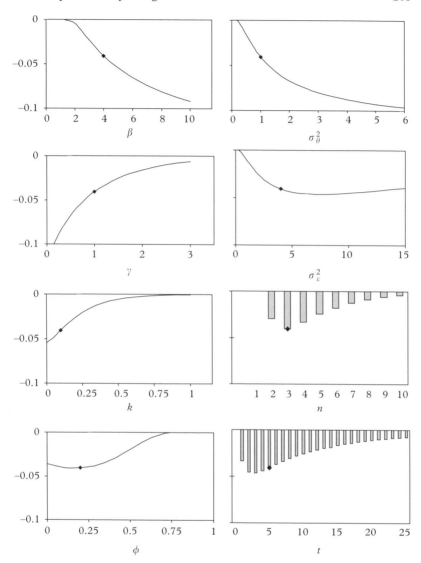

Figure 6.2 Expected quality of spin-off minus expected quality of parent. Diamond symbol on each graph indicates the baseline value used to generate all the other figures. The forms of the curves were verified for parameter ranges wider than those shown here

to rise as each parameter increases. Most notably, because this variable is most readily measured, the spin-offs with the lowest relative quality are those that form when the parent is of intermediate age. Spin-offs

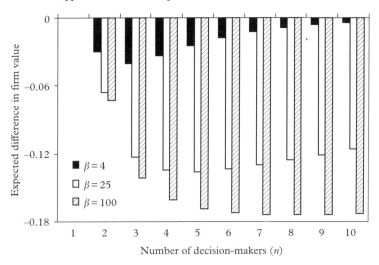

Figure 6.3   Expected quality of spin-off minus expected quality of parents

from both young and old parents are on average higher quality. These comparative statics are on the whole intuitive, and they are related to the non-monotonic relationship between these parameters and inequality in efficiency among firms explored by Jovanovic and Nyarko (1995) in their simpler setting.

The non-monotonic relationship between the relative quality of the spin-off and the number of decision-makers in the parent firm represents the trade-off between two opposing forces. On the one hand, the parent firm's decision is a weighted sum of n posteriors and hence the expected quality of the decision improves as n increases. On the other hand, the posterior of individual i is also a weighted sum of n posteriors, so that the average quality of any spin-off formed also increases in n. For small values of n, the former effect dominates, so the *relative* quality of the spin-off declines as n rises. For larger values of n the latter effect dominates. Of course, increasing n has a smaller effect on the quality of the spin-off the greater the degree of solipsism. Thus, one would expect that increasing β not only increases the average gap between the qualities of parent and spin-off, it also postpones to higher values of n the point at which the second effect dominates. Figure 6.3, which repeats the plot for n from Figure 6.2 with different values of β, confirms that this is the case. For our baseline value of $\beta = 4$, the expected quality difference is greatest at $n = 3$. For $\beta = 25$, it is greatest at $n = 5$, while for $\beta = 100$ it is greatest at $n = 8$. In the limit as $\beta \to \infty$, i's posterior does not benefit from the signals of his colleagues, and in this case the difference in relative quality asymptotically attains its greatest value as $n \to \infty$.

Our result that on average spin-offs will perform worse than their parents is a distinctive prediction of the model. Other theories envision spin-offs as either exploiting especially valuable ideas developed within their parent firm, pursuing ideas their parent rejected for parochial reasons or because of bureaucratic inertia, or competing on even terms with their parent based on knowledge gleaned from working for their parent. All of these theories suggest that on average spin-offs will perform as well if not better than their parent firms. Our model certainly leaves open the possibility that some spin-offs will outperform their parents, but predicts, on average, that this will not occur. Note that if there are technological spillovers, which seems inevitable, spin-offs could turn out to be socially productive even if, on average, they perform worse than their parents. This could help explain the observation by Klepper (2003) that spin-offs played an important role in advancing the technology of the U.S. automobile industry.

We can exploit the data collected in Klepper (2004b) on all automobile firms to compare the performance of spin-offs and their parents. For each firm, we can compute the number of years it produced automobiles, which is a kind of all-purpose measure of firm performance. By definition, a firm has to produce automobiles for some amount of time in order for it to be a parent, guaranteeing a minimal number of years of survival for parents. On the other hand, spin-offs could fail at any point. Consequently, it might be expected that a greater percentage of spin-offs would fail at young ages than their parents, but if spin-offs performed comparably with their parents then at older ages spin-offs and parents would have comparable survival rates. To test this, we construct Kaplan Meier survival curves for automobile spin-offs and their parents, which are reported in Figure 6.4.[15] For each age, the curves indicate the natural log of the percentage of spin-offs and parents surviving to that age (the slope of the curves at each age reflects the hazard of exit). As expected, spin-offs had lower survival rates than their parents at young ages. Consistent with our model, these lower survival rates persisted at older ages. For example, at age twenty the percent of survivors was nine percent for spin-offs and twenty-six percent for their parents, and at age forty the spin-off survival rate was three percent versus eleven percent

---

[15] There were 145 spin-offs and 88 firms that accounted for the 145 spin-offs. The best firms spawned more spin-offs [Klepper (2004b)], but we included only one observation for each parent so that our test would be conservative.

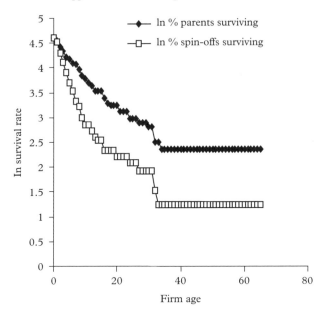

Figure 6.4   Survival curves for spin-offs and parents in the automobile industry

for their parents. No doubt there were many reasons parents survived longer than their spin-offs, but the results are encouraging regarding the model.

6.3.5   *Parent quality and the spin-off hazard*

A robust regularity identified in the empirical literature is that better-quality parents produce more spin-offs. It turns out that our abstract model predicts the opposite. A formal proof is difficult. We restrict attention to the special case in which $\phi_j = (1 - \phi_i)(1 - n)^{-1}$ for all $j \neq i$, and provide a numerical evaluation. Figure 6.5 provides representative plots of the hazard that individual i forms a spin-off for given values of the unknown parameter, θ, and the firm's decision, $\bar{\theta}_t$.[16] Panel A plots the case in which i has less decision-making influence than the remainder of the team, while panel B plots the case in which he has more influence. Both graphs are symmetric around an axis where the decision, $\bar{\theta}_t$, equals the unknown target, θ, and they have a minimum

---

[16] Appendix B provides the formal derivations. After suitable scaling of the axes, different sets of parameter values produce very similar graphs.

PANEL A:

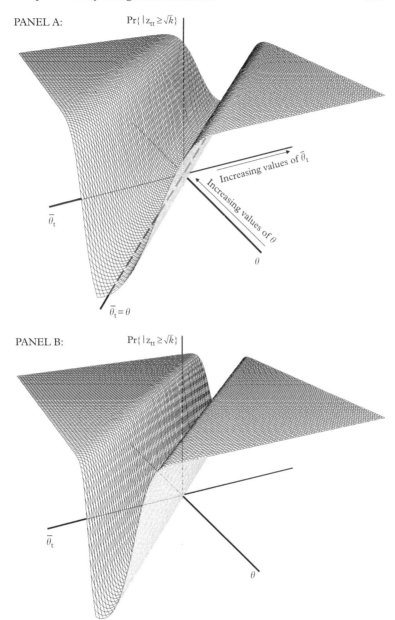

PANEL B:

Figure 6.5   Representative plots of probability that individual i chooses to form a spin-off, as a function of the unknown target value, θ, and the firm's decision, $\bar{\theta}_t$

along that axis. That is, the probability that individual i chooses to leave the parent firm is lowest for high-quality firms.[17]

PROPOSITION 5. *The hazard of spin-off formation is decreasing in firm quality.*

Factors other than disagreements appear to be the driving force behind the empirical finding that better-quality firms spawn more spin-offs. There is, however, a very simple extension to the model that can overturn Proposition 5. Assume that firm value is given by $v = -\zeta(\theta - x)^2$, where ζ is a parameter varying from firm to firm that measures the quality of the decision-making group, and assume that spin-offs inherit their parent's . Then, individual i will choose to form a spin-off the first instance that $(\theta_{it} - \bar{\theta}_t) \geq k/\zeta$. That is, an increase in ζ acts just like a reduction in the entry-cost, which is positively related to the spin-off hazard. Consequently, if ζ varies sufficiently across firms, the model would predict a positive relationship between firm quality and the spin-off hazard.

6.4   Discussion

Our model of spin-offs is based on the idea that disagreements naturally arise in the course of doing business, and under certain circumstances they will lead to spin-offs. The model is preliminary and rather abstract. Nonetheless, it already generates some distinctive predictions that resonate with prior findings, suggesting that this is a promising line of inquiry. First, the model predicts that the hazard rate of spin-offs initially rises with the age of the prospective parent firm, but eventually declines to zero. This accords with findings for autos [Klepper (2004b)] and lasers [Klepper and Sleeper (2005)]. Second, the model predicts that spin-offs are more likely in industries where there is considerable uncertainty about the target, which seems consistent with the prevalence of spin-offs in high-tech and younger industries [Garvin (1983)]. Third, the model predicts that spin-offs have on average lower quality than their parents and this accords with some preliminary evidence for autos reported in this paper.

The model also predicts that spin-offs are more likely when there is a strongly hierarchical structure in decision-making, as measured by the Herfindahl index of concentration of decision-making authority, and that individuals with less decision-making influence will be more likely to start spin-offs. These last two results seem to be at the heart of why

---

[17] Because this result holds for $\phi_i > n^{-1}$ and $\phi_i < n^{-1}$, it is noteworthy that parental quality does *not* matter in the symmetric case, where $\phi_i = n^{-1}$ (see appendix).

acquisitions increase the chance of spin-offs. They shift control of decision-making to the acquirer. If the acquirer has a distinctive "culture" then managers of the incumbent firm will have little influence on the decisions of the acquirer, raising the prospect of a culture clash in which employees leave to pursue their ideas in their own firms. This seems consistent with Mitton's (1990) observations about how acquisitions of San Diego biotech firms by non-biotech firms led to spin-offs.

Our model needs to be developed further to accommodate the full range of common patterns that we detailed about spin-offs. The model does not allow for heterogeneity across firms, hence it cannot address why better firms have more and better spin-offs. In fact, our model predicts that better firms have fewer spin-offs, while the relationship between parent and spin-off quality is ambiguous. We suspect the explanation lies in spin-offs inheriting an aspect of quality from their parents. We are undecided about whether this is because better firms are able to hire more able team members, some of whom subsequently form spin-offs, or whether employees of better firms are able to learn more. Perhaps both factors are in play, but at least one such mechanism needs to be built into the model.

We also need to draw out and test other distinctive implications of our model. One promising direction concerns the effect of spin-offs on industry performance. Most models of spin-offs suggest that spin-offs will harm their parents, which accords with Phillips' (2002) findings for Silicon Valley law firms. Our theory suggests a more benign view of spin-offs. Although, on average, they do worse than their parents, some spin-offs will inevitably do much better. Spin-offs therefore provide parents, and quite possibly the industry as a whole, the opportunity to observe the outcome of decisions that incumbents had chosen not to make. When such spillovers exist, all firms in the industry may learn much more quickly when spin-offs are prevalent. We suspect that this benign role of spin-offs may be quantitatively important (this seems to be the case for autos, where spin-offs played a major role in the development and diffusion of new technology), but this mechanism is absent from the model.

While there is much to do, our findings to date are encouraging about the model. We intend to structure the model further to accommodate the full range of findings about spin-offs and then use data from multiple industries to test its distinctive implications. If we want to sort out the welfare implications of spin-offs and appropriate public policies to deal with them, we will need to push all theories, including ours, much harder. Given how well spin-offs performed in the high-tech industries where they have been studied, the agenda seems well worth engaging.

6.5   Appendices

A. *Derivation of the distribution of spin-off quality relative to parent quality for $k > 0$.*

Let

$$y_{it} = \frac{t\sigma_\theta^2[1 + (\beta - 1)(1 - \phi_i)]\overline{S}_{it}}{(\beta + n - 1)t\sigma_\theta^2 + \beta\gamma\sigma_\varepsilon^2} \tag{A.1}$$

and

$$y_{jt} = \frac{t\sigma_\theta^2[(n - 1) + (\beta - 1)(1 - \phi_i)]}{(\beta + n - 1)t\sigma_\theta^2 + \beta\gamma\sigma_\varepsilon^2} \sum_{j\neq i} \overline{S}_{jt}. \tag{A.2}$$

Clearly, y_{it} and y_{jt} are independent and normally distributed. Using (A.1) and (A.2) in (6.5) and (6.7) yields

$$\left|\overline{\theta}_t - \theta_{it}\right| = \left|\frac{(\beta - 1)(1 - \phi_i)y_{jt}}{(n - 1) + (\beta - 1)(1 - \phi_i)} - \frac{(\beta - 1)(1 - \phi_i)y_{it}}{1 + (\beta - 1)\phi_i}\right|. \tag{A.3}$$

Hence, i's departure requires that or $y_{jt} \leq a_-$ or $y_{jt} \geq a_+$, where

$$a_- = \left(\frac{(n - 1) + (\beta - 1)(1 - \phi_i)}{1 + (\beta - 1)\phi_i}\right)y_{it} - \left(1 + \frac{n - 1}{(\beta - 1)(1 - \phi_i)}\right)\sqrt{k}$$

and

$$a_+ = \left(\frac{(n - 1) + (\beta - 1)(1 - \phi_i)}{1 + (\beta - 1)\phi_i}\right)y_{it} + \left(1 + \frac{n - 1}{(\beta - 1)(1 - \phi_i)}\right)\sqrt{k}.$$

We begin by conditioning on the unobserved θ. It is then easy to see that the expected difference in the values of parent and spin-off, conditional on i forming a spin-off, is

$$\begin{aligned}
\mu(\theta) &\equiv E\left[-(\theta - \overline{\theta}_t^2) + (\theta - \theta_{it}^2)\big||\overline{\theta}_t - \theta_{it}| \geq \sqrt{k}; \theta\right] \\
&= \int_{-\infty}^{\infty} \left\{\left[1 - \int_{a_-}^{a_+} d\Psi_j(y_{jt})\right]^{-1}\left[\int_{-\infty}^{a_-} h(y_{jt}, y_{it}; \theta)d\Psi_j(y_{jt})\right.\right. \\
&\quad \left.\left. + \int_{a_+}^{\infty} h(y_{jt}, y_{it}; \theta)d\Psi_j(y_{jt})\right]\right\}d\Psi_i(y_{it}),
\end{aligned} \tag{A.4}$$

where

$$h(y_{jt}, y_{it}; \theta) = (\theta - (y_{jt} + y_{it}))^2$$

$$- \left(\theta - \left(\frac{\beta y_{it}}{1 + (\beta - 1)\phi_i} + \frac{(n-1)y_{jt}}{(n-1) + (\beta - 1)(1 - \phi_i)} \right) \right)^2,$$

$$\Psi_i(y_{it}) \equiv N \left(\theta \eta_t (1 + (\beta - 1)\phi_i), \frac{\eta_t^2 \sigma_\varepsilon^2}{t} (1 + (\beta - 1)\phi_i)^2 \right),$$

and

$$\Psi_j(y_{jt}) \equiv N \left(\theta \eta_t ((n-1) + (\beta - 1)(1 - \phi_i)), \right.$$

$$\left. \frac{\eta_t^2 (n-1) \sigma_\varepsilon^2}{t} \left(1 + \frac{(\beta - 1)(1 - \phi_i)}{n-1} \right)^2 \right).$$

Finally, we take expectations over θ, giving

$$E\left[-(\theta - \bar{\theta}_t^2) + (\theta - \theta_{it}^2) \big| |\bar{\theta}_i - \theta_{it}| \geq \sqrt{k} \right] = \int_{-\infty}^{\infty} \mu(\theta) d\Psi(\theta), \qquad \text{(A.5)}$$

where θ is normally distributed with zero mean and variance σ_θ^2. Equation (A.5) must be evaluated numerically. Figure 6.2 plots the results. The calculations were conducted using Derive™ 6, on a Dell Dimension 4600 with 512MB RAM. Numerical approximations were carried out with accuracy to 10 significant digits.

B. *Derivation of the spin-off hazard as a function of parent quality.* We continue to focus on the probability that individual i chooses to form a spin-off at time t, and for ease of notation we assume that $\phi_j = (1 - \phi_i)(1 - n)^{-1}$ for all $j \neq i$. That is, we assume that the posteriors of all individuals except for i are weighted equally in arriving at the firm's decision.

Given a decision $\bar{\theta}_t$, the probability that i chooses to leave is given by

$$\Pr\{|z_{it}| > \sqrt{k}\} = 1 - \int_{a_t}^{b_t} dF(S_{it} | \bar{\theta}_t), \qquad \text{(B.1)}$$

where the limits,

$$a_t = \frac{\bar{\theta}_t}{\eta_t(n + \beta - 1)} - \frac{\sqrt{k}}{\eta_t(\beta - 1)}, \quad b_t = \frac{\bar{\theta}_t}{\eta_t(n + \beta - 1)} + \frac{\sqrt{k}}{\eta_t(\beta - 1)}$$

with

$$\eta_t = \frac{(\beta - 1)t\sigma_\theta^2}{(\beta + n - 1)t\sigma_\theta^2 + \beta\gamma\sigma_\varepsilon^2}$$

are obtained from a rearrangement of (6.8). By Bayes' rule, (B.1) can be written as

$$\Pr\{|\, z_{it}\,| > \sqrt{k}\} = 1 - \int_{a_t}^{b_t} \frac{\psi(S_{it}\,|\,\bar{\theta}_t)\psi(\bar{S}_{it})}{\psi(\bar{\theta}_t)}\,dS_{it} \qquad (B.2)$$

where, as the notation in (B.1) suggests, the densities turn out all to be Gaussian. The first unconditional density, $\psi(S_{it})$, is normal with mean θ and variance σ_ε^2/t. The decision, $\bar{\theta}_t$, is the weighted sum of n independent normals. Hence, $\psi(\bar{\theta}_t)$ is a normal density with mean $[(1 + (\beta-1)(1-\phi_i)/(n-1))(n-1) + (1 + (\beta-1)\phi_i)]\eta_t\theta$ and variance $[(1 + (\beta-1)(1-\phi_i)/(n-1))^2(n-1) + (1 + (\beta-1)\phi_i)^2]\eta_t^2\sigma_\varepsilon^2/t$. Finally, the conditional density, $\psi(\bar{\theta}_t|s_{it})$ is the convolution of $\tilde{n}1$ independent normals evaluated at $\bar{\theta}_t = \Sigma_{j\neq i}S_{jt} - S_{it}$. It is therefore also a normal density function, with mean $[(1 + (\beta-1)(1-\phi_i)/(n-1)(n-1) + [(1 + (\beta-1)\phi_i]\eta_t(\theta + S_{it})$ and variance $(n-1)h_t^2 S_e^2[1 + (\beta-1)(1-\phi_i)/(n-1)]^2/t$. There is no closed-form expression for (B.2), which is evaluated numerically in Figure 6.5 for the cases $\phi_i < n^{-1}$ and $\phi_i > n^{-1}$.

A special analytical result exists for the symmetric case $f_i = n^{-1}$. In this case, (B.2) simplifies to

$$\Pr\{|\, z_{it}\,| > \sqrt{k}\} = 1 - \text{erf}(y),$$

where erf is the error function, $2\pi^{-0.5}\int_0^y e^{-v^2}dv$, and

$$y = \frac{\sqrt{n(n-1)k}}{\sqrt{2}(n-1)(\beta-1)}\left(\frac{\gamma\beta\sigma_\varepsilon}{\sigma_\theta^2\sqrt{t}} + \frac{(\beta + n - 1)\sqrt{t}}{\sigma_\varepsilon}\right).$$

Neither $\bar{\theta}_t$ nor θ appear in the expression, so the hazard of a spin-off is independent of the quality of the parent.

References

Agarwal, R., Raj, E., Franco, A. M. and Sarkar, M. B. 2004. "Knowledge transfer through inheritance: Spinout generation, development, and survival," *Academy of Management Journal* 47(4): 501–22

Akerlof, G. A. 1982. "Labor contracts as partial gift exchange," *Quarterly Journal of Economics* 97: 543–69

Amador, M. and Landier, A. 2003. "Entrepreneurial pressure and innovation," Manuscript, MIT

Anderson, L. R and Holt, C. A. 1996. "Information cascades in the laboratory," *American Economic Review* 87(5): 847–62

Anton, J. J and Yao, D. A. 1995. "Start-ups, spin-offs, and internal projects," *Journal of Law, Economics, and Organization* 11: 362–78

Aroian, L. A. 1947. "The probability function of the product of two normally distributed variables," *The Annals of Mathematical Statistics* 18(2): 265–71

Aumann, R. J. 1976. "Agreeing to disagree," *Annals of Statistics* 4: 1236–39

Aumann, R. J. 1987. "Correlated equilibrium as expression of Bayesian rationality," *Econometrica* 55: 1–18

Aumann, R. J. 1988. "Common priors: reply to Gul." *Econometrica* 66(4): 929–38

Bankman, J. and Gilson, R. J. 1999. "Why start-ups?," *Stanford Law Review* 51: 289–308

Brittain, J. W. and Freeman, J. 1986. "Entrepreneurship in the semiconductor industry," Manuscript

Busenitz, L. W., and Barney, J. B. 1997. "Differences between entrepreneurs and managers in large organizations: Biases and heuristics in strategic decision-making," *Journal of Business Venturing* 12(1): 9–30

Cassiman, B. and Ueda, M. 2002. "Optimal project rejection and new firm start-ups," Manuscript, Universidad de Navarra

Carroll, G. R., Bigelow, L. S., Seidel, M-D. L., and Tsai, L. B. 1996. "The fates of de novo and de alio producers in the American automobile industry 1885–1981," *Strategic Management Journal* 17: 117–37

Chen, Qi and Wei, Jiang 2003. "Analysts' weighting of private and public information." Manuscript, Duke University

Christensen, C. M. 1993. "The rigid disk drive industry: a history of commercial and technological turbulence," *Business History Review* Winter, 531–88

Cooper, A. C., Woo, C. Y., and Dunkelberg, W. C. 1988. "Entreprenuers' perceived chances for success," *Journal of Business Venturing* 3: 97–108

Craig, C. C. 1936. "On the frequency function of xy," *The Annals of Mathematical Statistics* 7(1): 1–15

Daniel, K. D., Hirshleifer, D., and Subrahmanyam, A. 1998. "A theory of overconfidence, self-attribution, and security market under- and over-reactions," *Journal of Finance* 53(5): 1839–86

DeBondt, W. M. and Thaler, R. H. 1995. "Financial decision-making in markets and firms: a behavioral perspective," in R. Jarrow, et al. (eds.) *Finance*, Amsterdam: Elsevier

Franco, A. M. and Filson, D. 2000. "Knowledge diffusion through employee mobility," Federal Reserve Bank of Minneapolis Staff Report 272

Garvin, D. A. 1983. "Spin-offs and the new firm formation process," *California Management Review* 25(2): 3–20

Geanakoplos, J. D. and Polemarcharkis, H. M. 1982. "We can't disagree forever," *Journal of Economic Theory* 28: 192–200

Geroski, P. 1995. "What do we know about entry?," *International Journal of Industrial Organisation* 13: 421–40

Geroski, P. A., Mata, J., and Portugal, P. 2002. *"Founding conditions and the survival of new firms,"* Manuscript, Universidade Nova de Lisboa

Gompers, P., Lerner, J., and Scharfstein, D. 2003. "Entrepreneurial spawning: public corporations and the genesis of new ventures, 1986–1999", NBER working paper No. 9816

Gromb, D. and Scharfstein, D. 2002. "Entrepreneurship in equilibrium." NBER working paper No. 9001

Gul, F. 1988. "A comment on Aumann's Bayesian view," *Econometrica* 66(4): 923–27

Harrison, J. M. and Kreps, D. 1978. "Speculative investor behavior in a stock market with heterogeneous expectations," *Quarterly Journal of Economics* 92 (2): 323–36

Hellman, T. 2002. "When do employees become entrepreneurs?," Manuscript, Stanford University

Henderson, R. A. and Clark, K. B. 1990. "Architectural innovation: the reconfiguration of existing systems and the failure of established firms," *Administrative Science Quarterly* 35: 9–30

Holmes, T. J. and Schmitz, J. A. Jr. 1990. "A theory of entrepreneurship and its application to the study of business transfers," *Journal of Political Economy* 98(2): 265–94

Irigoyen, C. 2002. "Where do entrepreneurs come from?," Manuscript, University of Chicago

Jackson, T. (1998): *Inside Intel.* New York: Penguin Putnam

Jovanovic, B. and Yaw, N. 1995. "A Bayesian learning model fitted to a variety of empirical learning curves," Brookings Papers on Economic Activity: Microeconomics 1995(1): 247–305

King, A. A. and Tucci, C. L. 2002. "Incumbent entry into new market niches: the role of experience and managerial choice in the creation of dynamic capabilities." *Management Science,* 48(2): 171–86

Klepper, S. 2002a. "Firm survival and the evolution of oligopoly," *RAND Journal of Economics* 33(1): 37–69

Klepper, S. 2002b. "The capabilities of new firms and the evolution of the US automobile industry," *Industrial and Corporate Change* 11(4): 645–66

Klepper, S. 2003. "The organizing and financing of innovative companies in the evolution of the U.S. automobile industry," N. Lamoreaux and K. Sokoloff (eds.) *The Financing of Innovation,* Cambridge: MIT Press.

Klepper, S. 2004a. "Pre-entry experience and firm performance in the evolution of the U.S. automobile industry," Business Strategy Over the Industry Life Cycle, Advances in Strategic Management 21: 289–315

Klepper, S. 2004b. "Agglomeration through spinoffs: how Detroit became the capital of the U.S. automobile industry." Manuscript, Carnegie Mellon University

Klepper, S. and Simons, K. L. 2000. "Dominance by birthright: entry of prior radio producers and competitive ramifications in the US television receiver industry," *Strategic Management Journal* 21: 997–1016

Klepper, S. and Sleeper, S. D. 2005. "Entry by spinoffs", Management, 51: 1291–1306

Kunda, Z. 1987. "Motivated inference: self-serving generation and evaluation of causal theories," *Journal of Personality and Social Psychology* 53: 636–47

Lindholm, A. 1994. *The Economics of Technology-related Ownership Changes.* Ph.D. Thesis, Chalmers University of Technology

Lucas, R. E., Jr. 1978. "On the size distribution of business firms," *Bell Journal of Economics*, 9: 508–23

Malmendier, U., and Tate, G. 2002. "CEO overconfidence and corporate investment," Manuscript, Stanford University

Malmendier, U. and Tate, G. 2003. "Who makes acquisitions? CEO overconfidence and the market's reaction," Manuscript, Stanford University

McKelvey, R. D. and Page, T. 1986. "Common knowledge, consensus, and aggregate information," *Econometrica* 54: 109–28

Mitton, D. G. 1990. "Bring on the clones: a longitudinal study of the proliferation, development, and growth of the biotech industry in San Diego." in N. C. Churchill, W. D. Bygrave, J. A. Hornaday, D. F. Muzyka, K. H. Vesper, and W. E. Wetzel (eds.), *Frontiers of Entrepreneurship Research, 1990.* Babson College: Wellesley, MA, pp. 344–58

Morris, S. 1995. "The common prior assumption in economic theory," *Economics and Philosophy* 11: 227–53

Odean, T. 1998. "Volume, volatility, price and profit when all trades are above average," *Journal of Finance* 53(6): 1887–934

Pakes, A. and Nitzan, S. 1983. "Optimum contracts for research personnel, research employment, and the establishment of 'rival' enterprises," *Journal of Labor Economics* 1(4): 345–65

Phillips, D. J. 2002. "A genealogical approach to organizational life chances: the parent–progeny transfer," *Administrative Sciences Quarterly* 47(3): 474–506

Russo, J. E. and Schoemaker, P. J. 1992. "Managing overconfidence," *Sloan Management Review* 33: 7–17

Stuart, T. E. and Sorenson, O. 2003. "Liquidity events and the geographic distribution of entrepreneurial activity," *Administrative Sciences Quarterly* 48(2): 175–201

Thompson, P. 2005. "Selection and firm survival. Evidence from the shipbuilding industry, 1825–1914," *Review of Economics and Statistics* 87(1): 26–36

Tushman, M. L. and Anderson, P. 1986. "Technological discontinuities and organizational environments," *Administrative Science Quarterly* 31: 439–65

Van den Steen, E. J. 2001. *Essays on the Managerial Implications of Differing Priors.* Ph.D. Thesis, Stanford University

Walsh, S. T., Kirchhoff, B. A., and Boylan, R. L. 1996. "Founder backgrounds and entrepreneurial success: implications for core competence strategy applications to new ventures." in P. D. Reynolds, S. Birely, J. E. Butler, W. D. Bygrave, P. Davidson, W. B. Gartner, and P. P. McDougall (eds.), *Frontiers of Entrepreneurship Research, 1996.* Babson College: Wellesley, MA, pp. 146–54

Ware, R. and Lad, F. 2003. "Approximating the distribution for sums of products of normal variables," Manuscript: University of Queensland

Weinstein, N. D. 1980. "Unrealistic optimism about future life events," *Journal of Personality and Social Psychology* 39: 806–20

Wiggins, S. N. 1995. "Entrepreneurial enterprises, endogenous ownership, and the limits to firm size," *Economic Inquiry* 33: 54–69

Comments to Chapters 5 and 6

Luigi Orsenigo

Industrial dynamics is a relatively young subfield of industrial economics. Despite around two decades of rapid growth, still almost no textbook makes systematic reference to this burgeoning literature, let alone being organized around these themes. Ever since its inception, industrial dynamics has been growing through the "discovery" (or rediscovery) of empirical evidence quite at odds with the assumptions and predictions of the standard theory of competition (both as it regards the Structure–Conduct–Performance approach and the New IO paradigm). At the same time, theoretical analysis has systematically attempted at providing explanations to these results and ensuing "paradoxes," like, for example skewed distributions of firms' sizes, turbulence in the population of firms through high rates of entry, exit and changes in market shares, persistence of profits, and industries life cycles.

The two chapters by Dosi and Klepper and Thompson mark in my view not only significant contributions to this (still) new field but, more than this, a signal of the maturation of the (sub-) discipline, through the introduction of new techniques of analysis, concepts, discovery, or reconsideration of "stylized facts" as well as the investigation of some of the puzzles that had been noted earlier on but had not received systematic attention by subsequent research.

Dosi's chapter is essentially a review article, which summarizes previous work by the author himself and colleagues. He discusses some statistical properties of central phenomena of interest in industrial dynamics like firms' size distributions, firms' growth, profitability, productivity, and linkages among these variables. The primary contribution of this chapter lies – in my view – in the use of "descriptive" statistical methods which are certainly not new in other scientific disciplines (including some parts of economics) but relatively novel in industrial dynamics. This statistical analysis goes beyond the standard emphasis on average relations between the relevant variables and focuses instead on the identification of the deeper properties of the underlying distributions. Indeed, this evidence highlights a much richer heterogeneity and complex dynamic structure than it is usually detected in this field.

Typically, most of the theoretical and empirical literature dealing with industrial dynamics has assumed small idiosyncratic shocks as the main drivers of evolution and as the determinants of the observed distributions of, for example, firms' sizes. One may wonder if the straw man of an "additive" dynamics driven by small independent shocks is excessively caricatural. Yet, it is fair to recognize that a large body of literature in our field has indeed adhered to this representation, at least as a useful reference model.

Dosi's work shows instead that some important regularities – or stylized facts – that have attracted the attention of scholars in recent years do not actually appear to resist a deeper statistical scrutiny and/or they are likely to result from the aggregation of microentities that are strongly heterogeneous. Distributions of firms' size and related explanations based on some version of Gibrat's Law are the most prominent example offered in Dosi's paper.

These findings do not only or simply run against the notion of the "representative agent – which after all has long been dead in most (but not all) microeconomics for a long time – but suggest that:

(1) "strong and ugly" heterogeneity, which cannot be simply washed away and averaged out via processes of aggregation, is the hallmark of industrial evolution (Dosi and Kaniovski 1994).
(2) such heterogeneity exhibits strong degrees of persistence, with only weak tendencies to mean reversion. This observation is also reflected in the detection of higher than expected degrees of correlation in the structure of the stochastic processes (the "error term"), which hint at the existence of highly structured processes and lumpy events driving firms' behavior and performances.
(3) aggregate industry dynamics is often strongly influenced by the behavior of few outliers.
(4) as a consequence, the dynamic processes driving firms' growth and industry evolution are also likely to be far more structured and complex than those conventionally employed in theoretical and empirical models, involving strong nonlinearities, positive feedbacks, autocorrelation.

All in all, Dosi's chapter suggests that industrial evolution occurs in and generates a non-Gaussian (or Poissonian) world, which bears the typical signatures of "self-organization" and "complex systems" dynamics. The statistical and economic interpretation of these results is not yet – and admittedly – fully developed in Dosi's chapter. Certainly, these results are broadly in tune with evolutionary explanations of the dynamics of industrial structures and with competence-based theories of the firm, whereby evolution is driven by the coupled dynamics of processes of (market and other nonmarket) selection and learning by relatively inertial but innovative agents. But the chapter suggests further fascinating conjectures and preliminary pieces of evidence concerning the existence of different timescales between processes of learning (change at the level of the individual agent) and processes of market selection – which appear to work much more slowly and imperfectly than it is usually assumed in this stream of research.

Dosi's chapter raises also some remarkable puzzles. What can account for the purported absence of linkages between productivity and profitability on the one hand and growth on the other, despite evidence of positive relationships between productivity, innovativeness, and profitability? Why and in what way do market-selection processes seem to work so slowly, at least as compared with conventional wisdom?

Clearly, these results need further empirical corroboration, as well as solid theoretical explanations. As it regards the former, systematic extension of this methodology to other datasets (countries and industries) seems an exciting and necessary step forward. As it concerns the latter, various theoretical models in the evolutionary tradition (but also in less heterodox contributions) provide some background to the empirical findings of Dosi's paper. Clearly, a lot remains to be done, especially in the direction of more systematic exploration of the statistical properties of the data generated by those models as compared with those observed in the real world.

Finally, it is worth noting that Dosi's chapter does not address fundamental elements of industrial dynamics like entry and exit. This is most likely due to lack of information on these issues in the datasets referred to in this contribution. Yet, a deeper exploration of the properties of industrial demography along the methodological lines suggested by Dosi seems essential to our understanding of the patterns of industrial evolution.

Klepper and Thompson focus instead precisely on entry. Both from a methodological and a theoretical perspective, they take a rather different direction from Giovanni Dosi. They start from carefully gathered empirical evidence, but essentially they try to build a theoretical model that can explain this evidence. They focus on a seemingly much more specified phenomenon, that is, spin-offs from other firms. They develop a theory which can explain not only a major factor leading to the formation of spin-offs, namely disagreements within existing firms that lead frustrated employees to pursue their ideas by starting a new firm, but which can also account for various empirically observed phenomena linked to the emergence and performance of spin-offs. In this respect, the Klepper and Thompson's chapter is a nice example of the tradition of industrial dynamics (identify interesting and puzzling "stylized facts"; build a theory that can account for many of them at the same time, thus providing a strong empirical discipline to the theoretical argument). Theoretically, the chapter is placed in the neoclassical/behavioral approach, which introduces deviations from perfect rationality (as identified by the new behavioral economics), in an otherwise orthodox framework.

There are three main features of the Klepper and Thompson chapter that I find particularly interesting. First, they offer an interesting contribution to the theory of entry. Second, they introduce elements of the theory of the firm as an organization into a literature where the internal structure of firms has been seldom playing a major role (except, of course in the evolutionary tradition à la Nelson and Winter). Third, they use concepts ad results from the new behavioral economics to model individuals and firms.

Indeed, there are now substantial reasons to be skeptical of the standard "model" of entry, which essentially assumes an infinite queue of potential, fully rational entrants who will actually start producing as long as expected profits are positive.[1] A first set of reasons concerns the hypothesis that within an industry (or an economy) there will always be a number of potential entrants able and/or willing to take advantage of notional opportunities. Lack of capabilities, for example, may very well constrain the population of entrants. A second set of objections relates to the incentives, motivations, and behavior of such entrepreneurs. Third, more generally (and as Klepper and Thompson mention in their chapter), little is known about the processes that lead to the formation of the cohort of potential entrants. While a substantial body of literature has now explored important traits of entrepreneurs (in terms of skills, motivations, psychology, etc.) and some of the factors that are associated to subsequent superior postentry performance, still the idea that the population of potential entrants should not be defined exogenously

---

[1] In a somewhat different context, the same idea is central in the work by John Sutton (Sutton 1999) through the "non arbitrage principle", which essentially states that no profit opportunities will be foregone by potential entrants. But empirical evidence shows that actual and/or expected profits bear very little and typically statistically non significant relationships with observed rates of entry (Geroski, 1995).

but should be determined endogenously has not been systematically explored in the industrial dynamics literature.[2]

In this respect, the Klepper and Thompson chapter, by focusing attention on incumbent firms as important incubators of new entrepreneurs, constitutes a significant contribution in the direction of building a theory of endogenous entry. However, the objective of this chapter is more specific, that is, to provide a theoretical rationale for the emergence and above all persistence of disagreement among members of the parent firms that may lead to the formation of spin-offs. The issue is important if agents are rational. In this context, (persistent) disagreement cannot exist in equilibrium if agents hold common priors. The way out suggested by Klepper and Thomson is to assume "almost perfectly" rational agents. Moreover, the proposed model differs from other possible explanations based on strategic behaviors by opportunistic agents (which might be the natural approach in the new neoclassical theory of the firm). Here, agents are not playing strategically against other members of the organization but are genuinely interested in the value of the firm's activities.

Building on the results of the new behavioral economics, Klepper and Thompson rely on a modification of the (now) well-known concept of overconfidence, which they term as solipsism. The notion of overconfidence has been already applied to the theory of entry (see for instance, Camerer and Lovallo, 1999 and Dosi and Lovallo, 1997). However, Klepper and Thomson note that overconfidence – referring to individuals who underestimate the noise of any signals, whether their own or those inferred from their colleagues – presupposes and magnifies disagreement, but cannot explain it (in an almost rational world). Solipsism – defined instead as asymmetric weighting of private and nonprivate signals (in favor of the former) – is a necessary condition for disagreement. The model presented in this chapter is preliminary and rather abstract. Nonetheless, it is also elegant and ingenious and it generates some predictions that are in accordance with prior empirical evidence.

I would argue that the two chapters discussed here, despite their obvious profound differences in their subject, methodology, and ultimate approach, share some similarities, in their fundamental commitment to

---

[2] This issue has become particularly prominent in the field of economic geography, especially in relation to the studies concerning the formation and development of clusters of industrial and especially innovative activities. Most likely, it is not by chance that earlier papers by Steven Klepper on spin-offs are contemporary to his work on the emergence of Detroit as the capital city of the automobile industry (Klepper, 2002 and 2004).

(1) patient, time-consuming and humble collection of data and empirical evidence as a necessary departure point for subsequent analysis;

(2) introduction and use of increasingly sophisticated techniques of empirical and theoretical analysis for understanding and interpreting the evidence;

(3) empirically based theorizing, including representation of agents' behavior which departs from perfect rationality;

(4) emphasis on dynamics;

(5) use of evolutionary concepts as crucial interpretative tools.

Clearly, Dosi adopts a rather heterodox stance and takes all the previous five points as central tenets for the further development of evolutionary economics. The Klepper and Thompson chapter remains rooted into neoclassical economics and the authors are certainly not willing to depart too much from it. Still, they take on board at least some elements of points (3) and (5), while taking very seriously indeed points (1), (2) and (4). On my part, as much as I am committed to the evolutionary approach, I consider these two chapters as an exemplary and promising step toward the development of a "post-neoclassical" theory of industrial organization, where research and debate can be more fruitfully framed within the boundaries of points (1) to (5) than it is now.

References

Camerer, C. and Lovallo, D. 1999. "Overconfidence and excess entry: an experimental approach," *American Economic Review* 89(1): 306–18

Dosi, G. and Kaniovski, Y. 1994. "On 'badly behaved' dynamics," *Journal of Evolutionary Economics* 4: 93–123

Dosi, G. and Lovallo, D. 1997. "Rational entrepreneurs or optimistic martyrs? Some considerations on technological regimes, corporate entries and the evolutionary role of decision biases," in R. Garud, P. Nayyar, and Z. Shapiroz (eds.) *Technological Innovation: Oversights and Foresights.* Cambridge: Cambridge University Press, pp. 41–68

Geroski, P. 1995. "What do we know about entry?," *International Journal of Industrial Organization* 13: 421–40

Klepper, S. 2002. "The capabilities of new firms and the evolution of the US automobile industry," *Industrial and Corporate Change* 11(4): 645–66

Klepper, S. 2004. "Agglomeration through spin offs: how Detroit became the capital of the U.S. automobile industry," Manuscript, Carnegie Mellon University

Sutton, J. 1999. *Technology and Market Structure.* Cambridge MA: MIT Press

Part 4

Innovation and institutions

7   Schumpeterian innovation in institutions

Masahiko Aoki

Introduction: Schumpeter revisited on innovation

In the last decade or so a near-consensus seems to have emerged among
economists that "institutions matter" in making economic performance
differ across economies.[1] However, if institutions are nothing more than
codified laws, organizational entities and other such deliberately
designed human devices, why cannot a badly performing economy jet-
tison "ineffective" institutions and then emulate the "good" institutions
operating in a well-performing economy? Is it simply because politicians
are bad? But, if so, why do they survive? Is their persistence itself not an
institution? Alternatively, is it the case that good institutions cannot be
emulated because they are shaped in each economy in a path-dependent
manner? Does it then follow that there cannot be "innovation" in
institutions?

In order to be able to provide intelligible answers to these questions,
we need first to be clear about what is meant by the term "institution."
However, there does not seem to be an unambiguous consensus among
economists regarding this point. Certainly, it would be unproductive to
quarrel over a semantic definition of institutions as such. Any definition
would do, as long as it serves to answer a meaningful and analytical
question. And one of the objectives of this chapter is to understand a
mechanism of change in an "institutional" framework that may constrain or enable economic, political, and other social activities relevant
to economic performance and development.

In pursuing this purpose, this chapter first introduces a static conceptualization of an institution based on a game-theoretic equilibrium
notion. Some ardent students of the evolutionary approach may
immediately dismiss this approach as inappropriate for an earnest study
of institutional change of an inherently dynamic nature. But if we want
to understand how and why institutions change, we need to first clarify

---

[1] North (1990), World Bank (2002), Nelson and Sampat (2001).

what change we are dealing with. We must try to understand institutional change as a process of discontinuous and spontaneous shift of equilibrium. It may be remembered that Schumpeter, whose youthful hero was Walras,[2] began his classic, *The Theory of Economic Development* (1934: 64.n.), with a chapter entitled "The Circular Flow of Economic Life as Conditioned by Given Circumstances." It was essentially devoted to a characterization of "static equilibrium." The book then went on to distinguish the stationary process in which equilibrium only passively adapts to changes in external data (tastes, war, policy, etc.), on the one hand, and the kinds of economic change (development) "arising from within the system *which so displaces its equilibrium point that the new one cannot be reached from the old one by infinitesimal steps*" italics by the present author) on the other. Thus the notion of equilibrium was introduced to provide "necessary principles," as well as "conceptual devices," by which the object of his analysis (change in the latter kind of equilibrium) could be distinguished from that of Walras (the former). I submit that this chapter is essentially in accordance with Schumpeter's methodology and spirit.

Our unit of analysis is the primitive domain of the game people repeatedly play, rather than the entrepreneur, who was the focal point of Schumpeter's analysis. However, analogously to Schumpeter's interests in creative changes in the combination of productive factors initiated by the entrepreneur, we deal with ways through which games of different domains are interlinked, delinked, and newly linked through the process of institutional change. Relying on equilibrium analytical tools, we are able to identify three major modes of such linkages and the corresponding changes: embeddedness, complementarities, and bundling. The first two modes tend to provide the process of institutional change with more or less inertial, path-dependent characteristics, while the last one contributes to a discontinuous characteristic of the Schumpeterian type. In actual processes of institutional change, all three modes may interact and play respective roles in varying degrees. However, it is our hope that our approach will provide a clue to synthesizing the historical, path-dependent approach to institutional change with the Schumpeterian evolutionary approach.

The plan of this chapter is as follows. Section 7.1 lays out a game-theoretic equilibrium conceptualization of institutions. Section 7.2 identifies basic types of the primitive domains of games and discusses the above-mentioned three modes by which they can be interlinked

---

[2] He stated that "as an economist," he owed more to "the Walrasian conception and technique ... than to any other influence." (1937/1965: 65).

institutionally through the equilibrium play of the games. Section 7.3 then applies these conceptual devices to the dynamic context and identifies the corresponding mechanisms of institutional change, among which the Schumpeterian innovation of institutions is identified. Section 7.4 applies the developed frame to the emergence of the Silicon Valley clustering of entrepreneurial firms and clarifies its fundamental nature as an instance of Schumpeterian innovation in an institution of innovation. Section 7.5 concludes the chapter.

7.1   Using the game-theoretic frame for understanding institutions

Although we adopt a game-theoretic-like frame below for defining and understanding institutions, its concern is not expository refinement or analytical rigor. On the contrary, the exposition below remains simple and nontechnical (it involves few symbols and no proof!).[3] But the frame is adopted here because I strongly believe that the game-oriented frame of thinking is the most suitable for socially scientific subject matter.[4] Its essence is that every constituent member of any social domain has his or her own motivation (aside from how it is formed), and any social outcome (specifically an institution) can be understood as a result of the interplay among such members. In order to develop such a framework, we will first define the basic terms and concepts, such as domains and the game form that we will use below.

We will treat a game defined in a certain domain as a unit of analysis, somewhat analogously to the way transaction-cost economists treat a transaction. The domain of the game is composed of a set of agents (players) – either individuals or organizations – and sets of physically feasible actions open to the choice of each agent (that may be conditioned by his or her mental state and innate or acquired abilities) in successive periods. The set of actions chosen in one period by all the agents in the domain is known as their action profile. In the classical formulation of the game, an action profile determines the distribution of payoffs among the agents. We decompose the payoff functions into objective and subjective elements. Namely, given the external environments and historically determined states of the domain at the beginning of a period, an action profile in that period generates a physical

[3] See Aoki (2001) for a technical and detailed exposition and analysis.

[4] The analogy of the society with the game can be dated back to Adam Smith, who stated in *The Theory of Moral Sentiments* that "[I]n the great chess board of human society, every single piece has a principle of motion of its own, altogether different from that which the legislature might choose to impress upon it." (1759/1976: 234).

consequence (that also determines the initial state of the next period). We call the function (rule) that assigns a physical consequence at the end of the period for each action profile of that period, the consequence function. Various environmental factors, such as technology, "institutions" prevailing in other domains, as well as statutory laws and policy determined in the polity domain, parametrically define the form of the consequence function, which may be said to represent the *exogenous rules of the game*. A pair of a domain and an associated consequence function specify the game form.[5] The agents have respective preference orderings over the possible consequences. They strategically choose actions (or action plans over periods) so as to maximize their own satisfaction on the basis of expectations regarding others' possible choices and their consequences. An action (action plan) thus chosen by each agent is called his or her strategy.

Using this simple setup, we can readily distinguish three different views of institutions:

- *Institutions as players (=organizations) of the game*: Institutions may be identified with prominent organizations, such as "industry associations, technical societies, universities, courts, government agencies, legislatures, etc." (Nelson, 1995: 57)
- *Institutions as the exogenous rules (the consequence function) of the game*: This view is subscribed to prominently by North (1990) and Williamson (2000). It includes formal rules, such as constitutions, laws, and regulations, as well as informal rules, such as customs, in institutions thus defined. North draws a sharp distinction between the rules of the game and the players (organizations and political entrepreneurs) of the game. The latter act as agents of institutional change, that is, as rule-makers, in the polity, distinguished from the domain of the economic game (North, 1995). One well-known problem with this view is the enforceability question: Who enforces the rules? What incentives do the enforcers have to enforce or not to enforce? In short, who enforces the enforcer? An attempt to endogenize a solution of this problem leads us to the following view.
- *Institutions as the equilibrium outcome of the game*: By recognizing the enforcer (such as the court, the police, and Law Merchants) as a strategic player of the game, and by examining its equilibrium outcome, exogenous rules (such as the law) may be shown to be

---

[5] The reason we introduce the game form, in lieu of the three-tuple game formulation composed of the set of players, the sets of action choices, and the payoff functions, is to define the notion of exogenous rules of the game free from the subjective notion of utility. The terminology is due to Hurwicz (1996).

enforceable or not. Unenforceable exogenous rules may not be considered institutions. Also social norms, organizational conventions, and the like that may constrain the economic action choices of the agents in the economic transaction domain may also be understood as equilibrium outcomes of social-exchange and organizational-exchange domains. There are two approaches to this view: the evolutionary game approach based on some kind of evolutionary equilibrium notion (Schotter, 1981; Sugden, 1989; Young, 1998; Aoki, 2001 chapters 2.1 and 5; Bowles, 2003), and the repeated game approach based on the classical subgame-perfect equilibrium notion (Aoki, 2001 chapters 2.2, 6, 10–11; Greif, 1997; Dixit, 2003). In this chapter, we adopt the kind of equilibrium view that follows, since there can be many advantages to an equilibrium view beyond the enforceability question.[6]

An institution is a summary representation of the invariant and salient features of a (Nash) equilibrium path, internalized as shared beliefs among all the players in the domain, regarding the ways in which the game is being repeatedly played.

Three remarks are made regarding this conceptualization.

First, we leave the notion of equilibrium unspecified beyond Nash. As is well known, Nash equilibrium is the state in which no player is motivated to change his or her strategy, as long as other players remain committed to the current strategies. It essentially amounts to the state in which the current strategic profile is self-enforcing (for example, people refrain from breaking a law in the expectation that they will be punished for doing so). In general, there can be multiple Nash equilibria for a game that have a reasonable degree of freedom of choice. The attempts of game-theorists to refine the concept of equilibrium in order to reduce the multiplicity to a unique equilibrium have proved to be unsuccessful. But this should be regarded as a blessing in disguise for institutional economists because it indicates that a variety of institutions may be possible in the same environment. If so, how can a particular

---

[6] We do not deny the value of an attempt to develop a comprehensive view that may integrate the three different views and more. For example, Nelson and Sampat (2001) provide such an attempt. They propose defining institutions as prevailing "social technologies" and view the language of "routines," as developed in Nelson and Winter (1982), as appropriate to such a characterization. Although their exposition is not explicitly game-theoretic, this characterization may be thought of as including "the rules of the game" view, as well as a kind of equilibrium view, particularly when they refer to "the way things customarily are done." Further, they state that broad social and cultural values, norms, beliefs, and expectations are "behind the scenes in [their] formulation." Recently, Greif (2006) has been trying to develop a similar, comprehensive view of institutions that may integrate all three game-theoretic notions plus the classical cultural notion of institutions. His attempt is indeed ambitious and worthy of attention.

equilibrium (institution) be chosen from the many that are possible in one domain but not in others? Since game theory cannot provide an answer, a solution must be found in historical analysis. In other words, the game-theoretic approach to institutions needs to be inherently complemented by historical analysis (Greif, 1997, 2006). However, we need to be a little bit careful about characterizing the situation with the dictum "history matters." This can imply either that historical events and accidents condition the choice of the subsequent path out of the many that are possible (in that case a historical narrative has a unique value) or that there could be some generic mechanism operating that conditions institutional change in a path-dependent way (in that case a game-theoretic analysis may retain some value in interpreting history). We will argue that the latter possibility cannot be entirely dismissed.

Second, the conceptualization refers to a *summary representation* of equilibrium as an institution rather than an equilibrium as a whole. The meaning of this characterization may be captured analogously by referring to the theory of the informational efficiency of the price mechanism developed by Hayek (1945), Koopmans (1957), and Hurwicz (1960). At a Walrasian general equilibrium, every market participant does not need to know what technologies other producers are employing, nor what tastes other consumers entertain in order to derive maximum satisfaction of his or her own within the constraints of the resource endowment and technology that he or she faces. The only information that he or she needs to know is the equilibrium set of prices whose dimension is no more than the number of commodities in markets less than one, assuming one of them serves as the numeraire. This information constrains the agent's choice by defining his or her budget constraint, but also provides him or her with sufficient information to gain maximum satisfaction in the rigorous sense of a sufficient statistic.[7] Marginal changes in prices in adaptive responses to gradual changes in tastes and technologies belong to the realm of the stationary economy in the Schumpeterian sense, and do not have direct bearings on the issue of institutional change (we suppress such marginal change by assuming that an institution is a summary of invariant, salient features of a possibly continually moving equilibrium). The analogy with market prices clarifies one important dual role of institution: It acts as a device to summarize complex information and economize information-processing for the agents so that they are enabled to act effectively within the limits

---

[7] The sufficient statistic property in nonclassical environments, such as those that include externalities and increasing returns to scale, was investigated by Aoki (1971a) and (1971b).

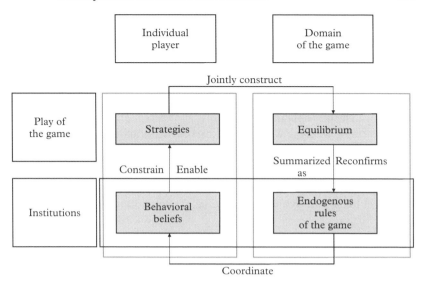

Figure 7.1 An institution: endogenous/exogenous, informative/constraining dualities

of their time and information-processing abilities, while it constrains their choices consistently with the over-all resource constraints (this constraining or enabling duality is represented by the upward arrow of the left side of Figure 7.1).

Third, the above conceptualization and Figure 7.1 also capture another important duality of an institution – the endogenous–exogenous duality. Namely, an institution as an equilibrium is produced and must be repeatedly reproduced as an endogenous outcome of the strategic interplays of all the agents in the domain (indicated by the upper right arrow in Figure 7.1), while it appears to be an external constraint (rules to be followed) for each agent making a choice (indicated by the left upward arrow).[8] In the former sense, an institution thus defined may be characterized as the *endogenous rules of the game*. Related to this duality is the internal–external duality. An institution appears as external constraints to each individual agent, something that is felt to exist beyond his or her control. But, on the other hand, it cannot become self-enforcing (i.e., an equilibrium), unless it becomes taken-for-granted by

[8] This duality characteristic was explicitly discussed in the sociology literature in Berger and Luckmann's (1966) ethnographic study on the construction of everyday social reality. Their proposition is thought to have a close affinity with the game-theoretic notion of equilibrium.

them. That is, the rules of the game can become institutions and sustained as such, only when they are deeply internalized by each individual, constituting part of his or her mind-set. In that sense, institutions may not be implemented or abandoned by the mere will of political elites or the enactment of laws, unless the mind-set (shared beliefs) of the people are changed accordingly. Thus we seem to need a deeper theory of institutional change than one that assumes, expects, or posits, that it is political elites who can change institutions, although they play certain roles in the process of institutional change. This recognition may become imperative only if we adopt the institution-as-equilibrium view.

7.2   The primitive domains and three modes of their linkages

We have introduced the concept of both the domain of the game and that of an institution as an equilibrium in it as basic conceptual devices. In order to introduce the operational notion of a Schumpeterian mechanism of institutional change, however, we need to take one more preparatory step of an analytical nature. As hinted in the Introduction, institutional change (innovation) a la Schumpeter needs to be distinguished from a gradual equilibrium change in the domain in response to data (exogenous rules of the game). According to Schumpeter, innovation involves the introduction of some kind of new combinations (in his case, productive factors), as well as a destruction of some old combinations that will emerge from within and consequently bring about a qualitative shift of equilibrium. With this perspective in mind, this section introduces concepts of linkages (combinations) of domains through plays of the game(s). It analytically distinguishes three different modes of linkages and subsequently presents that one of them may indeed be related to the Schumpeterian notion of combinations. In order to do so in a somewhat concrete and applicable manner, let us begin by identifying three primitive kinds of domains with some illustrations.

Two-person (symmetric) domain: One type of this simplest domain may be found in an *economic exchange* domain composed of two agents who possess respective goods that are exchangeable for mutual benefits. In the sense that exchange can take place only if the two agents agree to it and mutually fulfill their promises, they may be regarded as symmetrically situated and the exchange becomes essentially a contract. It is well known that if a game is repeatedly played in such a domain, then mutually beneficial contracts of exchange may be agreed upon and enforced over time through the reputation mechanism, or else a no-trade option may ensue. Another even more primitive type is the *commons* domain characterized by the presence of common resources whose use may be shared between the two (or more) agents. A *de facto* property rights arrangement may evolve between them as customary rights,[9] or else the tragedy of the commons may ensue.

Three-person asymmetric domain: The essential characteristics of games involving asymmetric players as in the polity or the organization may be generically analyzed by considering this type of primitive domain. Suppose that the primitive *political-exchange* domain is composed of the government and two private agents. The government is assumed to possess the power of unilaterally protecting or transgressing property rights (and or other human rights) of the private agents, while the latter may choose, respectively, to agree on coordinated resistance against potential abuse of power by the government, submit to it, or collude with the government at the cost of the other (there is no exit option for the private agents).[10] Even in this primitive domain, multiple equilibria may be identified, which suggests the diversity of the state as a form of political institution.[11] Likewise, the primitive *organizational-exchange* domain may be thought of as being composed of one superordinate and two subordinates. A different type of organizational architecture may be distinguished by the ways in which types of information-processing activities are asymmetrically distributed and governed between them as an equilibrium. In this type of primitive domain too, there can be multiple equilibria, indicating a diversity of organizational architecture-cum-governance arrangements, such as a hierarchy, a team, and a modularity.[12] More complex organizational architecture may be thought of as nested (equilibrium) bundling of the primitive organizational domains.

N-person symmetric (asymmetric) domain: One type of this domain may be found in the primitive *social-exchange* domain in which social symbols (languages, rituals, gestures, etc.) that directly affect the payoffs of recipient agents are exchanged at large.[13] Equilbria arising in this type of domain (in linkages with other types of domains) may be identified with norms enforced by the threat of ostracism from the domain, gradational

---

[9] See Demsetz (1967), Aoki (2001) Chapter 2.

[10] Weingast (1993), Aoki (2001) Chapter 6.

[11] Aoki (2001) Chapter 6 identifies several equilibria for a simple three-person political domain and associates a notion of a political institution to each of them such as the democratic state, corporatist state, developmental state, and bureau-pluralist state. Indeed, the English word "state" is said to be derived from the same Latin word from which the English word "stable" is derived.

[12] Aoki (2001) Chapters 4 and 5.

[13] Blau (1964/1998), Aoki(2001) Chapter 1.

rankings of prestige or social status among the agents, and so on.[14] Norms are taken as exogenous rules of the "economic exchange" game in the North-Williamson framework, but their evolution may be susceptible to game-theoretic analysis, which we will return to shortly.

The equilibria associated with the primitive domains are nothing more than prototypes of institutions. However, more realistic models of institutions (institutional arrangements) may be captured as equilibrium linkages (combinations) of those prototypes. We now introduce some generic modes of linkages of primitive domains through the equilibria of games.

Linked games: Games are "linked" if one or more agents choose strategies across more-than-one (primitive) domain in a coordinating manner. Because of the possible externalities (social surpluses) created by such linkages, a choice pattern that is unsustainable in a single domain in isolation may become sustainable. There are four types of games in this linkage.

Bundling of multiple domains by a single player internal to each domain. For example, a single employer can bundle multiple employment contracts with workers. Then an equilibrium may emerge that can elicit higher levels of effort from each worker than would be possible under a single contract by making the threat of terminating a contract more credible with a worker who shirks, and replacing him or her with another, provided that the employer can prevent collusion among the workers.[15] Economic-exchange domains thus bundled are then transformed into an organizational domain. A somewhat similar example is found in the financing of multiple entrepreneurial projects of similar types by a single financier. In spite of multiple financing costs, multiple contracts may become beneficial to the financier when they can broaden future options in the presence of high developmental uncertainty. They may also be able to elicit higher entrepreneurial effort by creating a tournament-like situation among entrepreneurs by promising to share a large prize with the most successful entrepreneur (this type of bundling is discussed in greater detail later in Section 7.4).[16]

Bundling of multiple domains by a third strategic party. Suppose that the reputation mechanism does not sustain honest exchanges (mutual contract compliances) between two anonymous traders, because they are not expected to meet again. However, if multiple domains of this sort are bundled with an intermediary who can disseminate information

---

[14] Coleman (1990), Aoki (2001) Chapters 2 and 8.

[15] Murdock (1996).

[16] This type of linked game was earlier analyzed in the context of the linkage between share-cropping and loan contracts in the Indian agrarian economy. See Bardhan (1977), Braverman and Stiglitz (1982).

regarding the past contractual compliances of agents across them, the reputation mechanism may be effectively replicated, provided that honest information processing and dissemination by the third party can be motivated by his or her own reputation concerns (note that we treat the third party as a strategic player). The Law Merchants (Milgrom, North, and Weingast, 1990), credit bureaus, escrow services, on-line certificate authorities, and auction-sites are examples of such third parties.

An N-person social-exchange domain embeds (multiple) economic-exchange domains, commons domains or nested organizational domains. Suppose that a sufficient surplus cannot be generated in a single primitive economic-exchange domain to sustain cooperation (enforce contracts) among the agents. However, suppose that all the agents from this domain, or a multiple of these domains, at the same time belong to a single social-exchange domain, where sufficient social surplus can be generated through social cooperation. In this situation, the agents may be able to enforce cooperation in the former domain(s) as well, by the threat of social punishment (ostracism from the social exchange domain) in the event of noncooperation. This linkage mechanism may be considered to correspond to the sociological notion of "social embeddedness" (Granovetter, 1985). He argued that agents in markets and organizations in the modern society generate trust and discourage malfeasance by being embedded in "concrete personal relations and structures (networks)," while the norms and values prevailing therein are shaped by them "in part for their own strategic reasons"(p. 57). The same mechanism may be workable for controlling the classical tragedy of the commons, or providing incentives to the members of the free open-software development community in the form of professional reputation.

Institutional Complementarities. In linked games agents coordinate their strategic choices across domains and generate a single institution (equilibrium). Alternatively, we can conceive of the possibility that, even if agents may not strategically coordinate their choices across domains, they regard an institution in another domain as a parameter and accordingly choose strategies in their own domain, and vice versa. In such situations, institutions evolving in those domains may become interdependent and mutually reinforcing. This intuition can be game-theoretically warranted. Suppose simply that x' and x" are two alternative institutions in domain X, while z' and z" are two alternative institutions in domain Z. Suppose that the utility difference $U(x') - U(x'')$ increases for all the agents in domain X (they do not need to have the same utility function), when z' rather than z" prevails in domain Z. By the same token, suppose that utility difference $V(z') - V(z'')$ increases for all the agents in domain Z, when x' rather than x" prevails in X. Then x' and z' (alt. x" and z") are said to complement each other. If these conditions known to be supermodular hold, then the equilibrium combination can be either (x', z') or (x", z").[17] Further, even if one of them is less efficient in terms of Pareto-ranking, it may still prevail as an equilibrium (an overall institutional arrangement), once it is achieved.

This is a powerful and useful analytical tool for institutional analysis. First, as just mentioned, it explains why there can be a variety of over-all institutional arrangements even if economies face the same types of domain characteristics (such as technologies, common markets connecting them) and why a suboptimal overall institutional arrangement can persist. Second, institutional complementarities are not conditional on consensus among agents in domain X regarding the absolute ranking of x' versus x" (and, by the same token regarding the absolute ranking of z' versus z" among agents in domain Z), but only regarding the direction of change in their relative rankings when the parameter changes [i.e., it is not required that $U(x') - U(x") > 0$]. Thus the emergence and sustenance of an overall institutional arrangement may become stable even if there is a conflict of interests among agents about the absolute preference for a component institution in isolation.

By using the supermodular analysis described earlier, we can provide a rigorous logical support to intuitive notions of complementarities such as those between the rule of law (the liberal state) and anonymous competitive markets, the main bank system and lifetime-employment as observed in Japan, the corporatist state and codetermination in the corporate governance domain in Germany.[18]

7.3   Three mechanisms of institutional change

As suggested in the introduction, we intend to identify institutional change with a kind of discontinuous shift of equilibrium. Change or shift should be also endogenous (from within) in the sense that it is not a passive adaptation to marginal changes in the "data" of the game, as expressed according to our terms in the form of the consequence function (exogenous rules of the game). If it is, the equilibrium shift would remain gradual and marginal, leaving its summary representation (i.e., the institution) almost intact. Thus, it can be suggested that institutional change needs to involve qualitatively new (innovative), experimental choices initiated by some (or all) agents, and their

[17] Topkis (1978), Milgrom and Roberts (1990).
[18] See Aoki (2001) Chapter 10.

subsequent stabilization, accompanied by the strategic adaptations by other agents to them. The conceptual devices introduced in the previous sections may suggest that such a possibility may arise when (1) a novel way of linking the same or different domains is initiated and established as an equilibrium; and/or (2) a momentum is created for making hitherto unobserved or suboptimal choices viable and subsequently established (even creating a new type of domain) by the complementary reinforcement of existing institutions in other domains, the accumulated stock of relevant competence, policy changes, and the like. These two types of mechanisms may operate conjointly in a complex manner in the actual process of institutional change, but we begin by treating them singly.

7.3.1 Schumpeterian dis-bundling and new-bundling

We have already referred to the bundling of multiple employment contracts. The historical emergence of the factory system can be considered in that light as an institutional innovation of entrepreneurs for coping with the incentive problem of the workers, rather than merely an adaptive response to technological change. On the other hand, the reverse trend from the integrated organizational architecture of the modern firm to the spin-offs of hitherto-internalized activities to outsourcing can be thought of as an institutional change through rebundling. Equilibrium shifts from internalized bundling to third-party-mediated bundling are also often found in the process of institutional change. For example, business groups organized by financing headquarters that often include a trading company may emerge in developing economies as a way of enforcing contracts when the rule of law has not been firmly established. However, as professional competency for designing, agreeing upon, and enforcing contracts becomes more available, the relative economic value of business groupings may gradually decline unless relational contracting generates intrinsic economic values of its own, for example, through the continual sharing of technological and/or market information among the partners and the like. However, the group may strive for survival on the basis of monopoly rents resulting from exclusive bundling. An institutional change from corporate grouping to more open contractual relationships governed by a third party (such as the court) will involve entrepreneurial challenges, as well as political, economic, and social resistance by the old group. It may not be realized simply by a legislator's pen stroke. In the next section, we present an interesting recent example of Schumpeterian innovation in an institution emerging from the creative destruction of the old organizational bundling of innovation activities.

7.3.2   *Overlapping social embeddedness*

Let us imagine that the choice set of the agent potentially includes all physically possible action choices (possibly of infinite dimensions), but that the agent activates only his or her small subset of finite dimensions as a "repertoire" of actual choices at any point of time.[19] In general, the agent may change this repertoire over time by adding new choice possibilities and deleting obsolete ones in response to the available technological change, development of skills, as well as changes in his or her own physical and mental state shaped by training, experience, perception of the external world, and so forth. However, the speed of this change may vary depending on the types of domain. Choice possibilities open to agents, and accordingly to the equilibrium strategy profile, may change relatively slowly in the social-exchange domain, while those in the organizational domain may be relatively faster, because it is susceptible to the conscious design that is only subject to constraints from complementary institutions (such as labor and capital markets). Thus, it may happen that the same types of choice profile in the social domain may link themselves with changing choice profiles elsewhere. In other words, the same types of social norm and/or customs may embed different domains over time in an overlapping manner.

For example, consider the introduction of exchange relations with outside merchants into a rural community in the precapitalist period where the social norms of cooperation in the management of the commons, such as the irrigation system, have prevailed. Social science disciplines, including economics, sociology, and anthropology, have tended to draw a sharp line between community relationships in the precapitalist period and market relationships thereafter, until the resurgence of institutional economics in the late-twentieth century. However, the presence of social norms as an institutional device to promote and sustain cooperation in the rural community may facilitate or deter its transition to a market economy. Under certain conditions, the presence of social norms may serve as a transitory mechanism of contract enforcement in the underdevelopment of the rule of law governing the domain of exchanges. It may do so by facilitating collective punishment on breaches of contracts by merchants from the outside, while restraining their own breaches through peer monitoring in order to preserve their

---

[19] The term is due to Dosi and Marengo 1994.

collective reputation to outside markets. Cases of such possibilities mainly drawn from East Asian economies are documented, and their theoretical implications were examined in Aoki and Hayami (2001). It may also happen in a period of industrial takeoffs that the interplay of strategic efforts by entrepreneurs and the unconscious sustenance of social customs among workers drawn from rural communities may reproduce a form of community norm in the industrial context.

7.3.3   Dynamic institutional complementarities

The concept of static institutional complementarities has a natural dynamic version formulated in the Momentum Theorem by Milgrom, Qian, and Roberts (1991). Liberally rephrased, it holds that, even if the initial level of human competence conducive to the support of potential institution x is low, the presence of complementary institutions elsewhere may amplify the impact of a policy intended to induce x, and that once a momentum is initiated, x may gradually evolve as a viable institution. Conversely, even if laws are introduced to induce institution x, the absence of complementary institutions may make its realization difficult.

One example can be given by the role that the institutional infrastructure of Hong Kong played in the transition of the Chinese Economy to a market economy. It is now well recognized that China's remarkable growth since its initiation of reform in the late 70s was largely driven by foreign direct investment and commodity exports. By 1977, however, the Chinese economy was virtually closed to the world market economy. How could such an economy attract massive foreign investments in spite of regulations over the remittance of investment returns to home economies? How could it be expected to entertain export contracts in the absence of effective rules of law governing domestic market exchanges? In fact, large portions of capital inflow and commodity exports were mediated through Hong Kong, where the legal infrastructure of contract enforcement was relatively better developed and the associated human competence in law, accounting, consulting, trading, and foreign languages was already in place or were acquired from abroad. Once the Chinese economy became involved in exchange relationships with outside markets, it was able to gradually develop its own competence. Further, in 2001, China entered the World Trade Organization and expressed its policy commitment to the rule of law by merging external and internal trade administrations. Through the complementary interactions of these developments, market-governance institutions could begin to shape themselves, although gradually.

As noted previously, the mechanisms of Schumpeterian re-bundling, overlapping social embeddedness, and dynamic institutional complementarities, although conceptually distinct, are realistically likely to operate simultaneously and in an interactive manner. While Schumpeterian re-bundling may bring more innovative elements into the process of institutional innovation by destroying old bundling, overlapping social embeddednes and dynamic institutional complementarities may impact it with past legacies. On the other hand, Schumpeterian re-bundling may need to take place in the milieu of complementary institutions or competence as well as by becoming embedded in persistent social norms. Further, overlapping social embeddedness and dynamic complementarities may not necessarily deter institutional innovation. On the contrary, as was suggested, under certain conditions the former may facilitate the transition into a new institutional arrangement, while the latter may aid the evolution of embryonic institutions. At this stage of theoretical development, we are still far from being able to construct a general theory of institutional change based on the theoretical analysis of the interplay among these mechanisms. However, the hope is that the notions of these mechanisms may be effectively applied to an analysis of the concrete processes of institutional change and contribute to an understanding of its nature. We turn to such an application in the following section.

7.4   Schumpeterian innovation in an institution of innovation

The evolution of the Silicon Valley clustering of small entrepreneurial firms in the last two decades of the twentieth century is considered one of most spectacular historical examples of Schumpeterian innovation in an institution in the defined sense. Often-unimpressive performances of attempts to emulate the Silicon Valley development elsewhere by means of public policy may have stemmed from a misunderstanding of the truly innovative nature of the phenomena. Its essence lay in spontaneously realizing an entirely new way of combining engineers' design activities and their outcomes in the development of a complex technological system of potentially high commercial value. Its emergence has largely destructed the value of the traditional way of developing such a system exclusively on a particular organizational domain (i.e., a particular firm).

This process is now known as modularization and its emergence was well documented by Baldwin and Clark (2000). Their story began with how the ambitious project of developing a breakthrough mainframe computer, IBM system/360, was internally organized by the then-dominant manufacturer. In order to make the complex task feasible, a design procedure was conceived in such a way that the whole task was decomposed into quasi-independent modules with only common design rules centrally created beforehand and made open. As long as these design rules were followed, the individual modular designs were made decentralized and self-closed. This design architecture was conceived and design rules were laid down by a few individuals such as Amdahl, Blaaauw, and Broocks. This much may be characterized only as a sophistication, even though of a high degree, of the traditional hierarchical combination of design tasks in a single organizational domain. A Schumeptarian innovation followed as its unintended outcome, however.

An unintended consequence of this design architecture was the possibility of organizationally quasi-independent modular design. That is, as long as the *ex ante* specified design rules are followed, the design of each module can be "hidden" from the others in the sense that they can be performed independently of, and without intervention by, the design processes of other modules.[20] Then, once the design rules are made "open," there may be no technical reason why the modular designs need to be combined within a hierarchical order on a particular organizational domain. Thus, not a small number of engineers who had been engaged in the modular design of system/360 within IBM began to exit IBM after its commercialization. They started their own firms to capture profits on the improved modular design. Naturally, IBM tried to hinder the dissolution of their rigidly hierarchical design architecture to protect their monopoly rents. Alas, however, an one-century-old California law made the postemployment covenants not to compete unenforceable (Gilson, 1999).

Yet, the massive exits were only a half story. A spontaneous new combination started to emerge endogenously through the participation of exiting engineers. Some of them became entrepreneurs, financed by themselves, families, friends, and the like, while others became financial intermediaries (venture capitalists) with their engineering knowledge and capacity for judging and imagining a future technological road map as assets. The emergence of venture capital financing became a complementary vehicle for realizing a new combination of design tasks. Its revolutionary feature lies in that venture capitalists initially finance

---

[20] This is parallel to the software design procedure for large complex programming known as object-oriented.

multiple development projects of a similar nature, proceed to sequentially select the better prospects for additional financing and deny refinancing to the losers, and finally bring the winner to the initial public offering (IPO) market or arrange its acquisition by an established firm to share the returns with it. Thus the design of each module of a complex system is now made, not by a single internal unit on a particular organizational domain, but competitively by multiple entrepreneurs, and the final system can be composed through the *ex post* bundling of the best designed modules that are forthcoming.

Although the duplication of developmental financing and entrepreneurial efforts are involved, this type of combination can yield, under certain conditions, two kinds of economic values that were not possible with marginal improvements of the old types of combinations: option values and externalities generated by a tournament-like competition among module-designing entrepreneurs. Let us examine first each of those value sources.

For simplicity's sake, let us imagine for a while that in each modular design the level of effort by each entrepreneur is constant and observable and that the engineering uncertainty can be resolved at the end of the development process so that the best design can then be identified. This is the situation assumed by Baldwin and Clark (2000). Namely, they regard the result of modular designs as "real options." They show that the greater the number of parallel experiments, the greater the value of real options, which they call the "value of substitution." However, there is a cost to each experiment. Assuming the constant cost of an experiment and equating it with the marginal diminishing value, the optimal number of experiments can be determined. Applying the option value theory in a straightforward manner, it holds that the greater the uncertainty involved in modular design, the greater the optimal number of experiments to be financed.

In reality, the cost of an experiment is not constant, as the level of design effort can be controlled by the entrepreneur after the initial financing contract. The value of modular design will depend on the effort level of the module designer. However, as the level of design effort cannot be precisely observed from outside, the level of effort will be undersupplied in the case where only one entrepreneur is financed. It is because the utility-maximizing entrepreneur will equate the marginal cost of his or her own effort with its marginal value times his or her expected share in it, which is less than one. Will the situation be altered when the tournament-like competition is managed by the venture capitalist who evaluates the potential value of each modular design with some imprecision? Yes, it will, albeit in a second-best manner, because the marginal individual value of extra effort by the entrepreneur is now composed of three parts: (1) the expected marginal value of his effort when he wins the tournament; (2) the expected total value when he wins multiplied by the marginal increase in the expected probability of his winning the tournament, plus (3) the marginal expected value associated with the market uncertainty and/or the venture capitalist's accuracy when selecting a winner that the individual entrepreneur cannot control, each multiplied by his share in the final modular value (Aoki and Takizawa, 2002). In other words, the tournament-like competition managed by the venture capitalist can create the kinds of externality that can be shared by the venture capitalist and the winning entrepreneur. We can derive a theoretical proposition that, if the total value of a final product system is expected to be very high, and if the venture capitalist's judgment in the selection of tournament winners is believed by the competing entrepreneurs to be precise, then there exists an optimal number by which the same type of modular design can be financed.

Thus the possibility of the supply of venture capital competence is complementary to tournament-like competition. It is analogous to the situation where, unless the referee is believed to be fair and able, the players of the sports game may not be motivated to play earnestly. However, the availability of such a referee cannot be taken for granted. She needs to be well versed in the rules of the game and have been trained to use good judgmental skills. Also she herself needs to be motivated to be unbiased and neutral. Professional reputation may be one possible source for such motivation in a sports game. One of the major reasons why the clustering of the Silicon Valley types has not emerged so easily in Continental Europe and Japan until recently in spite of promotional public policy and the potential supply of risk capital may lie in the shortage of credible venture capitalists. In these regions venture capital firms were first usually set up by financial institutions. Their managers were not necessarily equipped to make proper judgments on staged-financing. As noted, the evolution of the Silicon Valley clustering was greatly aided by the ample supply of knowledgeable, skillful, and imaginative venture capitalists who had once been employed by large established organizations, such as IBM, Bell Labs, and Xerox. In this sense, the modularization in the initial design of the IBM/system 360 and the emergence of Silicon Valley clustering is interlinked in a path-dependent manner.

Thus the duplicated modular design and *ex post* bundling of selected modules that emerged in Silicon Valley may be considered a truly Schumpeterian innovation in an institution of innovation. However, besides the problem of a supply of venture capital competence, the production of values inherent to this unique bundling are not unconditional. Most importantly, they depend on the possibility of decomposing the design of a complex system into quasi-independent design modules, mutually related only through well-specified design rules. This roughly corresponds to the case where the product of each design module is made technologically and attribute-wise independent (not complement) of the others and consistent only with common interface rules.[21] This condition is more likely to be satisfied by digital-based communications and information-related systems, but not necessarily by complex mechanical or chemical systems.

The above-mentioned proposition on the inherent values of the Silicon Valley clustering is derived from a static model. However, it may not be the case that the re-bundling of modules is possible only when design rules are *ex ante* fixed and modules are combined accordingly. Once independent and competitive modular designs are institutionalized, then ways to rebundle them may also evolve. Not only can improved modular designs be substituted into an existing system, but the system design itself may also be improved, and even new systems may be invented through additions, subtractions, imports, nesting, porting, and other possible module operations, as Baldwin and Clark described. This is certainly the way the Internet has been evolving as an innovative mass communication system.

But such development also requires evolutive adaptation and specification, rather than hierarchical and *ex ante* specification, of the design rules themselves. This can be done possibly through intensive communications, partly competitively and partly cooperatively, among agents with the competence to draw a road map for future technological development. It is not done on a single organizational domain such as IBM now, but through intense and professional communications among independent agents, including venture capitalists, leading firms in niche markets, academia, standard-setting industrial associations, and so on. At the early stage of development of a potential technological system, communications are often informal because the knowledge exchanged is still at a formative stage; often it remains tacit and communicable only on a face-to-face basis. Standard-setting also is often evolutive. Knowledge exchanged in this situation is therefore not proprietary. Yet there are intensive interactions and communications among those agents and entrepreneurs in Silicon Valley, making clustering one of the distinguishing features of the region (Saxenian, 1994). Indeed, the sharing of generic knowledge relevant to the potential direction of total

---

[21] Aoki (2001) Chapters 4 and 11.

technology on one hand, and the encapsulation of the processing of potentially proprietary information on the other, appear to be inseparable dualistic characteristics of the Silicon Valley clustering.

However, if generic knowledge is potentially capable of generating proprietary knowledge, why do the agents exchange it without explicit pecuniary compensation? If they benefit from access to such knowledge, why do they not try to free-ride on the supply of knowledge by others while keeping their knowledge secret? Further, there may be differentials among agents in their ability to generate and disseminate potentially useful knowledge. Why are able agents willing to part with their knowledge to others without necessarily being directly reciprocated? Apparently, what is operating here is something reminiscent of the social embeddedness of communications in the professional community (the commons domain). Namely, they are able to gain higher status and esteem within the community by making best efforts in disseminating their superior knowledge. The opportunity costs of cooperative contributions incurred by them are compensated for by a larger amount of intangible social capital that is allotted to them, such as social esteem and an enhanced professional reputation. Further, reputation-building may eventually be useful for gaining access to potentially profitable projects as investors and financiers, and other pecuniary opportunities in the future.[24]

7.5   Concluding remarks

In this chapter I have been trying to show that, contrary to the conventional view, an equilibrium view of an institution is not necessarily inconsistent with the evolutionary approach in the tradition of Schumpeter. On the contrary, the game-theoretic equilibrium view of an institution can apply the essential point that Schumpeter made regarding the nature of innovation, that is, the creative destruction, to the study of institutions. We have extended the Schumpeterian original idea regarding the combinations of production factors to those (bundling) of games, and examined how discontinuous equilibrium shifts in this regard can be identified with institutional change. Combined with other types of equilibrium shift dynamics, such as overlapping social embeddedness and dynamic institutional complementarities (the momentum theorem), this approach may hopefully suggest a new fruitful research strategy for analyzing the nature of institutional change, the topic so vital to understanding present-day economic development. This chapter has

---

[24] Lerner and Tirole (2002).

provided one example of this analytical direction by examining the innovative nature of the Silicon Valley clustering as an emergent new institution of innovation. However, the applicability of this approach may not be limited to it, as suggested by the many casual references in the text and footnotes.

References

Aoki, M. 1971a. "Two Planning Processes for an Economy with Production Externalities," *International Economic Review* 12: 403–14

Aoki, M. 1971b. "An Investment Planning Process for an Economy with Increasing Returns," *Review of Economic Studies*, 38: 273–80. Reprinted in G. Heal (ed.) *The Economics of Increasing Returns*, International Library of Critical Writings in Economics, London: Edward Elgar

Aoki, M. 2001. *Toward a Comparative Institutional Analysis*, Cambridge, MA: MIT Press

Aoki, M., and Hayami, Y. (eds.) (2001), *Communities and Markets in Economic Development*, Oxford: Oxford University Press

Aoki, M. and Takizawa, H. 2002. "Information, Incentives, and Option Value: The Silicon-Valley Model," *Journal of Comparative Economics* 30: 759–86

Baldwin, C. Y., and Clark, K. B. 2000. *Design Rules, Vol. 1: Power of Modularity*. Cambridge, MA: MIT Press

Bardhan, P. 1977. "Variations in Forms of Tenancy in a Peasant Economy," *Journal of Development Economics* 4: 105–18

Berger, P. and Luckmann, T. 1966. *The Social Construction of Reality: A Treatise in the Sociology of Knowledge*. New York: Doubleday Anchor

Blau, P. 1964/1998. *Exchange and Power in Social Life,* with new introduction. Brunswick, NJ: Transaction Publishers

Bowles, S. 2003. *Microeconomics*. Princeton, NJ: Princeton University Press

Braverman, A. and Stiglitz, J. 1982. "Sharecropping and the Interlinking of Agrarian Markets," *American Economic Review*, 72: 695–715

Coleman, J. 1990. *Foundations of Social Theory*. Cambridge MA: Harvard University Press

Demsetz, H. 1967. "Towards a Theory of Property Rights," *American Economic Review* 57: 347–59

Dixit, A. 2003. *Lawlessness*. Princeton, NJ: Princeton University Press

Dosi, G. and Marengo, L. 1994. "Some Elements of an Evolutionary Theory of Organizational Competence," in England (ed.), *Evolutionary Concepts in Contemporary Economics*. Ann Arbor MI: University of Michigan Press: 157–78

Gilson, R. J. 1999. "The Legal Infrastructure of High Technology Industrial Districts: Silicon Valley, Route 128, and Covenants Not to Compete," *N.Y University Law Review* 74: 575–629

Greif, A. 1997. "Microtheory and Recent Developments in the Study of Economic Institutions Through Economic History," in Kreps and Wallis (eds.), *Advances in Economics and Econometrics: Theory and Applications 2*. Cambridge, UK: Cambridge University Press: 79–113

Greif, A. 2006. *Institutions and the Path to the Modern Economy*: Lessons from Medieval Traders' Coalition. Cambridge, UK: Cambridge University Press

Granovetter, M. 1985. "Economic Action and Social Structure: The Problem of Embeddedness," *American Journal of Sociology* 91: 480–510

Hayek, F. A. 1945. "The Use of Knowledge in Society," *American Economic Review* 35: 519–30

Hurwicz, L. 1960. "On the Dimensional Requirements for Non-wasteful Resource Allocation Systems," in K. J. Arrow, S. Karlin, and P. Suppes (eds.), *The Mathematical Methods in the Social Sciences 1959*. Stanford CA: Stanford University Press. Reprinted in K. J. Arrow and L. Hurwicz (1977), *Studies in Resource Allocation Processes*. Cambridge, UK. and New York: Cambridge University Press, pp. 413–24

Hurwicz, L. 1996. "Institutions as Families of Game Forms," *Japanese Economic Review* 47: 13–132

Koopmans, T. 1957. *Three Essays on the State of Economic Sciences*. New York: McGraw Hill

Lerner, J. and Tirole, J. 2002. "Some Simple Economics of Open Source," *Journal of Industrial Economics* 50: 197–234

Milgrom, P., North, D., and Weingast, B. 1990. "The Role of Institutions in the Revival of Trade: The Law Merchant, Private Judges, and the Champagne Fairs," *Economics and Politics* 2: 1–23

Milgrom, P., Qian, Y., and Roberts, J. 1991. "Complementarities, Momentum, and the Evolution of Modern Manufacturing," *American Economic Review*. 81 (AER Papers and Proceedings): 84–8

Milgrom, P. and Roberts, J. 1990. "Rationalizability, Learning, and Equilibrium in Games with Strategic Complementarities," *Econometrica* 59: 1255–77

Murdock, K. 1996. *The Role of Institutions in Enforcing Implicit Contracts: Analysis of Corporate and Government Policy*. Ph.D. dissertation submitted to Stanford University

Nelson, R. 1995. "Recent Evolutionary Theorizing about Economic Change," *Journal of Economic Literature* 33: 48–90

Nelson, R. and Sampat, B. N. 2001. "Making Sense of Institutions as a Factor Shaping Economic Performance," *Journal of Economic Behavior and Organizations* 44: 31–54

Nelson, R. and Winter, S. 1982. *An Evolutionary Theory of Economic Change*. Cambridge, MA: Harvard University Press

North, D. 1990. *Institutions, Institutional Change and Economic Performance*. Cambridge, U.K. and New York: Cambridge University Press

North, D. 1995. "Five Propositions about Institutional Change," in J. Knight and I. Sened (eds.), *Explaining Social Institutions*. Ann Arbor, MI: University of Michigan Press: 15–26

Saxenian, A. 1994. *Regional Advantage: Culture and Competition in Silicon Valley and Route 128*. Cambridge, MA: Harvard University Press

Schotter, A. 1981. *The Economic Theory of Social Institutions*. Cambridge, UK and New York: Cambridge University Press

Schumpeter, J. 1912/1934. *The Theory of Economic Development.* Cambridge, MA: Harvard University Press. English translation of *Theorie der Wirtschaftlichen Entwicklung*

Schumpeter, J. A. 1937/1965. "Preface to Japanese Edition of *Theorie der Wirtschaftlichen Entwicklung*" Reprinted from Japanese translation by Nakayama, I. and Tobata, S. (Iwanami Shoten) in R. V. Clemence (ed.), *Essays on Entrepreneurs, Innovations Business Cycles and the Evolution of Capitalism.* New Jersey NJ: Transaction Publishers, pp. 165–8

Smith, A. 1759/1976. *The Theory of Moral Sentiments.* Oxford and New York: Oxford University Press

Sugden, R. 1989. "Spontaneous Order," *Journal of Economic Perspectives* 3: 85–97

Topkis, D. 1978. "Minimizing a Submodular Function on a Lattice," *Operations Research* 26: 305–21

Weingast, B. 1993. "Constitutions as Governance Structures: The Political Foundations of Secure Markets," *Journal of Institutional and Theoretical Economics* 149: 286–311

Williamson, O. E. 2000 "The New Institutional Economics: Taking Stock, Looking Ahead," *Journal of Economic Literature* 38: 595–613

World Bank 2002. *World Development Report 2002: Building Institutions for Markets.* New York: Oxford University Press

Young, H. P. 1998. *Individual Strategy and Social Structure: An Evolutionary Theory of Institutions.* Princeton, NJ: Princeton University Press

8 Innovation and Europe's academic institutions – second thoughts about embracing the Bayh–Dole regime

Paul A. David

8.1 Innovation and universities' role in commercializing research results: should Europe be imitating America's Bayh–Dole experiment?

8.1.1 Introduction

To address the complex issue of the evolving role of universities in technological innovation poses a challenge of truly daunting proportions, especially in a brief presentation such as this. The institutional history of the university is one that is marked by both remarkable continuities and innovations in response to shifting societal expectations and pressures, in the course of which there has been a cumulative broadening of the "missions" that these academic organizations have embraced. Reflecting on this process unavoidably raises questions about the degree to which new roles for the university are compatible with the performance of historical functions, and whether, when there are tensions and conflicts the ancient should yield to the imperatives of the modern. My intention is to address this generic problem of institutional adaptation and survival as it manifests itself today in a role the Europe's universities are being asked

This chapter is an extension of my paper "New Science, New Industry – and New Institutions?: Second Thoughts on Innovation and Europe's Universities," contributed to the Conference organized by the Accademia Nazionale dei Lincei and the Fondazione Edison on *New Science, New Industry* held in Rome, Italy on October 13–14 2004. It draws, in turn upon material in "Innovation in the Past and Future of Europe's Universities," at lecture to the Coimbra General Assembly of the Historical European Universities, convened in Siena, Italy, April 14–16 2004, and my "Lectio Magistralis" delivered at the University of Torino, May 12 2003. I wish to express my gratitude to each of the institutions that have invited me to speak on this important subject, to the participants in those events from whose comments and suggestions this work has benefited, and to Professors Cristiano Antonelli (Torino), Ugo Pagano (Siena), and Albert Quadrino Curzio (Accademia Nazionale) for their respective roles in making those events stimulating, informative and memorable.

to take in the development and commercialization of faculty research as a basis for industrial renewal and economic growth.

Most readers of this chapter will readily appreciate that the relationship between fundamental advances in scientific understanding and technological innovation is complicated and multivalent, and uncertain. It involves the structure economic and organizational incentives for discovery and invention, entrepreneurship and finance, and for the formation of managerial expertise and workforce skills, and the diffusion of new processes and products. It is characterized by a multiplicity of expectational effects and dynamic feedbacks that interconnect all of the foregoing processes. Complicated as that is, it is the context within which one needs to explicitly consider the roles played by the institutions of higher education that are involved in training and research, in order to bring the discussion to bear on contemporary policy issues about long-term economic growth. Most salient among these at present are the issues arising from the expectations about the contributions that Europe's universities and kindred public institutions should be making as part of broader strategies for enhancing the region's international economic competitiveness – specifically those announced in 2000 at the Meeting of EU Council of Ministers in Lisbon, and subsequently elaborated at the Barcelona Meeting in 2002.

To keep the discussion within manageable bounds, I will focus on the nexus of issues raised by the Commission of the European Communities' *Communication* (of February 2003) on "the role of the universities in the Europe of knowledge".[1] That document assessed Europe's critical needs in the epoch of "knowledge-driven economic growth" and the means to meeting those needs. Beyond its specifics, I regard the general thrust of that text to be both influential and emblematic of the wider stream of thinking that has been shaping the policies of both EU agencies and the ministries of national governments on science, technology and innovation in recent years.

The EC's *Communication* identifies the university as the institution uniquely suited to meeting Europe's needs to become more effective in generating and exploiting science-based innovation, and it goes on to call for debate on the means by which the conditions prevailing among the region's universities can be changed – in order to better satisfy the requirements of the new societal role for which the Commission view those institutions to have been destined. Underlying that belief about

---

[1] European Commission, *Communication from the Commission: The role of the university in the Europe of knowledge*, COM(2003) 58 final. Brussels: Commission of the European Communities (5 February), 2003.

Europe's path to a brighter economic future, as far as I can discern, is an arresting assessment of the present situation regarding the region's R&D capabilities – namely, that in the EU the leading institutions of higher education possess the potential to be more effective at generating commercially successful technological innovation than are the mass of business firms comprising the economy's private sector.

At the same time, along with other recent pronouncements by representatives of the governments of the member states, the Commission's *Communication* finds fault with the universities' researchers and administrators for failing to make the realization of their "innovation-potential" an institutional priority. From this it follows that what is needed is a program of institutional reform and reorientation that would release this supposed (latent) potential of the research universities, and thereby fulfill two of Europe's key requirements for achieving faster productivity growth and greater international competitiveness. One is to be able to pay for deepening of human capital formation via an expansion of public education and training at the tertiary level. The second is to substantially raise the share of EU gross domestic product invested in R&D by roughly half again, that is to say, from the current two-percentage-point level to the three-percentage-point level that was set by the Lisbon Meeting as a strategic "target" for the year 2010.[2]

I am persuaded that both university adaptations and other institutional innovations are both possible and desirable as steps toward reinvigorating the performance of the so-called "European Research Area" (ERA). Nonetheless, it is equally clear that the premises upon which the EC's proposed programs of university redirection and reform are grounded, and consequently the basic economic logic of this aspect of the innovation strategy for Europe, should be resubmitted for a more careful, indeed, more skeptical examination that they have generally been receiving. But, to firmly establish that these doubts are in no way rooted in hostility to institutional change, I want to approach that argument a little indirectly – taking a few minutes to envisage with you the dawning of a new, innovative epoch in the development of higher education institutions for Europe. I shall set out this vision in compressed and appropriately futuristic terms, rather than in the historical narrative style that is more expected of me, and which, in truth, is my more natural voice.

---

[2] The persisting weak macroeconomic performance of the major industrial countries, particular those with substantial manufacturing sectors and R&D-intensive industries recently has forced a more realistic public "revision" of the R&D investment target for the EU private sector of two percent of GDP, and consequently a deferral of the date at which the total (public and private) R&D target investment rate would be attained.

8.2   An innovative epoch in the European university system

8.2.1   *Institutional creativity envisaged*

Communities of scholars and students from distant parts of the continent have assembled collectively to form centers of learning of unprecedented size, and by adapting preexisting organizational forms to create novel governance structures have given rise to numerous new and *more efficient* nodes for knowledge creation and knowledge dissemination. The resulting radically new institutional model lends itself readily for imitation, and soon is being replicated widely throughout the entire region of Western Europe, and eventually far beyond. Moreover, the academic life carried on within these new organizations is *infused with much infectious dedication, which makes possible the highest intellectual standards*, especially in those institutions that succeed in their struggles to free themselves from the repressive constraints imposed by various regulatory authorities. A movement is under way, with the active encouragement and the sanctions granted by a pan-European authority, to bring into being *an international university system . . . a European-wide academic commonwealth which would transcend race and provincialism in the collective pursuit and dissemination of learning.*

The profound departure from previous traditions and formal arrangements in higher education just depicted has been conceived during a period of intense and eclectic intellectual activity, featuring efforts to synthesize old and new systems of thought. This has required mastering and further developing of what for many of the participants is a powerful but unaccustomed set of analytical tools. The resulting new analytical mode finds many transdisciplinary applications, and has *advanced with such meteoric pace that it is displacing the established, classical norms of education.* It is destined to achieve intellectual dominance as *the centerpiece of the university curriculum*, thus marking *a radical transition* in pedagogy: from an educational *system founded on a passive attachment to an inherited culture*, to one in which an investigative and *questioning approach to all sources of knowledge is in the ascendant.*

These dramatic alterations in the cognitive and institutional structures of higher education in Europe are nonetheless firmly *rooted in utilitarian soil.* They are responses to the need to harness the expanding intellectual forces of the era to the increasingly demanding knowledge requirements of the surrounding society and economy. While pursuit of advanced inquiry by an intellectual elite – as a contribution to fundamental understanding of an ordered universe and the place of humankind within it – is held to be one of the universities' perennial functions, these institutions also are seen *as service agencies catering for a hierarchy of social needs.* The surrounding societies, with which these organizations soon develop increasingly strong symbiotic relationships, *have at their disposal only limited finances for the purposes of higher education. Returns of a concrete nature are expected ... from investment in university concerns. Scarce resources are not made available for the subsistence of ivory towers.* Instead, the society of this era expects its universities *to be vocational institutions responding to vocational needs.* Reciprocally, new arrangements are introduced by other institutions to facilitate the provision of finance for the continuing education of certain cadres of specialized workers who engage in knowledge-intensive service activities that the community at large deems to be particularly important. Thus, by mobilizing and rationally deploying Europe's intellectual and pedagogical resources on a hitherto unprecedented scale, the new system manages – despite the constraints of its situation – to *meet the professional expectations of a broad spectrum of contemporary society;* and its constituent institutions are judged to have achieved this *without becoming the monopolistic agencies of any one privileged section of the community.*

It sounds quite good, does it not? Indeed, it really was good. As many in this audience already must have surmised, the inspiration for the foregoing recitation has not come from my reading of the recent EC "Communication" on the future role of the universities. Rather, the preceding evocation of the emergence of a vibrant, innovating, and socially responsive university system reflects the assessment by modern historians of the rise the Europe's medieval universities in the epoch stretching from the twelfth to the fifteenth century – an epoch which saw the founding of the venerable institutions of Bologna and Padua, Paris and Montepellier, Oxford and Cambridge, Prague and many others. Indeed, I constructed my text by piecing together direct quotations and close paraphrasings of the works of Professor A. B. Cobban, a leading British historian of the origins and early development of the remarkably successful medieval institutional innovations.[3]

8.2.2   The "vision" deciphered

I make no apologies for the benign deception was entailed in relating modern historians' views of salient developments in the early medieval university scene, masked in the language of our contemporary

---

[3] The portions of the text in the preceding section at appear in italics are direct quotations drawn from Cobban (1988) Ch.1, which provides citations to numerous sources – including recent historical studies by other others.

discussions of university research and training policies in Europe. "The medieval university" has been receiving an unwarranted amount of bad press in recent times, in Britain, most notably in the public utterances of the Labor Government's former Minister of Education and Skills, Charles Clarke (who later went on to head the Home Office). In the spring of 2003, shortly after the circulation of the EC's *Communication*, Mr. Clarke urged Britain's universities to think more about how they benefit the economy and opined that "a medieval concept of a community of scholars seeking truth is not in itself a justification for the state to put money into that. We might do it at, say, a level of one hundredth of what we do now and have *one university of medieval seekers after truth*... as an adornment to our society."[4] Thus, there is some value simply in reminding ourselves from time to time, and reminding the makers of public policy, too, that to propose that the university community should serve the "utilitarian" needs of society hardly is a modern idea. The medieval roots of this remarkable institutional form, as Professor Cobban says, were firmly planted "in utilitarian soil." Greater awareness of that fact would be useful, especially in drawing attention to the difference between the principle of "service" – about which there can be no real debate – and the question of how best the modern university can serve the societies on which it ultimately must rely for its support.

Nevertheless, for the benefit of those who wish to complete their decoding of my "vision," I should quickly identify the most salient among the historical details to which I was alluding, before coming back to the future vision of the university's role in "the Europe of knowledge" – as the EC would have it.

The medieval idea of the *studium generale* was indeed a major institutional innovation. That was the term by which the medieval universities were first described, distinguishing them from *studium particulare* – institutions offering instruction in the arts by local scholars for local students. Although thought to require endorsement of the Pope or the Holy Roman Emperor, the status of *studium generale* was of a customary rather than legal nature until the late-thirteenth and fourteenth centuries, when Italian jurists devised the term *studium generale ex consetudine* and applied it to long-established centers such as Paris, Bologna, Montpellier, Padua, Oxford, and Cambridge.

---

[4] *The Guardian*, May 10, 2003, p. 3, reports these statements as taken from a transcript, released by the Department of Education and Skills, of the Minister's remarks made earlier in that week at a gathering at University College, Worcester. The emphasis in the quoted statement has been added.

By adapting pre-existing organizational forms – notably, corporate guild organization – the masters and students created novel governance structures for a rapidly increasing number of these new and *more efficient nodes* – or as we economists would say, agglomerations or "clusters" that generated economies of scale and scope for knowledge creation and knowledge dissemination. Only by accident did the Latin term *universitas – which in common usage denoted several types of corporate bodies, for example, craft guilds or municipal councils* – come to be specifically associated with university institutions. Before the fifteenth century *universitas* referred to guilds of students (as in the case of Bologna) or of the masters (in case of Paris), – that is to the personnel of the university rather than the university structure as a whole.

The "New Logic" of Aristotle, rediscovered and made available in Latin (c. 1150 – 1250), emerged as the characteristic analytical mode associated with these novel institutions of learning, *advancing with such meteoric pace that it soon displaced established classical norms of education and found many transdisciplinary applications*. Logic, or dialectic, was the indispensable instrument for deep penetration of all branches of learning – including theology, law, medicine, the natural sciences, and grammar; it soon achieved intellectual dominance as *the centerpiece of the new university curriculum, the quintessence of all that was forward-looking and creative*. Logic's rise thus marked *a radical transition* in pedagogy: from an educational *system founded on a passive attachment to an inherited culture* (namely that associated with studies of classical literature), to one that was committed to an investigative and *questioning approach to all sources of knowledge*.

These alterations in the cognitive and institutional structures of higher education in Europe were *rooted in utilitarian soil*; their respective societies expected its universities *to be vocational institutions responding to vocational needs*. The *studium generale* featured a regime offering graduate training – in at least one of the "superior faculties" of law (canon or civil or both), theology, and medicine. Training in logic and the art of disputation was the prescribed general preparation for most professional activities. Allied to knowledge of the relevant procedures, dialectical training could be applied successfully over a wide range of intricate administrative, litigious, educational, and diplomatic affairs.

Reciprocally, new arrangements were introduced by other institutions (namely, the Church) in order to *facilitate the provision of finance for the continuing education of certain cadres in specialized knowledge-intensive service activities* whose work was held to be socially important. The specific financing arrangements to which I thus referred were those permitting beneficed members of the clergy to receive the incomes of their benefices whilst absent from their parish and attending university as students or teachers.

By the thirteenth century a movement was underway with the encouragement and sanction of pan-European authorities (namely, that of the Pope and the Holy Roman Emperor) which was aimed at forming *an international university system. ...a European-wide academic commonwealth which would transcend race and provincialism in the collective pursuit and dissemination of learning.* This advanced the theoretical notion of the *ius ubique docendi* – the legal right of a graduate of one university to teach in another without undergoing examination. By the fourteenth century possession of this right by holders of a master's degree from the institution in question was a cardinal legal hallmark of the status of *studium generale* – and so normally was included in the institution's founding charters. In practice, however, university particularism (and job protection for the masters) tended to prevail against the supranational (and leveling) implications of the *ius ubique docendi*.

8.2.3   The evolving legacy of medieval institutional innovation

The dream of the *ius ubique docendi* is one that continues today to haunt the EC when it contemplates the advantages of creating a truly integrated labor market for science and engineering researchers that would embrace the entire ERA. Of course, that was not the only piece of business left "unfinished" by the medieval institutional innovators. Those who followed the foregoing recitation closely will have noticed the glaring absence – from both the cryptic and the decoded versions of my "vision" of the medieval university system – of any explicit references to scientific or technological research, indeed to organized research activities of all kinds. Resistance to the new mechanical philosophy of the sixteenth and seventeenth centuries by the entrenched university faculties meant that the fusion of mathematics with experimentalism which created the epistemological foundations of the Scientific Revolution was not carried forward within that institutional setting. Even though many of the great names associated with the movement (including Copernicus and Galileo) held university posts at some points in their careers, those who pioneered in the emerging experimental and observational sciences managed first to insinuate their new methods and style of discourse into the proceedings of the independent humanistic academies that had flourished during the late Renaissance era. At the end of the sixteenth century they had begun to form more specialized scientific societies, such as della Porta's *Accademia Secretorum Naturae* (founded in Napoli, 1589), the *Accademia dei Cacciatore* (in Venezia, 1596) and, of course, Federico Cesi's *Accademia dei Lincei* (Roma, 1603). Only subsequently, toward the latter part of the seventeenth century did mechanical philosophy – and the associated behavioral precepts of the equally novel social structure of "open science" become institutionalized under state patronage in the Royal Society of London (1662), and the Parisian *Académie royale des Sciences* (1666).

There ensued a further delay of almost two centuries duration before "research" as we know it became established alongside teaching as a central activity of the faculties of Europe's higher education institutions – on the model of Wilhem von Humbolt's proposed reform of Germany's universities. That development, and the subsequent importation of the Humboltian model into American universities during the two closing decades of the nineteenth century, belongs to the longer story into whose details it will be neither possible nor necessary for me to enter on this occasion. It is worth noting, nevertheless, that this second great institutional innovation led relatively swiftly to the experiment of closely coupling of graduate instruction and faculty research activities, a pedagogical shift that was fully embraced in the United States university setting from the closing decades of the nineteenth century onwards. There it proved to be so effective a means of forming large cadres of productive researchers in the physical, engineering, and life sciences that today one observes it being taken up by countries (including France and Japan) where science and engineering research formerly had been conducted largely in government laboratories and public institutes that were formally isolated from their universities' instructional activities.

With the progressive integration of advanced instruction with research in the universities, the institutionalization of open science throughout the modern world – albeit to a different degree in different places – was reinforced and its normative structure was imparted to successive cohorts of academics and public sector researchers. Generation after generation of graduate students were thereby exposed to and inculcated with the ethos of open science that became more and more clearly articulated in the democratic societies of the West from the late-nineteenth century onwards. This was a potent means of reinforcing the informal behavioral norms of cooperation in pursuit of knowledge, meritocratic universalism, and uninhibited intellectual skepticism. It promoted conditions in which the responsibility of faculty researchers and teachers to impart their knowledge fully to graduate students was well aligned with the open pursuit of scientific enquiry, rather than trammeled by secrecy, restrictions on the usage new research methods, and potential conflicts arising from organizations' and individuals' ambitions to create economically valuable intellectual property from which they would benefit directly. Furthermore, it reinforced and sustained the ability of the universities in many parts of the work to continue to function effectively as open nodes in an international information network that transmitted, received, and validated claims to discoveries and inventions that represented advances in knowledge, and upon which further advances in knowledge could be based. In this way, the universalist promise of the original, medieval European institutional innovation came to be realized throughout a far more extensive and culturally diverse domain: the global domain of the Republic of Science.[5]

The ethos of open science – as much as any formal institutional regulations designed to avoid conflicts of interest and the misuse of public resources for purely private gain – formed an effective bulwark in the democratic societies against the more subtle distorting pressures that commercial and political interests could bring to bear upon the conduct of university-based research and the reporting of its outcomes. Upon this and kindred fragile structures of institutionalized behavior came to rest the public trust that once was more or less automatically accorded to "disinterested" academic research; and consequently to the reliability of universities as the loci of enquiries that could, more than any other, impartially ascertain and report on the meaning and implications of new discoveries and devices for human knowledge, individual well-being, and the vitality of society.

8.3   Back to the future: the quest for the "wealth-creating" university reconsidered

These qualities, achievements, and potentialities of Europe's universities should be regarded as unique societal assets that would in all likelihood be placed at risk by a concerted effort to develop commercially oriented "knowledge management enterprises" within those institutions. Yet that is what appears to be contemplated today in Britain, where the Board of Trade and Industry speaks of the need to expand a "third stream" – in addition to the traditional channels of teaching and research – through which the university can contribute to national and regional "wealth creation," specifically by creating and exploiting intellectual property rights, by offering the services of its faculty as consultants to private firms, by introducing specialized degree courses tailored to the needs of industrial sponsors who would be able to select candidates for instruction, by developing "distance learning" services that could be marketed to the public both at home and overseas via the Internet, and so forth.

---

[5] On the economics of the ethos and institutions of open science, see Dasgupta and David (1994) David (2003).

Such a program constitutes perhaps the leading edge of reformist policy initiatives intended to reanimate the universities with an entrepreneurial spirit of "wealth creation." Certainly it envisages a much wider array of university profit-seeking activities than those proposed for discussion and debate by the EC's (2003) *Communication*. The latter, more conservatively, focuses attention upon the remaining changes in national regulations that would be needed not only to enable the universities to patent discoveries and inventions resulting from publicly funded research, and also provide stronger incentives for those working in public research organizations to engage in more applied, commercially oriented projects – by permitting them to share with their institutions the income derived from the exploitation of their findings. In this regard, the position presented by the Commission is hardly a radical one; it reflects policy initiatives that already were being actively considered and in some cases have been implemented by a number of Organization for Economic cooperation and Development (OECD) member nations, apparently in emulation of the experiment undertaken by the United States under the term of the Bayh–Dole Act and the Stevenson–Wydler Act, legislation passed by the Congress in 1980. The immediate effect of those Acts was to simply codify the terms on which institutions conducting federally sponsored research could seek intellectual property rights in the results.[6] Their ostensible purpose at the time was to facilitate the commercial application of such inventions by permitting the universities to own and readily license the patents secured on them, especially to small and medium-sized enterprises that were thought to be an important source of "job creation."[7] Only during the mid-1980s, during the height of the "international competitiveness crisis" in the United States, did the reorientation of public and private R&D toward "technology applications" as a basis for revived commercial innovation emerge as the *ex post* rationale for this legislative experiment.

The latter rationale has inspired subsequent initiatives elsewhere in the industrialized world, and beyond, to effect institutional change in universities and other public research organizations, although not in

[6] The Bayh–Dole Act was passed as Pub. L No. 96–517, Section 6(a) 3015, 3019–28 and codified as amended at 34 U.S.C. Sections 200–212 (1994); the Stevenson–Wydler Technology Innovation Act of 1980, Publ. L. No. 96–480, 94 Stat. 2311–2320 (codified as amended at 15 U.S.C., Sections 3701–2714) pertained to the assignment of title to federal research institutes and national laboratories, and they complemented, whereas Bayh–Dole pertained to vesting title to IPR in universities.

[7] The restrictions in the 1980 Act, which limited licensing to small–and medium–sized firms, were relaxed under the Reagan Administration by Executive Memorandum in 1983, and the Bayh–Dole Act was subsequently amended to remove the time limits on exclusive licenses to large corporations.

every instance by recourse to the same legislative measures.[8] In Italy, for example, legislation was adopted in 2001 to shift ownership of intellectual property based upon university research from the institution to individual researchers, whereas in Japanese universities the allocation of ownership of intellectual property rights (IPR) from publicly funded research is now being determined by a committee in each institution, and these bodies on occasion award title to individual researchers. In Britain there has been a trend to transfer ownership rights to patents (and electronic copyrights) on publicly supported research results from the funding agencies to the universities, and a variety of arrangements exist among the institutions for distributing royalties between individual researchers and their institutions. Professors in the university systems of Germany and Sweden long held ownership of intellectual property resulting from the activities of their laboratories, and while debate about shifting ownership to the university continues in Sweden, recent legislation in Germany has shifted the locus of ownership from the individual to the institution. In each of the countries mentioned, as well as in France, governments have encouraged the formation of external "technology licensing organizations." These may be affiliated with a given university – as is the norm in Britain, and in the United States – or be entirely independent entities. Overall, there has been an evident international movement toward engaging university faculty researchers in "patentable" research, and the involvement of their respective institutions in the ownership and licensing of intellectual property based upon publicly funded research results.

The scale on which these institutional innovations are being promoted is a matter for concern – not only in my view, but in that of other economists and science and technology analysts.[9] In order for these policy initiatives to succeed there must be a significant reorientation of university-based research, pushing it away from areas in which these institutions have a demonstrable comparative advantage. Within the familiar context of academic, "open science" norms and governance structures, the comparative advantage of university-based researchers' lies in conducting fundamental, exploratory enquiries that in many cases will turn out to have laid the foundations for subsequently fruitful investment in applications-oriented R&D. It is also the case that in some new, frontier areas of science physical effects are discovered and new research techniques are devised that quite readily can be translated into

---

[8] On international emulation, see OECD (2002), and the discussion by Mowery and Sampat (2006).

[9] See for example Trune and Goslin, (1998); Eisenberg Dreyfuss and H. First, (eds), (2001); Nelson (2001); Pavitt, Stockholm, 21–23 November (2002); Walsh, Arora and Cohen (2003); Mowery et al. (2004)

devices which provide prototypes for valuable commercial innovations – even before the fundamental underlying phenomena are thoroughly understood. One may think in this connection of contemporary fields as diverse as proteomics, nanotechnology, or the connection between advanced number theory and cryptographic algorithms; just as the exploratory, academic research of an earlier era in molecular chemistry, solid state physics, and photo-optics rather unexpectedly opened pathways for the industrial development of new synthetic materials, microwave devices and transistor technologies, and lasers.

Yet, such passages from exploratory science to commercially profitable R&D as a rule do not occur in a tightly coupled, highly predictable fashion that attracts the attention of research-intensive companies whose managers seek identifiable and dependable payoffs streams from portfolios of new product-development projects. Moreover, for university administrators to encourage (or even permit) political leaders to entertain the hope that the energies of their faculties and students could be harnesses to yield accelerated productivity growth, showers of better-quality products, enlarged export earnings, and local job creation – all within the brief time frame that will make a difference in the coming elections – is not merely deceptive. It is quite reckless in risking the almost certain disappointment of unrealistic expectations, and so may bring in its train public disaffection and damage to the university.

I believe it would be irresponsible to remain silent in the face of these possibilities and simply hope for the best. One must seriously question whether the prospects of gain can justify the potential costs of redirecting the energies of Europe's university communities in ways that surely complicate, even if they do not gravely jeopardize, their ability to perform the social functions that traditionally have brought these institutions public respect, material support, and a considerable measure of insulation from political inference in the conduct of their special educational and research missions.

Proceeding in this vein, however, there are only three main classes of questions that I can hope to answer on this occasion:

(1) Is there really a problem arising from a failure of European scientific expertise in the academic sphere to respond to industry's innovation needs, a problem for which the proposed redirection of university research activities toward commercial goals would be at least an important part of the solution?

(2) Does the example of the United States' experience with the Bayh–Dole Act (1980) indicate that by imitating this institutional innovation the EU's member countries can expect to stimulate

university researchers to develop and patent technological innovations that will provide that basis for new industrial products, private investment, and job creation?

(3) Is there an empirical evidence to support the expectation that by becoming better at "knowledge management" and accumulating intellectual property rights on the basis of the research of their faculty and students, Europe's universities individually and collectively will be able to contribute significantly to defraying the rising costs of public sector science and university education?

To come to the point immediately, my answer to these questions is simply "No":

No – the problem of the innovation gap in Europe, if it exists, is not attributable to the supposed failure on the part of the professoriate to patent inventions and discoveries in which industry would take an interest. Such statistical data as is recently being produced confirms a different picture: academic researchers in Europe are active in patenting, but the title to the intellectual property in their inventions tends – in contrast to the United States situation – to be assigned to industrial firms rather than to their universities.

No – the Bayh–Dole regime is not an appropriate model for emulation: its apparently positive effects upon the rise of science-based technological innovation and university patenting activity in the United States during the past two decades have been widely misperceived in European policy circles. Other factors, which may not be operative elsewhere under currently prevailing conditions, played a major role in those ostensibly positive developments during the 1980s and 1990s. On the other hand, the Bayh–Dole legislation interacted with features of the American university and legal systems to produce a number of unintended consequences that have been quiet perverse in their effects upon some areas of scientific and technological research and the social benefits derived therefore, as well as creating new and costly issues with which universities have been obliged to contend.

No – it is thoroughly misleading to suppose on the basis of the only reasonably well-documented experience with a regime of extensive university patenting activity – namely the recent American case, that the fiscal burden upon taxpayers of supporting research and human capital formation in institutions of higher education can be significantly reduced by creating institutional incentives for faculty to disclose potentially patentable inventions, and for university technology managers to file for patent protection and then seek to exploit whatever intellectual property rights the institution thereby acquires.

It should be appreciated that the questions I have just posed are not so simple when considered in all their ramifications. Consequently, the evidence and arguments that is available to support my rough conclusions in each instance are more intricate, and more subject to important qualifications that the foregoing un-nuanced assertions would suggest. A brief elaboration of my answers under the three headings will have to suffice to indicate the nature of the complicating issues, without – I hope – reducing the persuasiveness of my argument that Europe should be building new elements of an organizational infrastructure for science-based innovation, rather than setting new and inappropriate tasks for its existing academic institutions.

8.4   Towards "evidence-based policies" for science and technology in the ERA

8.4.1   Is there a problem, and where does it lie?

Is the problem of the Europe's so-called "innovation gap" – the alleged failure to fully apply the region's scientific research capabilities to the generation of profitable innovations – one that should be laid first at the door of the universities? We know that the proximate source of the shortfall in Europe's relative R&D expenditure level vis-à-vis other industrial nation's like Japan and the United States is not the lower rate of public sector research investment, but, rather, the comparative shortfall in private business investment in R&D as proportion of GDP.

Economists can think of two reasons why the private R&D rate is low: either the supply of potential innovations is very restricted, or the demand for inventions is weak for reasons having to do with market conditions, or financing costs, or lack of expertise on the part of industrial managers in perceiving existing opportunities, or all of these deficiencies. It is difficult to clearly disentangle the two main sets of forces, but it is possible to dispel the mistaken impression that researchers at Europe's universities are not inventive, or fail to generate inventions that are relevant to the needs of industry. Recent empirical studies show that there is a big difference between institutional patenting of inventions by universities in Europe and successful involvement of university faculty researchers in patentable inventions that are taken up by industry. For example, during 1978–1999 there were only 40 patents filed by Italian universities at the European Patent Office (EPO), whereas during the same period the EPO issued 1,475 patents – mostly in the areas of biotechnology, drugs and organic chemistry – to

Italian university faculty.[10] In the case of a single French institution, the University Louis Pasteur in Strasbourg, during 1993–2000 the EPO issued 463 faculty members to members of the faculty (mainly in the fields of genetics, biology and physics), but only Sixty-two patents to the university itself. Of course, it is well known that until the recent change of the law in Germany, the professoriate received and could retain the rights to all patents – some 1800 of which were issued to them between 1986 and 2000, principally for inventions in scientific instruments, telecommunications, and biotechnology and pharmaceuticals.[11]

What would it mean, then, to seek to fix the European innovation gap by administrative measures designed to raise the rates of university-originated applications for patents?

• It would displace some part of the assignment of patents on faculty inventions to industrial firms, without necessarily increasing the total flow of patentable inventions arising from university research.
• There is nothing to assure that the resulting shift in the initial ownership of patent rights from firms to universities would enhance the value of patent portfolios in the economy. Indeed, the outcome might well work in the opposite direction because university patent holdings would provide government authorities with a convenient "indicator" of institutions' comparative performance in meeting "targets" for commercially relevant research. The trouble with the use of patents for monitoring the universities is that their value to the institution in negotiations for research funding from government ministries may well make its administrators loath to part with them; by contrast, a private firm is less likely to consider the symbolic value that the patent has in some administrative transaction, and so will be ready for the rights assigned to them by their employees to be exploited by other firms, for whom their market-based value is higher.
• It would oblige firms that are interested in further developing those inventions and making them the basis for new products and new processes to negotiate for patent (or copyright) licenses with university technology managers, and it is likely to complicate some of the directly consultative relationships for knowledge transfer that

---

[10] See M. Balconi, S. Breschi, and F. Lissoni, "Il transferimento di conoscenze technologiche deall'universita all'industria in Italia: Nuova evidenze sui brevetti di patenit dei docenti," in A. Bonacorse, (ed.), *Il sistema della ricera pubblica in Italia*, Milano: Franco Angeli, 2003; M. Balconi, S. Breschi, and F. Lissoni, "Networks of inventors and the role of academia: An exploration of Italian patent data," *Research Policy*, 2004.

[11] The French and German statistics, as well as data for Belgium and Sweden are presented by Geuna and Nesta (2006).

would otherwise be concluded between faculty researchers and the companies to which they assign the patents on their inventions.

• The views of large R&D-intensive corporations in the United States regarding the experience in trying to negotiate with universities over the intellectual property rights arising from collaborative research should be instructive in the foregoing connections.[12] As one may see they are hardly flattering about the performance of university technology management offices.

8.4.2   Does the Bayh–Dole regime offer a suitable model for international adoption?

Ideas for European institutional reform and regeneration along those lines clearly have been inspired by perceptions of vigorous university–industry research partnerships, rising patenting activity, and the flourishing of academic entrepreneurship in the United States during the two closing decades of the past century. As those years followed immediately upon the date of the passage of the Bayh–Dole Act (1980), the latter has been accepted as a model for emulation. This is dangerously simplistic. Closer examination of the available record leads one to doubt that the Bayh–Dole regime offers an appropriate paradigm for the ERA, and, indeed, a growing number of observers of the U.S. university scene have recently voiced doubts about the wisdom of the experiment.

The Act initially was justified as a measure that would promote the transfer of university inventions to the business sector for further development and job creation, and therefore contained provisions that allowed universities to license their patents only to small-and-medium size firms; it was not intended as a stimulus to university-applied research. Originally, universities were restricted to granting exclusive licenses on their patents to small businesses, on the grounds that such rights would be needed to enable them obtain an attractive rate of return on their investment in developing new products; the encouragement of small business at the time was very much driven by government concerns for job creation and the then fashionable belief that small business formation was disproportionately responsible for generating increases in employment. Only in the course of the 1980s, amidst growing worries about competition from Japanese firm's penetration of the U.S. domestic and foreign markets hitherto held by large manufacturing companies, did the rationale for the institutional experiment undergo a transformation. The transfer of technological discoveries from universities to the private

---

[12] The following is drawn from Hertzfeld, Link, and Vonortas (2006).

sector became subservient to the stimulation of university-research- based businesses in the new scientific fields where – it was hoped – the United States was less likely to face immediate challenges from either Europe or the new industrial nations.[13]

The rise of university patenting and start-ups are traceable in significant part to factors other than to U.S. congressional initiative, particularly to the antecedent emergence of biotechnology and new foundational breakthroughs in the biomedical sciences more generally. The available data show that university patenting was rising in the 1970s, in advance of the Bayh–Dole Act, and in significant measure the impetus for the drafting of the legislation derived from the concern on the part of a small number of universities active in the biomedical research area about a possible reversal of the policies of the funding agencies that had enabled them to secure patent rights under individually negotiated Intellectual Property Agreements (the so-called called "IPA's").

A number of factors quite distinct from the legislative innovation of the Bayh–Dole Act underlay the emergence of university research as a driver of patented inventions in the United States. What the quantitative evidence shows, first, is that the rapid growth of patent citations to scientific papers in the United States was not an unprecedented development, having begun during the 1970s. The trend certainly has become more salient since the mid-1980s, an eight-fold increase having occurred in the number of such citations in a random sample of utility patents during 1987–1997. But, rather than being a reflection of an across-the-board tightening of the connection between advances in academic science and technological invention, it reflected a number of features that were more specific to the participation of university researchers in the biotechnology revolution:[14]

- A 3.4-fold rise in the number of university–industry research centers during 1985–1995 brought firms' researchers into closer contact with academic research publications: the "general propensity" of patent applications of all kinds (mainly by firms) to cite scientific publications surged in the 1980s.
- A 3.5-fold increase in patenting by research universities during 1985–1995 contributed disproportionately: university-assigned patents (in the aggregate and in every major technical field) cite scientific papers more frequently than other patents.
- The rise of biotechnology – promoted by the shift of federal R&D funding toward the life sciences – is the main factor behind the

[13] On this background, see Sampat (2006).
[14] The following draws on Branstetter (2004); Sampat (2006).

aggregate trend of patent citations to scientific publications: drug and medicine patents are 260 times more likely to cite science than mechanical patents; biomedical research paper are thirty-eight times more likely to be cited than biology papers.

Thus, with regard to the Commission's strategic vision of the existence of an easy path to renewed industrial innovativeness via university research, the burden of evidence on the factors underlying the rising trend of patent citations to university science suggests that this strategy could be tantamount to "betting the farm on the future of biotech." Moreover, even were one to suppose that the concentration of citations in the 'bioscience–biotech nexus' reflects the actual underlying distribution of knowledge "spill overs," rather than the peculiarities of citation practices in this particular research area, the clearest positive lesson to be drawn from the U.S. data points to a rather different policy than the emulation of the Bayh–Dole regime. The massive shift of U.S. public funding toward the life sciences laid foundations for expanded industry R&D expenditures in the biomedical field, and a rising stream of product innovations. That "payoff," however, required matching increases in levels of private sector investment. Perhaps the right lesson for Europe to draw from this experience, therefore, is to emulate the United States focused public funding approach – in a newly emerging area of science, and to prepare its private sector firms to take advantage of the expected "spill overs."

8.4.3   Could the exploitation of intellectual property really offset universities' costs?

The results of universities' attempts to exploit IPR, whether by licensing or by faculty "start-ups," is likely to bring significant financial gains for only a very few institutions at best, whereas the increased administrative problems and the private and social costs almost inevitably will be quite widespread and represent an added distraction (and expense) that will deflect resources from the performance of the institutions, main social missions.

The plain truth is that most of the offices of technology licensing (OTLs) at American universities do not produce enough revenue to cover their own costs. Only for a very few institutions is it likely that the net income from their intellectual property rights will be substantial enough to materially contribute to solution of the universities' funding problems. There is a pronounced skew in the distribution of patent income receivers, as there is in the distribution of public R&D funding. In 1993, for example, fifty percent of public R&D funding for university-based research in the United States went to the top twenty-five percent of the 200 research universities. The skew in the returns to patenting are more pronounced than that: just three institutions (the University of California, Stanford, and Columbia) received one-third of all the royalties earned by U.S. university patent licenses in 1995. The top ten royalty-earners garnered far more than two-thirds of university patent licensing revenues, whereas roughly forty-five percent of the institutions with OTLs received no royalty income at all in 1997.[15]

Nor have the institutions that subsidize the operations of technology licensing offices been willing to bear the adverse publicity and, in the case of state universities, pay the possible political costs of shutting them down. What president of a state university wants to explain to the institution's politically appointed Regents, and ultimately to the legislators, that her budget cannot afford to go on paying for patenting inventions that might be of interest to local businesses, and might be the basis for regional job growth – just because there has not been any noticeable revenue from any of the past patents its technology managers have managed to obtain? Patenting for profit is a lottery. The business of lotteries thrives on hope. It is politically costly to deny hope, even when doing so would favor the public interest.

Besides, the political economy of university patenting has made the collective commitment to this institutional experiment more and more difficult to reverse. Even if a core of university administrators at leading institutions became convinced that the Bayh–Dole regime requires significant reforms, they would need to contend with vigorous public opposition in defense of the *status quo*. Another unintended consequence of the federal legislation has been the fostering of a new profession, and the building of a new professional organization: the Association of University Technology Managers(AUTM). With its nearly 3000 members, newsletter, and conference program, the AUTM now constitutes a well-organized and vocal professional lobby – a vested interest whose very survival is dependent upon the perpetuation of university patenting activity.

And so we have had all this for in exchange for $1.4 billion in annual licensing revenues,[16] which amounts to well less than a tenth of U.S. expenditures on university research, and roughly one-hundredth of the costs of operating the nation's 200-odd research universities.

If major revenue-generating patents have not been induced by the promise of returns to academic inventors and their institutions, programs modeled upon the U.S. Bayh–Dole Act – which the EC's "Communication" much admires – what then is their effect? From the economist's

---

[15] See Mowery et al. (2004), for further statistics.   [16] See AUTM (2003).

perspective the patenting outcomes of research that would in any case have been undertaken under public or private foundation sponsorship is tantamount to permitting universities to levy a tax on users of the new knowledge. The tax falls first upon the business firms that wish to invest in exploiting those discoveries and inventions, but, by granting exclusive licenses, the universities can sell private parties a chance to collect the tax in the form of monopoly profits (rents) extracted from the ultimate users of their new, knowledge-based goods and services.[17]

The sale of monopoly rights to private parties was utilized by Europe's kings and princes assiduously as a mechanism of financing various purposes of the state – which, in the political theory of the day, generally were not distinguishable from their own purposes. But that took place in the epoch before the modern state acquired its extensive fiscal powers. Quite apart from the political troubles that historically ensued on more than one occasion from a sovereign's grants of such rights to favorites, and the high leakage of revenues gathered by "tax-farming," this means of arranging what are in effect transfer payments has long been eschewed for sound economic reasons. As a government device *for financing university activities*, even the most visible successes of the Bayh–Dole university-patenting regime stand indicted as involving a doubly inefficient allocation of society's resources: first, because monopoly pricing imposes a restraint on the use made of new knowledge, and second, because in the situation considered, the university research was publicly subsidized and its outcomes cannot reasonably be ascribed the effects of prospects of eventual financial rewards deriving from commercial exploitation of the results.

8.5   Developing institutional innovations for innovative Europe

The proposal for today's public universities (along with the state-subsidized private institutions of higher education) to help support themselves financially by owning and exploiting intellectual property is a bad idea. It is a misuse of the economic rationale for the system of intellectual property protection, namely that the granting of legally enforceable monopoly rights is justifiable as a means of providing incentives for undertaking investment in intellectually creative activities.

---

[17] One might cite as particularly appropriate the supporting statement by Niels Reimers, formerly the director of the Office of Technology Licensing at Stanford University, and in a sense the modern founder of the profession of university technology managers, to the effect that university patenting is simply "a tax" whenever the ability to grant an exclusive license for commercial development of the innovation does not induce further, complementary R&D investment from industry. See Riemers (1987).

The modern university form, having fused pedagogy and research, has been privileged and supported by tax-paying members of society at large because it provides a home and haven, and a social institution with distinctive internal incentives and norms governing the activities of individuals who independently are motivated to engage in creative activities.

Asking, let alone demanding, that those responsible for university administration to attend to the profit-potentialities of their faculties, has turned out to be pernicious in its unintended side effects. If pursued rigorously, it is likely to be destructive of the very qualities for which these institutions rightly have come to be admired and maintained by open societies – however grudging and inadequate their support may be at the present time. Although not in the habit of appealing to Papal Authority to reinforce my economic arguments, the occasion calls for an exception: I would direct your attention to the following thoughtful passage in a letter to the apostolic nuncio in Poland on March 25, 2002, in which John Paul II decried the "overriding financial interests" that had become particularly manifest in the areas of biomedical and pharmaceutical research:[18]

"[T]he pre-eminence of the profit motive in conducting scientific research ultimately means that science is deprived of its epistemological character, according to which its primary goal is discovery of the truth. The risk is that when research takes a utilitarian turn, its speculative dimension, which is the inner dynamic of man's intellectual journey, will be diminished or stifled.

Moreover, as has been amply demonstrated by the experience of U.S. universities with institutionalized technology licensing under the terms permitted by the Bayh–Dole legislation, it offers no realistic solution to the problems of university finance. At best, and for a very few institutions, direct and indirect licensing of patents returns only a small portion of the costs of research performed by university personnel, whereas, for the overwhelming majority the activities of their technology licensing offices represent a net financial burden.

8.5.1   Hopeful monsters and plain monsters

The Bayh–Dole regime in the United States has developed into something rather different than that which its creators intended. They argued that socially useful innovations often could not be derived immediately from publicly funded discoveries and inventions, but that the additional R&D investments which were needed would be forthcoming from business firms if only they could be assigned exclusive rights to exploit

---

[18] This passage is reproduced as quoted in Horton (2004).

those university research findings that proved to be patentable. Installing the profit-making impulse into the body of the research university was not the original intention, although that outcome – surely the creature of a Frankenstein experiment, if ever such a thing could be conceived of in the area of institutional reform – has been celebrated by some enthusiasts for the emerging "entrepreneurial university." There was never a reason to believe that throwing open the doors of university offices and laboratories to commercial entrepreneurship was a proposition entirely different in its ability to yield unwanted consequences than was another particularly American higher education innovation – the idea of having universities meet the cost of their athletic programs (and why not operating expenses in general?) by the commercial exploitation of admission and media broadcast rights to college football, basketball, and other sports events.

Earlier in this chapter I alluded to the European medieval university innovation in metaphoric, evolutionary terms, as one of those "hopeful monsters" – a mutant form of the Cathedral schools, so to speak, which turned out to be not only viable but marvelously adaptable, and socially productive. Nevertheless, as widely as this innovation has propagated itself, the individual organizations carrying the germ of the "university idea" remain fragile bequests from the past, and history has shown that whole populations of such institutions are terribly vulnerable to shocks from alterations in the political climate, as well as to adverse trends in their economic environment. The proposed transformation of the university into a knowledge-management business would undoubtedly constitute a further innovative enterprise. Even if it is advocated with the best of intentions by political leaders and their policy advisors, we would do well to protect this remarkable institutional heritage from pressures to embrace new and potentially self-debilitating missions.

Innovation as a human activity is good. It carries risks, but modern societies are right to give it encouragement, for without the novelty that regenerates diversity, the possibilities of learning and selecting new social and cultural ways eventually would be exhausted. By the same token, we must be prepared to accept the reality that even the best-intended innovations may turn out not be "good enough." Society must be ready to experiment, and even to experiment in more radical ways than thinking up new purposes to be tackled with familiar devices. But it must also recognize and act decisively on the difference between a "hopeful monster," and an attempted hybrid such as the proprietary research university, which can be seen to be "a hopeless monster."

It is not just a matter of preserving the cultural legacies that are "the historical universities of Europe," and freeing those who wish to work therein from the distractions and tensions of managing, or trying to manage multiple and mutually conflicting missions. Another purpose is served by my skeptical examination of European policy makers' too-ready surmise that the United States has discovered the secret of universal institutional "best practice" in the organization of its innovation system. Not only is the Bayh–Dole regime a dubious paradigm for Europe to emulate – a growing number of my fellow economists now argue that the legislation and the system it has spawned stands in need of significant reform in the very place where it has become entrenched. What Europe needs, in my view, and what Europe has to offer the knowledge society and the knowledge-driven economy, is a new surge of institutional innovation, complementing its universities and institutions of higher education with novel organizations that are better suited to fostering the generation of commercially successful innovations based upon the results of publicly supported research.

If latter is accepted as a truly important and enduring societal priority, then the attention of creative people and the necessary public resources should be liberated from the distraction of tinkering with inherited institutional forms that are ill-suited for that mission, however well they serve society in other vital respects. It is clearly a job for talented individuals with a wide variety of public and private sector experience with various aspects of the worlds of R&D-based enterprise and "open science" throughout western Europe – and why not also in the accession states of the EU?

It is not as though there were no experiments from which to learn, and on which it might be possible to build: there are 'bridge institutions' like the Fraunhofer Gesellschaft, independent research consortia operating under sponsorship of business firms and public foundations, such as the Inter-university Microelectronic Centre (IMEC, founded 1984 by the Flemish government), regional incubators linked to universities and to research parks. But rather than being peripheral, and rather marginal, the development of novel institutional forms and procedures – to populate the organizational terrain situated between the university, the state agency, and the business corporation – now should be brought to the center of the stage, promoted, and accordingly resourced. It calls for a commensurately serious response if the ERA is to become more than a wishful conceptualization – a suitably symbolic gesture with which to usher in a new millennium – that the European Commission was able to persuade the national leaders of the EU member states to embrace on the occasion of their meeting in Lisbon in October of 2000. In my vision of the future, the creative task of enriching the institutional infrastructure for scientific and technological advance in a way that protects and sustains the vital heritage of the universities within that structure is the critical challenge that should and can be met by an "innovative Europe."

8.6   A summing up

This chapter has been framed as a response to trends encouraged by the February 2003 *Communication on the role of the universities in the Europe of knowledge*, issued by the Commission of the European Communities. That thought-provoking document assessed Europe's critical needs in the epoch of "knowledge-driven economic growth," and identified the university as the institution uniquely suited to meeting those needs. It called for debate on the means by which the conditions of European universities can be changed to satisfy the requirements of the new societal role for which the Commission believes them to be destined. Reduced to its essence, this presented a view of Europe's institutions of higher education as possessing the potential to be more effective than its industry at the business of technological innovation. But it also faulted the university researchers and administrators for failing to make the realization of that potential a priority. What is being advocated, therefore, is tantamount to a program of institutional reforms intended to mobilize that capability in order to meet a dual societal problem: financing the rising costs of public education and research, and enlarging the share of EU gross domestic product that is devoted to public and private investment in R&D.

This approach to fostering what the Commission referred to as "a Europe of knowledge" aims to "harness" the energies of university professors, students, and administrators for a new and highly instrumental goal, the advancement of knowledge for national and regional "wealth creation." But the likely costs, as well as the promised benefits of this proposal deserve more careful consideration than they have been receiving from enthusiasts for the grand goal. With regard to the costs, it is apparent that many of the features of universities that have rendered them particularly effective when called upon to perform in their historical societal role as "nodes" in the international dissemination of knowledge – and, since Humboldt, as generators of fundamental advances in scientific understanding, might have to be sacrificed in order to effectively carry through the institutional reforms suggested by the EC's *Communication*. Within the familiar context of academic, "open science" norms and governance structures, the comparative advantage of university-based researchers lies in conducting inquiries that may provide the foundations for valuable commercial innovations. But the best way to do this is precisely not the closely managed, tightly coupled search for discoveries and inventions that fires the imaginations of many political leaders, policy-advisors – and financially hard-pressed university administrators – who are seeking predictable and readily identifiable near-term payoffs.

Turning to the supposed benefits, it is equally apparent that the EC's *Communication* (and many similar policy pronouncements of national government ministries) have failed to show that there is an adequate evidentiary basis for supposing that the envisaged societal gains will be substantial enough to justify attempting to transform Europe's most prestigious academic institutions into "knowledge-management enterprises." It is not plausible to suppose that more than a few among Europe's research universities would, by exploiting the intellectual property created by the people who study and work there, be enabled to contribute materially to the costs of their own upkeep. Ideas for European institutional reform and regeneration along those lines clearly have been inspired by perceptions of vigorous university-industry research partnerships, rising patenting activity and the flourishing of academic entrepreneurship in the United States during the two closing decades of the past century. As those years followed immediately upon the date of the passage of the Bayh–Dole Act (1980), the latter has been accepted as a model for emulation. This has been seen to be dangerously simplistic in several respects.

First, closer examination of the available record leads one to doubt that the Bayh–Dole regime offers an appropriate paradigm for the ERA. The rise of university patenting and start-ups are traceable in significant part to factors other than the U.S. congressional initiative, and particularly the advances in biomedical knowledge driven by the rise of massive public research funding predating 1980. Secondly, the universities' attempts to exploit IPR, whether by licensing or by faculty "start-ups," has brought significant financial gains for only a very few U.S. institutions, whereas the increased administrative problems, and the private and social costs of patenting – especially in the biomedical areas – are widely felt. Third, there have been unforeseen and somewhat perverse consequences of this institutional experiment. The highly decentralized approach of the Bayh–Dole Act, in giving every university and public research institute the responsibility for securing and exploiting its intellectual property portfolio, has imposed significant "learning costs" on the system as a whole and brought into existence a new professional group – university technology managers – who have personal and collective interests in the perpetuation of these arrangements. Concomitantly, there are few if any large, R&D-intensive firms in the United States that now express general enthusiasm for the Bayh–Dole regime, and, many of their executives now speak in very critical terms about the performance of most of the universities' technology licensing offices.

In sum, then, European policy-makers concerned with the scientific and technological foundations for business innovation and economic growth should be considering reforms and revitalizing measures that build upon the region's own rich and diverse institutional foundations, rather than risking doing damage to them by blindly imitating a dubious American experiment.

References

AUTM 2003. *Licensing Survey: FY 2002.* Norwalk, CT: Association of University Technology Manager Manager, Inc

Balconi, M., Breschi, S., and Lissoni, F. 2003. "IL transferimento di conoscenze technologiche deall'universita all' Industria in Itaila: Nuova evidenze sui bervetti di patenit dei docenti", in A. Bonacorse (ed.), *ll sistema della ricera pubblica in Italia*, Milano: Franco Angeli

Balconi, M., Breschi, S., and Lissoni, F. 2004. "Networks of Inventors and Their Role of Academia: An Exploration of Patent Data," *Research Policy*, 33(1): pp. 127–145.

Branstetter, L. 2004. "Exploring the Link Between Academic Science and Industrial Innovation: The Case of California's Research Universities," Columbia University Business School Working Paper, forthcoming in *Annales d'Economie et de Statistique.*

Cobban, A. B. 1988. *Europe's Medieval Universities.* Oxford: Oxford University Press

Dasgupta, P. and David, P. A. 1994. "Towers a New Economics of Science," *Research Policy* 23: 487–521

David, P. A. 2003. "The Economic Logic of 'Open Secience' and the Balance Between Private Property Rights and the Public Domain in Scientific Data and Information," in J. M. Esau and P. F. Uhlir (eds.), *The Role of Scientific and Technical Data and Information in the Public Domain.* Washington, D. C.: National Academies Press

Eisenberg, R. S. 2001. Bargaining over the Transfer of Proprietary Research Tools: Is This Market Emerging or Failing?," in D. I. Zimmerman, R. C. Dreyfuss, and H. First (eds.), *Expanding the Bounds of Intellectual Property: Innovation Policy for the Knowledge Society.* Oxford: Oxford University Press

European Commision 2003. *Communication from the Commission: The Role of the University in the Europe of Knowledge*, COM (2003) 58 final. Brussels: Commission of the European Commmunities (5 February), 2003

Geuna, A. and Nesta, L. 2006. "University Patenting and Its Effects on Academic Research: The Emerging European" *Research Policy* 35 (June–July) [Special Issue on Property and the Pursuit of Knowledge: IPR Issues Affecting Scientific Research, P. A. David and B. H. Hall (eds.)]

Hertzfeld, H. R., Link, A. N., and Vonortas, N. S. 2006. "Intellectual Property Production Mechanisms in Research Partnerships" *Research Policy* 35 (June–July) [Special Issue on Property and the Pursuit of Knowledge: IPR Issues Affecting Scientific Research, P. A. David and B. H. Hall(eds.)]

Horton, Richard 2004. "The Dawn of McScience," *The New York Review of Books* 1.1(4): 7

Mowery, D. C. and Sampat, B. 2003. "The Bayh–Dole Act of 1980 and University Industry Technology Transfer: A Model for other OECD Governments?," *Journal of Technology Transfer*, 30(1): pp. 115–127

Mowery, D. C., Nelson, R. R., Sampat, B. N., and Siedonis, A. A. 2004. *Ivory Tower and Industrial Innovation: University–Industry Technology Transfer Before and After the Bayh–Dole Act*. Standford, C.A: Standford University Press

Nelson, R. R. 2001. "Observations on the Post-Bayh–Dole rise of patenting American University," *Journal of Technology Transfer* 26: 1319

OECD 2002. *Benchmarking Science-Industry Relationships*. Paris: OECD

Pavitt, K. 2002. "Changing Patterns of Usefulness of University Research: Opportunities and dangers." Brighton: SPRU, University of Sussex: Prepard for the Noble Symposium 123: Science and Industry in the Twentieth Century, Stockholm, 21–3 November

Riemers, N. 1987. "Tiger by the Tail," *Chemtech* 17 (8): 464–71

Sampat, B. N. 2004. "Patening and U. S. Academic Research in the Twentieth Century: The World Before and After Bayh–Dole" *Research Policy* (35) (June–July) [Special Issue on Property and the Pursuit of Knowledge: IPR Issues Affecting Scientific Research , P. A. David and B. H. Hall (eds.)]. Trune, D. and Goslin, L. 1998. "University Technology Transfer Programs: A Profit/Loss Analysis," *Technological Forecasting and Social Change:* 197–204

Walsh, J. P., Arora, A., and Cohen, W. M. 2003. "Research Tools Patenting and Licensing and Biomedical Innovation," in W. M. Chen and S. Merril (eds.), *The Patent System in the Knowledge-Based Economy*. Washington, D.C.: National Academies Press

Comments to Chapters 7 and 8:

Institutions and innovation

Bengt-Å ke Lundvall

Institutions

It is obvious that Schumpeter saw 'institutions' as being of fundamental importance for understanding innovation and economic development. This is clear both from his early work on capitalist development and his work on *Capitalism, Socialism and Democracy*. In *Theory of Economic Development*, the distribution of entrepreneurial spirit and talent in the population as well as the design of the banking system are important dimensions of the institutional setup. In *Capitalism, Socialism and Democracy* his rather pessimistic view on the future of capitalism refers to changes in the institutional framework that had undermined the fundamental role of the bourgeois family.

Some of the first attempts to link explicitly the economic literature on institutions to processes of innovation were by my colleague Björn Johnson (1988 and 1992). One of his basic points was that the uncertainty that characterizes innovation makes it even more necessary to include institutions in the analysis when the focus is on innovation and innovation systems. Rational choice referring to well-defined alternatives cannot explain what comes out of a process where outcomes are by definition unknown. Therefore institutions understood as rules, norms and habits are crucial for the outcome of what individuals decide and do in relation to innovation.

Later Johnson and Edquist developed the distinction between organizations and institutions in relation to innovation and innovation systems (Edquist and Johnson 1997). While specific organizations may be seen as 'incarnations' or 'containers' of institutions – such as patent offices incarnating intellectual property rights or as universities housing a specific knowledge production mode – they should not be defined as institutions.

Institutions as rules of the game

That institutions matter for innovation is thus not a controversial issue and from the very beginning the literature on 'innovation systems' takes this as fundamental starting point (Freeman 1982, Lundvall 1985). To understand how and why institutions change, sometimes incrementally and sometimes radically, is a more challenging task and this is what Masahiko Aoki sets out as ambition in his chapter. In the introduction to the chapter he raises interesting normative questions similar to the ones posed by the literature on national innovation systems. How can it be that there are constellations of institutions in one country that appear to promote performance that cannot be easily transferred to another? But his main focus is on how to explain *radical innovation in institutions*.

Aoki develops his ideas with explicit reference to Schumpeter's distinction between circular flow and incremental change on the one hand and innovation as discontinuous change and new combination on the other. He adds his own methodological preferences: 'I strongly believe that the game-oriented frame of thinking is the most suitable for socially scientific subject matter'. On this background he argues that institutions may be seen as Nash-equilibria (defined as games where individual agents stick to the rules as long as all the other players follow them).[1]

The game theory presented by Aoki is more complex than standard game theory. Different games are played in different domains and the same players may be engaged simultaneously in the different domains with important linkage and spillover effects. When there are 'institutional complementarities' certain rules in one domain may be viable only because there are other rules in a different domain that make them attractive for players. The ruling constellation may not be optimal but, according to Aoki, it may still reflect a Nash-equilibrium.

Using these concepts Aoki identifies three mechanisms that may give rise to and shape the Schumpeterian innovation in institutions.

(1) Schumpeterian bundling and disbundling
(2) Overlapping social embeddedness
(3) Dynamic institutional complementarities

On the basis of the definitions given in the text it is tempting to rephrase these mechanisms so that they refer to well-known phenomena.

---

[1] While it is clear that Schumpeter uses 'circular flow' as abstraction and as a method to link economic processes to *general equilibrium where everything is at rest* it is not clear if the equilibrium concept used by Aoki refers to general or partial equilibrium. Since he links games in different domains to each other I assume that he refers to something in between general and partial equilibrium.

Most examples on bundling and disbundling seem to fall under the heading of change in economic organization and more specifically *vertical integration and disintegration*.

Overlapping social embeddedness refers to *institutional drag and its opposite*. It is about norms inherited from the past and that may either hamper or support the emergence of new economic institutions/new ways of organizing the economy.

Dynamic institutional complementarities refer to *institutional learning*. To begin with agents may not be able to fully exploit the advantages of a new type of bundling of activities, but if there are supporting institutions they may obtain the necessary competencies.

As we read Aoki what is at the core of institutional innovation is what he calls Schumpeterian bundling and disbundling while the prevailing forms respectively of social embeddedness and dynamic institutional complementarities explain the success and direction of a potential institutional innovation.

Some critical remarks

Aoki brings into his analysis important aspects of institutional dynamics. The idea that institutions in different domains may support each other in terms of viability and that change in one domain may spill over into institutional change in another is certainly useful. The idea that old institutions may match or not match new emerging economic structure is also important. Similar ideas but presented without reference to game theory or equilibrium states are strongly present in the works by Christopher Freeman on the historical development of innovation systems (Freeman 2002).

All these ideas are worth developing further. The unique insights of Aoki when it comes to systematic comparison of the Japanese and the United States institutional setups gives him an excellent basis for this line of research. It is less obvious why it is necessary and useful to define institutions as *equilibrium rules of the game*. I would argue that in this context it is a *misleading and information-poor concept*.

Kenneth Arrow has coined the phrase that 'you cannot buy trust and if you could buy it – it would have no value whatsoever'. This elegant aphorism covers the wider insight that socio-economic systems based *exclusively* on agents' strategic behaviour are not sustainable. One reason that institutions can offer stability and reduce complexity in a rapidly changing world is that they are internalized and not much reflected upon. Therefore it is especially problematic to apply a game-theoretic approach, with its inherent emphasis on strategic behaviour, to institutions. If agents were constantly calculating the costs and benefits of sticking to the old rules, the basic function of institutions could not be fulfilled. We therefore do not believe that 'game-oriented frame of thinking is the most suitable' for this subject matter – at least not without solid support from history and broader social theory.

Neither do I see the equilibrium as it is used as being especially helpful in analysing change in existing institutions. According to Aoki, there is equilibrium when a set of rules is followed by agents and therefore remains unchanged. But the absence of change may cover over very different underlying situations. One dimension that may be added is the distinction between *harmony and tension*. Another distinction has to do with the *degree of internalization and explicitation* of norms and conventions. Both dimensions may change over time also when there is 'rest at the surface', and these underlying changes may be more interesting than the fact that there is 'rest' on the surface.

The fact that there is no earthquake in the Bay Area this year does not rule out the fact that tension has been growing in the underground and that seismologic analysis may even be able to forecast that an earthquake will arrive one of the next coming years. Similarly, the basic institutions of the socialist regimes in Eastern Europe remained 'in equilibrium' until they suddenly broke down, but at the end of the period tension was much higher than in the seventies and eighties.

For long historical periods the subordination of women, a factor of major importance for economic development, were internalized and not reflected upon. This has changed radically over the last century but to a very different degree in different parts of the world. An internalized institution may be brought explicitly on the agenda while it still remains in 'rest'.

To put under the same equilibrium heading a situation where tension is high and radical change is on the agenda and a situation where there is little tension and full internalization of norms might not be very useful.

The Silicon Valley story

Aoki uses the Silicon Valley to illustrate his analytical concepts. The major feature of the story is that the key to the success of the region is a combination of technological features making modularization feasible, spin-offs from major firms such as IBM and especially highly competent venture capital. Modularisation of technology and venture capital gives rise to tournaments with strong and just competition re-enforcing the manifold parallel efforts of entrepreneurs and designers to find excellent solutions.

This seems to be the main story but Aoki supplements it with a side story about 'intensive communication'. Tacit knowledge is exchanged face to face in informal ways. And it is said that 'Indeed, the sharing of generic knowledge relevant to the potential direction of the total technology on the one hand, and the encapsulation of the processing of the potentially proprietary information on the other, appears to be inseparable dualistic characteristics of Silicon Valley clustering'.

At first sight the two stories seem to bring forward games where the rules tend to counteract and undermine each other. Why should entrepreneurs and designers working very hard to win tournaments where 'winner takes it all' be willing to share tacit knowledge among themselves knowing that this is 'potentially proprietary information'? Rather than equilibrium we would expect tension to grow and the system to break down. Aoki brings in a series of extra arguments to explain this second part of the story relating to 'the commons domain' and to how players may attain social status by sharing, and so forth. But he does not explain why the two sides of the dualism do not undermine each other.

I believe that a different story is easier to reconcile with the history of Silicon Valley and at the core of this story is 'interactive learning'. To make modularization possible social capital (here I will define it as 'the willingness and capability of citizens to make commitments to each other, collaborate with each other and trust each other in processes of exchange *and* interactive learning') was as important as was the technical opportunities to split up the production process in separate steps (see Lundvall 2006).

Modularization has opened up more interfaces than before where a multitude of diverse agents can meet as users and producers in common efforts to develop new products. This deepening of the development of division of labour enriches the processes of interactive learning and has a major positive impact on the speed and diversity of the innovation process.

What makes Silicon Valley stand out is not that entrepreneurs and designers work harder and make bigger efforts (actually the ones I have met seem rather 'laid back') than elsewhere but rather the speed- of learning and competence-building at all stages of the production process. According to this story the sharing of tacit knowledge is part of the main story and not going against the overall logic of the system.

But as all stories, neither the first nor the second tells the whole truth, and a combination of the two may be getting closer to the whole truth. Aoki is right in arguing that intensified competition is an important factor driving growth in the region but it is a new kind of learning-based competition where becoming smarter is more important than working harder. And in order to become smarter people need to interact and share knowledge with others.

Bringing Bayh–Dohle to Europe – a case of pervert international institutional learning?

If we define institutions as rules, norms and routines, universities are not institutions but organizations. But universities carry with them institutions as genes inherited from the past since they impose rules of behaviour for those operating within them. Today universities are generally recognized as important elements in the overall innovation system. But what role they should play is controversial. Today there is a growing pressure among European policy-makers to make universities more 'market-oriented' and more directly useful for industry. More specifically, it is assumed that making universities more active when it comes to establishing and exploiting patents and other forms of intellectual property rights would benefit innovation and knowledge-based growth in Europe while at the same time create a new source of funding for university research.

Paul David's contribution gives a critical reflection on the current tendency to move in this direction in Europe and to import some of the features of the US academic system. He shows first that the basic assumption that European universities do not contribute to innovation is not well founded. Academic researchers do contribute substantially but the patents are registered not in the name of the university. Second he shows that the impact of the Bayh–Dole act is overestimated. Much of the United States-surge in patenting with reference to science came because more public funds were allocated to new areas such as bio-technology and pharmaceuticals, and if those fields are excluded the change was incremental rather than radical. Finally, referring again to the United States experience, he shows that it is an illusion that university research could find a major source of finance in the business of patenting and licensing.

The costs of the move toward 'entrepreneurial universities' supporting a new class of administrators who push for useful results by instituting benchmarking of researchers' patenting are substantial according to David. It means that the classical function of universities as sites where open science thrives comes under threat. Long-term research with vague objectives but with great potential to produce radically new knowledge is substituted for by more short-term profit oriented research surrounded by secrecy.

The chapter is brief and the policy proposals shaping an alternative strategy are even briefer. Basically David proposes to develop new institutions that can bridge between universities and industry. In this context he refers to Fraunhofer Gesellschaft and other similar organizations that have as their major task to transform scientific knowledge into technological opportunities for industry.

The role of universities in the learning economy

I have a lot of sympathy for the argument put forward, and as always when ideas are presented by Paul David the analysis is enriched by historical insight that makes current debates look somewhat shallow and futile. But I miss some reflections on why the policy community in Europe has become so eager to integrate universities as market actors. Following Paul David's arguments you might get the impression that we are confronted with massive collective ignorance, misinformation and lack of wisdom in policy circles. (Actually he invokes elements of public rational choice explanation when referring to the interests pursued by a new class of university bureaucrats.) This might not be the full story and neglecting what lies behind the current movement may actually weaken the criticism of what is going on.

There is of course no room for a full and consistent analysis here but let me list quickly some of my own assumptions regarding what lies behind the current urge to commercialize university research (see also Lundvall 2002):

1 In the most rapidly expanding scientific fields the distance from academic research to commercial use is much shorter than before, and biotechnology, the extreme case, has set a kind of standard for all university research (this point is acknowledged by David).

2 Competition in this but also in other science-based fields has increasingly become an issue of *speed*. To move quickly from scientific breakthrough to industrial application has become more important than before.

3 Scientific knowledge has become an important input for a wider population of firms. In high-income economies even small- and medium-sized firms operating in more traditional sectors may become more active in terms of innovation when they get better access to scientific knowledge.

4 New analytical concepts such as knowledge production of type Mode II, Triple Helix and National Systems of Innovation have focused

attention on the linkages within the system including the linkage between universities and industry.

Can we infer from these changes that the innovation system as a whole would benefit from universities becoming more active in terms of taking out and licensing patents to firms? I believe that the direct effects on innovation are as ambiguous as set out in David's analysis. Neither is it in the interest of science-based firms to initiate this kind of reform. An interesting case from Denmark is that the very person in the ministry of research who was instrumental in introducing the patenting rights for universities after becoming R&D-director of a major Danish company has made his regrets widely known. He complains that the access to new research ideas for business has become much more difficult than it was before the reform.

Walras+ as strategy for Europe

Therefore, we need to introduce other factors to explain the current urge to commercialize university research and look at the indirect effects of the reforms. The general ideological offensive in favour of markets and against taxes combined with the assumption that private actors always do rational things while public ones serve primarily their own self-interests has resulted in a strategy that might be referred to as Walras+, where the plus refers to innovation policy (cf. the seventies when European trade Unions tried to launch Keynes+ as strategical concept). The knowledge-based economy is combined with traditional assumptions about micro-economic institutions and processes. Since private firms do what is best for them, and for the system as a whole, the parts of the innovation system that policy-makers can legitimately attempt to govern and reform is the public sector. According to this perspective to make universities more market-oriented is a way to make them more rational.

And the indirect long-term effects of importing Bay-Dohle may very well be a speed-up of innovation, and at least it will make universities more malleable and easy to collaborate with for business. Introducing a commercial logic among scholars at universities will gradually shift their attention toward application and make them quicker to respond to commercial opportunities. The logic of open science will be moved toward protecting ideas and bringing the academic culture closer to what you find in profit-oriented firms. In this sense the patenting reforms may be seen as a response to the speed-up of change characterizing the learning economy.

But the long-term indirect effects may also be the most serious. As Paul David points out the long term efforts in science with less immediate objectives have proved most 'productive' in a historical perspective and I believe that he is right in warning against undermining this kind of research as well as against the possible break down of the norms of open science. The function of universities to be places where the public could trust scholars to be 'disinterested' and report about new insights without giving too much attention to the ruling political and economic interests would also be fundamentally changed.

In a different context I have argued that the background assumption behind much of the current thinking about reform – that we are in a knowledge based economy – should lead reformers to consider the need for institutions that take on the role of assessing the quality of knowledge (Lundvall 2002). To make the point more in more vivid terms I have drawn the parallel to the almost general acceptance today that central banks not only should have autonomy from the political sphere but also should not be involved in ordinary profit-making activities.

It is striking that this respect for those who are seen as guardians of the reliability of money as medium is not transferred to the guardians of knowledge. Serious deflationary crisis of knowledge could be started by private business (cf. ENRON) by public authorities (attempts to cover up major political mistakes and corruption) but it could also come through university scholars producing results according to the financial strength of the firms paying for the research, or environmental gurus supported by oil companies who argue that the global warming is not yet a proven phenomenon. In an economy based upon knowledge such deflationary crises may create damages difficult to repair.

A final consideration that is missing in Paul David's paper – it is understandable because it is short and focused on a specific issue – is that defending the current system against the Bay-Dohle kind of reform does not imply that the current university systems should be left to go on without reform. Neither does it imply that more efforts in the direction of applied research is necessarily bad for the quality of basic research. There is much low-quality research and teaching going on at universities and there are cases where the internal mechanisms do not work to correct the situation. In such cases one way to upgrade the quality may be to expose universities to external collaboration with users. To diversify the organization of the university so that the same scholar over a longer period may move between activities that are more or less close to, respectively, basic science and practical use of knowledge may actually be the best solution to the dilemma posed by the learning economy (Lundvall 2002).

Conclusion

While Aoki introduces important elements for understanding how institutional innovation may come about, David evokes an area where there is a strong need to find new institutional forms that combine the classical function of universities with the new context. While Aoki points to the difficulty in transplanting institutional forms from one context to another in general terms, David warns against a specific case of naïve transplantation policy.

One aspect of Europe that needs to be taken more into account is that it remains a continent composed of national institutional systems at rather different levels of development. The European Commission has a tendency to overlook this simple fact and often comes with standard solutions for Europe as a whole, for instance, by assuming that there are 'best-practice' institutions that can and should be diffused to all national systems in Europe regardless of 'social embeddedness'.

This is a serious mistake both in terms of analysis and political method. Only by starting with an analysis of the national institutional formations can a common European strategy aiming a convergence and complementarity become successful (Lorenz and Lundvall 2006). While it might be of interest to make it possible for scholars and students to move more freely across national boarders, there is no good reason to assume that it would benefit Europe if universities in France, the United Kingdom, Denmark and Greece became identical in terms of organization and research strategy.

It is also fundamentally important to realize that the learning economy calls for dramatic institutional and organizational change *also within the private sector*. A major problem is that many of the firms do not have the internal capability to absorb scientific knowledge. The diffusion of good organizational practices matching the new context is slow and there is a need to rethink entrepreneurship so that it does not remain an ideological excuse for promoting old-fashioned and conservative family businesses. As strategy for institutional reform inspired by a Walrus+ philosophy Walras+ is insufficient.

References

Edquist, C. and Johnson, B. 1997. 'Institutions and Organizations in Systems of Innovation', in Edquist (ed.), *Systems of Innovation – Technologies, Institutions and Organizations*. London, Pinter.

Freeman, C. 1982. 'Technological Infrastructure and International Competitiveness', Draft paper submitted to the OECD Ad hoc-group on Science, Technology and Competitiveness, August 1982, mimeo. Now in

Freeman, C. 2004. 'Technological Infrastructure and International Competitiveness', *Industrial and Corporate Change* 13(3): 541–69

Freeman, C. 2002. 'Continental, national and sub-national innovation system – complementarity and economic growth', Research Policy, 3: 191–211.

Johnson, B. 1988. 'An Institutional Approach to The Small Country Problem', in Freeman and Lundvall (eds.), *Small Countries Facing the Technological Revolution*. London and New York: Pinter Publishers

Johnson, B. 1992. 'Institutional Learning', in B.-Å. Lundvall, (ed.), *National Innovation Systems: Towards a Theory of Innovation and Interactive Learning*. London: Pinter Publishers

Lorenz, E. and Lundvall, B.-Å (eds.) 2006. *How Europe's Economics Learn: Coordinating Competing Models*, Oxford: Oxford University Press

Lundvall, B.-Å. 1985. *Product Innovation and User-Producer Interaction*. Aalborg: Aalborg University Press

Lundvall, B.-Å. 2002. 'The University in the Learning Economy', *DRUID Working Papers*, No. 6, 2002. ISBN: 87-7873-122-4

Lundvall, B.-Å. 2006. 'Interactive learning, social capital and economic performance', in D. Foray and B. Kahin (eds.) *Advancing Knowledge and the Knowledge Economy*. Washington January 10–11, 2005, organized by Cambridge, MA: MIT Press

Part 5

Innovation, firms' organization, and business strategies

9 Bringing selection back into our evolutionary theories of innovation

Daniel A. Levinthal

Introduction

Organization innovation, from an evolutionary perspective, involves at least three distinct challenges: problems of competence, problems of variety, and problems of selection. Both capabilities and selection are inherently organizational phenomena. However, while the field has developed a deep commitment to the first of these two propositions (Nelson and Winter, 1982), we are only beginning to fully appreciate the latter. After a brief review of the challenges of competence and variety, I consider some of the intellectual challenges that we face in incorporating issues of selection in our models and analyses of organizational innovation.

A few basic issues are highlighted. First, heterogeneity in selection criteria across organization is a function of the demand environments in which firms operate and that organizations, failing to engage in selective intervention, inevitably restrict the variety of selection criteria imposed on innovative initiatives. Second, selection forces internal and external to the firm operate not only on fully realized development efforts, but on intermediate forms. Selectable traits that are favored by intermediate selection need not be associated with properties of initiatives that, if allowed to reach that ultimate realization, are associated with high levels of technical or economic performance. This issue of the need for intermediate selection is applied to provide a critical perspective on recent interest in the use of real options as an analytical solution to the problem of making investments in the face of uncertain futures. The notion of search is central in behavioral theories of the firm (March and Simon, 1958; Cyert and March, 1963) and in evolutionary arguments regarding technical advance (Nelson and Winter, 1982). However, search has largely been construed in terms of variety generation and the specification of a latent alternative set. Search, however, also necessitates the screening and selection amongst those alternatives identified.

294   *Daniel A. Levinthal*

While there are some exceptions (Nelson, 1982), the literature, starting with Simon's (1955) original contribution on bounded rationality, has tended to ignore this other facet of search processes.

9.1   Problems of competence

I think that it is fair to say that the greatest emphasis has been placed on the problem of competence. Firms have been recognized as being bound by their path-dependent trajectory of capability development. Thus, firms' technical capabilities are, on the one hand, a source of tremendous strength, providing the basis for competitive advantage. At the same time, these firm-level technological trajectories are also highly constraining. Capabilities in the electromechanics sphere may not prove of much use in electronics, nor may chemical-based drug-discovery efforts provide a strong platform for efforts at rational drug design. In this spirit, the problem of firm innovation and, in particular the specter of displacement of leading firms through a Schumpeterian dynamic, has been primarily viewed as a problem of capabilities – the problem of once highly valued technical capabilities rendered less valuable through innovation and changes in the underlying science base.

There are important footnotes, or qualifications, to this basic capability argument in the literature. One of the most important of these is the idea of complementary capabilities introduced by Teece (1987) and developed by, among others, Mitchell (1989) and Tripsas (1997). The argument that the set of relevant capabilities for firm performance are not restricted merely to technical ones, but a range of nontechnical capabilities in manufacturing, marketing, and distribution may buffer a firm from changes in its relative technical competence. In addition, scholars are recognizing that technologies are typically not discrete entities but are often embedded in larger technical systems.

Recognizing technologies as being elements of broader technical systems is critical to Henderson and Clark's (1990) arguments regarding architectural change. Consistent with Simon's arguments regarding the adaptive dynamics of nearly decomposable systems, Henderson and Clark note a striking robustness of firms in the ability of photolithographic industry to respond to quite substantial changes in the technical base of individual components. However, seemingly modest changes in technology that impacted the manner in which components of the technical system interacted had rather severe impacts on the capabilities and market leadership of these firms. The work in recent years by Brusoni and Prencipe together, and in conjunction with their former colleague Keith Pavitt, have substantially advanced our understanding of the distinct issues associated with the development of technical systems and the role that near-decomposability and modularity play in that context (Brusoni and Prencipe, 2001; Brusoni, Prencipe, and Pavitt, 2001).

This provides a very brief, and clearly incomplete, overview of this, what is arguably the dominant perspective on the problem of technical change, within the evolutionary economics tradition. It constitutes a rich and powerful line of argument, one that clearly accounts for a substantial portion of the observed variation in changes in technical leadership across time and, as well, reflects the "supply-side" focus of the field. This "supply-side" focus is, I think, a natural by-product of the fact that one of the important motivating agendas for this effort was a desire to understand the heterogeneity in production capabilities (Nelson and Winter, 1982). However, as I argue subsequently, as we "bring selection back into" our discussions of technological innovation, demand-side considerations take a more prominent role.

9.2 Problems of variety

The literature on organizational learning (Levitt and March, 1988) is certainly sensitive to the problem of capabilities. Indeed, this literature has long identified the issue of competence traps. However, the capability "problem" takes a somewhat different form in this literature. The argument is that firms are simultaneously learning *what* actions to take at the same time that they are developing *competence* at particular actions. Rapid learning with respect to competence will tend to make, other things being equal, the current actions increasingly attractive. The joint effect of rapid learning as to what constitutes appropriate action in conjunction with rapid competence learning is to lead organizations to lock into a set of actions that may lead to relatively modest performance compared with other, latent possibilities.

Slow learning, particularly with respect to preferences over what constitutes desired action, is suggested as one mechanism to mitigate the competence trap. The virtue of slow learning in this regard is that it preserves variety and thereby sustains experimentation. More generally, the problem of variety can be expressed as the tendency for learning processes to be myopic and to tend toward excessive reliance on exploitation at the expense of possible exploratory efforts (Levinthal and March, 1993). Turnover, slow rates of adaptation, and noisy inferences regarding what constitutes superior performance are all mechanisms to sustain experimentation in the face of such biases towards exploitation.

9.3   Problems of selection

Variety, however, is clearly not sufficient for innovation. To take Kanter's (1988) imagery of "letting a thousand flowers bloom" such diversity in blooming will not be of consequence if the organization only has one type of lawnmower, or less metaphorically, one type of screening criteria. Innovation within organizations requires resources. In turn, sustaining diversity requires ongoing resource commitments to a diverse set of emergent efforts. It is to these questions of selection and resource allocation that I think the evolutionary economics tradition has under-attended.

9.2.1   *Heterogeneity in selection criteria*

Learning processes are feedback-driven. As a result, the particular context in which one operates influences the feedback received. I interpret Christensen's work on the disk-drive industry in this light (1997). One can take a bundle of performance characteristics regarding cost, processing capabilities, weight, and power consumption and get very different responses in terms of perceived value depending upon which customer constituency one asks. The desk-top user community responded with a shrug of their collective shoulders when offered drives that were smaller and lighter, while the emerging community of laptop producers responded with enthusiasm for such possibilities.

Now, the fact that firms and the products they produce compete in heterogeneous demand environments is an issue that has been of long-standing interest to our colleagues in marketing. However, marketing research tends to suffer the opposite problem of evolutionary modeling. They offer lots of representations and tools to engage heterogeneity of demand, techniques of conjoint modeling and the like, but they tend to operate on an implicit assumption of enormous plasticity in the range of what the firm is capable of producing. The marketing challenge is understanding the appropriate degree of bitterness of a beer, and perhaps what the desired images are associated with a product, but there is no question of brewing, of how one might actually produce the beer with the desired attributes. For those of us interested in processes of learning, heterogeneity in demand context not only says something about desired positioning, but also about what sort of capability might emerge.

Adner and Levinthal (2001) develop this sort of argument in an examination of the dynamics of process and product innovation. Product innovations expand the set of customers for whom the technology meets some minimal threshold of functionality, while process innovations expand markets by lowering the cost of goods or services. Heterogeneity in this work is simply represented by variation in customer's functionality requirements and their willingness to pay. These attributes need not be negatively correlated. For instance, early adopters of xerography, firms involved in typesetting for whom xerography allowed them more readily to create a master copy, had very low functionality requirements regarding ease of use and reliability relative to the mass market *and* a very high willingness to pay. Adner (2002) develops this modeling apparatus further to consider the possibility of divergent technical trajectories and applied this structure to the evolution of the disk-drive industry.

In a similar spirit, heterogeneous demand environments have been a central feature of the history-friendly modeling work of Malerba et al. (1999). The U.S. Defense Department provided a very different basis of feedback for the early computer and semiconductor market than industrial users and, in turn, resulted in firms that served these distinct constituencies with very different evolutionary paths for their capabilities. In further extensions of this research, we see that the presence or absence of upstream or downstream suppliers, a different facet of the environment, also had profound impacts on firms' evolutionary paths.

From a more proactive, incentive-based perspective, the presence of submarkets plays a central role in Sutton's work on industry evolution (1998). When consumers place a high value on meeting their idiosyncratic preferences, this leads to relatively fragmented industry such as flowmeters; in the absence of such value, there is an escalation of investment along a particular trajectory and concentrated industry structures such as aircraft.

Levinthal (1998) examined the evolution of wireless technology and argued that the critical junctures in that technology evolution were speciation events in which the existing wireless technology at that time was ported to new domains of application. From Marconi adopting Hertz's laboratory equipment to provide the initial wireless telegraphy, to the application of the subsequent advances in wireless telegraphy, and in particular the innovation of a continuous wave transmitter by a Westinghouse engineer to radio broadcasting, existing technologies were reapplied to new application domains with dramatic commercial and technological consequences. These "revolutions" in communication technology were remarkably incremental. An existing technology was ported to a new application domain, a Schumpeterian recombination. Once shifted to a new domain, a distinct, and in many cases rapid, path of lineage development ensued driven both by the distinct selection, or performance, criteria of the new domain, as well as the potential financial resources available in the distinct application domains. While Hertz had to do with makeshift lab equipment, Marconi had the backing of the British Admiralty and later a public company. Radio broadcasting is initiated by a Westinghouse engineer as a hobby, but is quickly adopted by Sarnoff at Radio Corporation of America (*RCA*), RCA having been previously founded to pursue wireless telegraphy.

9.2.2   *Selection in organizations and the Iron Law of Hierarchy*

Keith Pavitt made a convincing case that large enterprises can sustain an enormous diversity with respect to their technical competence (Pavitt, 1998). However, I suggest that a different sort of diversity is more problematic, that is the diversity of perspectives as to what constitutes useful endeavors for the firm; in particular, diversity with respect to the selection criteria associated with the firm's resource-allocation pro-cesses. Underlying this difficulty of organizations sustaining a diversity of selection criteria is the tendency for resources to be allocated by a singular authority structure within an organization. Thus, while a large organization may have sufficient resources to make multiple "bets," those individuals who control resource-allocation decisions are unlikely to be of multiple minds. While there may well be considerable diversity of opinion within the organization, there is typically a dominant political coalition, and the perspective of this ruling group will tend to drive the resource-allocation decisions.

Contrast this characterization with a population of organizations. Even if individual organizations make a singular "bet" with regard to a given opportunity, there may be tremendous diversity across the population of organizations. While there may be some pressure to conform to the perspective of other, respected organizations, individual organizations may receive highly differentiated feedback from their environment, and this distinct feedback may lead them to different views of the same business opportunity. Indeed, the motivation of entrepreneurs to leave their previous organization may stem as much from their inability to convince their previous firm to pursue an opportunity that they feel has tremendous promise as it is associated with an incentive to appropriate for themselves the returns associated with the pursuit of the opportunity.

Conceptually, a single firm could engage in, using Williamson's (1985) terminology, selective intervention and replicate the virtues of a population of independent organizations. However, doing so will not be sequentially rational (Selton, 1972). The corporate office will have a point of view about the appropriate direction for the firm and the relative promise of individual initiatives. When faced with a given funding decision, it cannot commit to "throw away" its belief structure.

The formation of a new, distinct organization, freed from any authority structure from the corporate office is one clear solution to this challenge. It may be possible to design commitment devices that restrict the impulses that make selective intervention not sequentially rational. One mechanism is to abdicate budgetary authority. This is often seen on a small scale in which a corporation may allow a modest percentage of a subunit's operating budget to be used at the subunit's full discretion. 3M has received attention for instituting such a role at the level of individual managers, who are free to spend a portion of their time pursuing whatever initiatives that they perceive to be valuable. The limitation of such an approach is that successful initiatives may not be financially self-sufficient and, as a result, may require external funding. One is then back in the position of having to convince some central authority of the merits of the particular initiative. Thus, the inherent hierarchy of organizations (Michels, 1915) constrains the variety that a single organization, independent of its size, can sustain.

9.2.3 Problems of intermediate selection along development journeys

A fundamental problem for selection processes is that selection is occurring over a "moving target." Indeed, as clear from evolutionary arguments, selection can only be intelligent if there is a high degree of stability over what is being selected. However, in the innovation context, it is inevitable and quite appropriate that elements of selection occur even when development processes are far from complete. Firms need to make interim judgments as to whether to continue to commit resources to a technology or product-development effort and cannot afford to wait for their full fruition or failure. Similarly, capital markets, particularly markets for venture capital, need to make interim evaluations as to whether a given concern is worthy of further resources.

One glaring example, in my view, within the management literature of the neglect of the problem of intermediate selection is the burgeoning literature and enthusiasm among academics and practitioners for the tool of real options as a solution for the problem of how firms should manage their uncertain futures, particularly with regard to technological uncertainty. As Adner and Levinthal (2004) argue, real options are not quite the panacea that its proponents tend to suggest. The basic real options argument as applied to problems of strategic management has the following basic structure: the world is uncertain; therefore we will make lots of modestly sized "bets" and, as future states are revealed, the firm will exercise those options that now appear attractive, having positioned itself to do so as a result of its earlier investments. One of the basic concerns that Adner and Levinthal (2004) pose is how will the firm know in this metaphoric "stage 2" which investments are attractive to strike or not. Unlike financial options, for which opening the pages of the Financial Times might suffice, real options on technology provide no such clarity.

Indeed, the typical early-stage innovative effort results in a partial failure, or put more optimistically, a partial success. Deadlines for technical hurdles are not quite met, but some substantial progress is made. Potential users have not reacted with unabashed enthusiasm for the product, but it appears that some modification of the feature set may result in a product with considerable appeal. If this is the modal outcome, what is the implication for managerial action and subsequent resource commitments? In the same spirit, critical to the logic of real options that enhances the value of initial "bets" on risking technologies is that exit and the termination of initiatives is a real possibility. However, analogous to Popper's arguments regarding hypothesis testing that we can only prove hypotheses false but can never prove them to be true, any innovative effort cannot prove the impossibility of future success (Adner and Levinthal, 2004). Rather, one observes a failure of the current embodiment of the technology to meet certain technical standards or satisfy the needs of a particular set of consumers. Such failure does not rule out the possibility that future incarnations of the technology might meet such standards, perhaps by pursuing somewhat different approaches, or that the firm might be able to identify a different user community that would respond more positively to the technology.

Adner and Levinthal (2004) argue that to preserve the analytical logic of real options, a firm would have to put tight boundaries around the scope of an innovative effort, boundaries concerning technical approaches, markets to which the product is to be sold into, and perhaps temporal boundaries. However, imposing such boundaries has enormous potential costs; they deprive the firm of exploiting the unanticipated discoveries of possibilities that is common in innovative efforts. Thus, real options may certainly be applicable to situations of well-defined risk, where there is uncertainty over known possible states of the world, but are deeply problematic in the face of Knightian uncertainty.

While Adner and Levinthal (2004) address the limitations of rationalistic approaches to allocating resources to innovative efforts, Levinthal and Posen (2005) examine the dual processes of firm learning and population selection. First, a striking observation when you look at the literature on organizational learning is the extent to which this work examines learning issues in, essentially, selection-free environments. Formal models of firm learning tend to have the structure of seeding a population of organizations with diverse learning strategies or organizational structures and then observe the variance in performance among the population after some large number of learning trials. However, these nominally process-oriented modeling efforts tend to ignore the path to these performance asymptotes. Imagine that learning does not take place in the benign petri dish of a simulation model, but in a competitive environment in which survival until the end of the period of observation cannot be taken for granted. What then are the implications for the desirability of alternative learning strategies?

First, once learning dynamics are placed in a context of selection pressures, the meaning of what is a high-performing learning strategy becomes nontrivial. Is a good strategy one that generates high expected performance conditional on survival? This is implicitly the criteria of the business press, which extols the virtues of dramatic gambles that paid off well. Alternatively, is a good strategy one that leads to a higher probability of survival? Finally, it is important to emphasize again that the criteria used in most models of organizational learning is the average performance of alternative strategies assuming that *all* organizations survive.

Development paths are subject to more intelligent intermediate selection, that is selection prior to the full realization of their potential, to the extent that the correlation in the performance of development efforts across time is relatively high. That is, to the extent that early success is suggestive of ultimate success, then intermediate selection can operate effectively. Development approaches, however, are likely to vary in their degree of correlation across time. The particular contrast that Levinthal and Posen (2005) explore is between development efforts in which initial efforts focus on one facet of the overall development effort, which we term the technical development subproblem, and efforts in which the full business system of technology, manufacturing, and marketing is jointly searched. Exploring subproblems has the virtue that it leads to rapid early performance gains and therefore is more likely to survive early screening efforts. However, such a search effort that initially attempts to optimize a particular subsystem will tend to lead to lower correlation in performance across time than an integrated development effort. Thus, while the focused strategy leads to higher survival rates from early screening efforts, the filtering process over this strategy is far less intelligent than selection in the context of integrated development efforts. As a result, in this analysis, integrated search strategies lead to higher average performance *conditional* on survival, even though the average performance under this search strategy in the absence of selection is inferior.

A further implication of this argument is that introducing survival concerns turns on its head the now established view of managing the dynamics of exploration and exploitation. The standard result from search models is that in early stages one should engage in exploration so as to learn more about the set of possible actions and then, after some knowledge has developed, to engage in more exploitative behavior. However, again, these analyses do not concern themselves with survival. Young, small, vulnerable firms have an acute survival problem. They need to exploit whatever modicum of wisdom they have about the world if they are to survive. Exploration, we suggest, is for the richer, more established firm; indeed, this is a notion, suggested by the *Behavioral Theory of the Firm* (Cyert and March, 1963) with the notion of slack search.

9.2.4   *Bringing evaluation into our models of search*[1]

Two points are focal in Simon's argument regarding bounded rationality, and they have served as central building blocks of behavioral economics since then. One is that only a subset of the entire space of alternatives might be considered in a given choice setting. Furthermore, decision-makers may be confronted with a sequential unfolding of these possible alternatives, even among the limited set considered. Second, he postulated that these alternatives are evaluated by a simple discrete value function that distinguishes between satisfactory and unsatisfactory outcomes. In this sense, Simon substituted for the usual objective function of economic theory an additional constraint of what constitutes a feasible solution to the choice problem. The value function becomes no different than the requirement that, say in the context of the decision to purchase a home, that the home have the requisite number of bedrooms.

What is less salient, though considered in the original discussion, is how actors are to evaluate the proposed solutions or alternatives. How do we know whether the various feasibility constraints are satisfied or not? Simon notes that there may be uncertainty as to whether a particular alternative may yield a state of nature that is in the satisfactory set or not, but the text suggests that this indeterminacy may be resolved by identifying a new alternative that does not suffer this risk.

---

[1] This section draws from Gavetti and Levinthal (2000) and Levinthal (2002).

Yet, this discussion points to an important lacuna in this early work and subsequent development of behavioral economic theories of individuals and firms. While ideas of search are central in behavioral theories of the firm, the mechanisms by which these alternatives are evaluated are less clearly developed.[2] Typical models of adaptive search have the following characteristics. Some space of possible alternatives is sampled. The realization from this "draw" is then compared either with the current status quo action or in other cases with an aspiration level. When the space of alternatives constitutes attributes such as prices, the model does not seem to require any elaboration. However, consider other possible spaces of alternatives, such as the space of possible new production technologies for a factory or the space of possible spouses. When presented with a new alternative from one of these sorts of "spaces," how is one to recognize a satisfactory solution when one is confronted with one?

Quick inspection of a possible spouse or a production plan may reveal certain proposed alternatives to be unsatisfactory, and some basic feasibility constraints may be revealed to be violated. However, the satisfaction of other constraints may not be so self-evident. How will the workforce respond to the production process? How reliable will the process prove? Similarly, will this proposed spouse prove to be an enjoyable companion upon repeated dining experiences, and will he or she prove reasonably tolerant of your array of annoying habits?

The evaluation of proposed alternatives is a relatively undeveloped facet of the behavioral theory of the firm. To provide some structure with which to consider such issues, it is useful to distinguish between two sorts of evaluation mechanism: the distinction of "online" and "off-line" evaluation (Gavetti and Levinthal, 2000). Online evaluation refers to those settings in which evaluation can only take place by actual trial of the proposed alternative, whereas off-line indicates the ability to assess value in the absence of such a trial. As with many dichotomies, this one is both informative and misleading. The distinction is clearly important. Some possibilities are evaluated by thinking, by imaging possible futures should that alternative (spouse, production process, car, etc.) be adopted. Sometimes this thinking is supported with various tools of analytical reasoning such as spreadsheets and yellow-pads.

However, the dichotomy is also quite misleading. There is an enormous gray area between these two poles and most evaluation processes

---

[2] An important exception is Nelson (1982), which considers the role knowledge plays in both guiding R&D search efforts and in facilitating the "testing" of possible new techniques.

occur somewhere in this intermediate zone. New production processes need not require shutting down the firm's entire operations and substituting the proposed process. One plant may serve as a test, while the prior technology is exploited in the remaining plants. In cases of more incremental changes, only a single line or shift of the production process may serve to provide an experiential basis on which to evaluate the proposal.

In other cases, an "artificial" environment is created in order to evaluate a proposed alternative that does not introduce the risk associated with a full commitment. The natural examples of this in the spousal problem are first-dates and weekends in the country. In the case of production processes, a pilot plant may be established. The pilot plant operates at a smaller scale than the ultimate substantiation of the alternative would imply, but again it allows a detailed examination of feasibility at lower cost and lower risk than full adoption.

A particular type of artificial environment, wind tunnels to test the performance of new aircraft, offers some additional insights as the boundaries of on- and off-line evaluation. Wind tunnels allow engineers to test loft and drag in a variety of conditions of a possible airframe. However, windtunnels have substitute modes of evaluation. One, of course, is to engage in the enormous financial commitment of the full development of a working prototype and the human risk posed to the pilot of such a craft. The other route is cognitive: to build computer models that simulate the performance of proposed designs. As knowledge of the underlying material and aeronautical engineering improves, off-line evaluation can substitute for more online forms of evaluation. But note that this is really a matter of degree. The computer simulation in some form creates its own kind of experience base. It is simply a lower-cost artificial world than ones that involve bending metal, such as wind tunnels, or pouring concrete, such as pilot plants.

A different sort of experience is the experience of others (March, Sproull, and Tamuz, 1991; Miner and Hanuschild, 1995). This sort of experience has the virtue that trial does not require the disruption of one's own activities and, furthermore, that the set of alternatives that are being explored at a given time are vast. It's weakness lies, of course, in the inferential difficulties that it poses. How much do I learn by watching a woman with another man what she would be like as a wife? Or, perhaps less daunting, how much does some other plant's experience tell me about my likely success with a new production technology? We are probably more comfortable with generalizing in the latter case, but they may in part stem from the fact that more of us have experience in being spouses than in being plant managers and are more keenly

aware of the idiosyncratic features of such relationships than of production processes.

In some sense, the issue of on- or off-line search becomes less a
categorical distinction than factors that influence the cost, risk, and
possibly accuracy of the evaluation process. Online search often entails a
particular sort of cost, that of the opportunity cost of not making use of
established options. It is this opportunity cost that underlies the tension
in the now oft-cited exploration, exploitation trade-off. In turn, the
degree to which current operations need be disrupted by the need to
evaluate a proposed alternative influences how painful that trade-off is.

Neighborhood search in the context of experiential, or online, evaluation of alternatives, however, has a distinct virtue. Neighborhood
search provides an effective, though not necessarily optimal, balance for
the need to explore alternative bases of action at the same time one
exploits current wisdom about the world (Gavetti and Levinthal, 2000).
The need for this balance between exploration and exploitation depends
entirely on whether the evaluation process of proposed alternatives is
on- or off-line. Thus, the wisdom of a particular sampling strategy
is intimately connected to the form of evaluation of those samples that is
possible. Many of our discussions of search processes have suffered by
not sufficiently disentangling these two features of search processes.

9.3   Conclusion

Evolutionary economics generally treats selection operating at the level
of the overall business enterprise. A first step in opening up our
understanding of selection is the recognition that economic environments, even within a focal industry, are heterogeneous and therefore
may create divergent market feedback and selection pressures. This step
has been taken, although its full consequences have probably not been
fully developed. However, the recognition that selection is hierarchical
with feedback from the market mediated by organizational processes is
far less developed.

Perhaps the most fundamental fact about business organizations is
that they comprise an aggregate unit by which a vast set of underlying
activities is allocated payoffs by an economy. Firms receive profits and
losses. Individual initiatives within firms, writing business plans,
investing in capital equipment, or arguing with a colleague do not. A
critical role of the firm is therefore the internalization of the effects of
environmental selection. An evolutionary theory of the firm acting in
markets clearly needs to engage this issue of a hierarchy of selection.
This brief sketch of some of the issues and research opportunities that

I see is, I hope, suggestive; suggestive meaning that the sketch is surely incomplete and, at points, perhaps mistaken, but also perhaps suggestive in the more positive sense of encouraging some movement and redirection to a possibly neglected aspect of our evolutionary theorizing.

References

Adner, R. 2002. "When are technologies disruptive? A demand-based view of the emergence of new technologies," *Strategic Management Journal* 23: 667–88

Adner, R. and Levinthal, D. 2001. "Technology evolution and demand heterogeneity: implications for product and process innovation," *Management Science* 47: 611–28

Adner, R. and Levinthal, D. 2004. "What is *not* a real option: considering boundaries for the application of real options to business strategy," *Academy of Management Review* 29: 74–85

Brusoni, S. and Prencipe, A. 2001. "Unpacking the black box of modularity: technologies, products and organizations," *Industrial and Corporate Change* 10: 179–205

Brusoni, S., Prencipe, A. and Pavitt, K.Brusoni, S., Prencipe, A. and Pavitt, K. 2001. "Knowledge specialization, organization coupling, and the boundaries of the firm: why do firms know more than they make?", *Administrative Science Quarterly* 46: 597–625

Christensen, C. 1997. *The Innovator's Dilemma.* Boston, MA: Harvard University Press

Cyert, R. and March, J. 1963. *A Behavioral Theory of the Firm.* Engelwood NJ: Prentice-Hall

Gavetti, G. and Levinthal, D. 2000. "Looking forward and looking backward: cognitive and experiential search," *Administrative Science Quarterly* 45: 113–37

Henderson, R. M. and Clark, K. B. 1990. "Architectural innovation: the reconfiguration of existing product technologies and the failure of established firms," *Administrative Science Quarterly* 35: 9–30

Kanter, R. M. 1988. "When a thousand flowers bloom: Structural, Collective, and Social Conditions for Innovation in Organization," in *Research in Organizational Behavior* 10: 169–211

Levinthal, D. 1998. "The slow pace of rapid technological change: gradualism and punctuation in technological change," *Industrial and Corporate Change* 7: 217–47

Levinthal, D. 2002. "Cognition and models of adaptive learning," in March, J. and Augier. M. (eds.), *Economics of Change, Choice, and Structure: Essays in the Memory of Richard M. Cyert.* Cheltenhave, U.K: Edward Elgar Publishing, Ltd

Levinthal, D. and March, J. 1993. "The myopia of learning," *Strategic Management Journal* 14: 95–112

Levinthal, D. and Posen, H. 2005. "Myopia of Selection: does organizational adaptation limit the efficacy of population selection? Working Paper, Wharton School, University of Pennsylvania

Levitt, B. and March, J. G. 1988. "Organizational learning," *Annual Review of Sociology* 14: 319–40

Malerba, F., Nelson, R., Orsenigo, L., and Winter, S. 1999. "History-friendly" models of industry evolution: the computer industry," *Industrial and Corporate Change* 8: 3–40

March, J. G. and Simon, H. A. 1958. *Organizations*. New York: John Wiley & Sons

March, J. G., Sproull, L. S., and Tamuz, M. 1991. "Learning from samples of one and fewer," *Organization Science* 2: 1–13

Michels, R. 1915. *Political Parties*. New York, NY: Hearst Library Co.

Miner, A. and Haunschild, P. 1995. "Population level learning" in Cummings, L. L. and Staw B. M. (eds.), *Research in Organizational Behavior*. Greenwich, CT: JAI Press, pp. 115–66

Mitchell, W. 1989. "Whether and when? Probability and timing of incumbents' entry into emerging industry sub-fields," *Administrative Science Quarterly*, 34: 208–30

Nelson, R. 1982. "The role of knowledge in R&D efficiency," *Quarterly Journal of Economics* 97: 453–70

Nelson, R. and Winter, S. 1982. *An Evolutionary Theory of Economic Change*. Cambridge, MA: Harvard University Press

Pavitt, K. 1998. "Technologies, products and organization in the innovating firm: What Adam Smith tells us and Joseph Schumpeter doesn't," *Industrial and Corporate Change* 7: 433–52

Simon, H. A. 1955. "A behavioral model of rational choice," *American Economic Review* 69: 99–118

Sutton, J. 1998. *Technology and Market Structure*. Cambridge, MA: MIT Press

Teece, D. 1987. "Capturing value from technological innovation," in Teece, D. (ed.), *The Competitive Challenge*. New York, NY: Harper & Row

Tripsas, M. 1997. "Unraveling the process of creative destruction: complementary assets and incumbent survival in the typesetting industry," *Strategic Management Journal* 18: 119–42

Williamson, O. E. 1985. *The Economic Institutions of Capitalism*. New York, NY: Free Press

10 From leadership to management: mobilizing knowledge for innovation in strategic alliances

Yves L. Doz, Andrea Cuomo, and Julie Wrazel

Introduction

Barriers to co-innovation between firms often arise from the contradiction between the uncertainty, ambiguity, and unforeseeable sharing of risks and benefits that characterize co-innovation and the need to specify the exchange between the parties, and be able to write a complete contract. Opportunities for significant value creation that would require co-innovation may thus go unexploited.

This chapter addresses these issues from a Schumpeterian innovation perspective, by showing how two firms – Hewlett Packard (HP) and STMicroelectronics (ST) – developed an innovative approach toward convergent interests and relational contracting in the development and manufacture of inkjet printer cartridges.

In addition to the central problem of uncertainty and ambiguity, these two firms also faced the problem of bringing together new knowledge from a plurality of locations, and therefore, across geographic, organizational, and contextual boundaries.

In particular, their innovation thus had to overcome at least four different barriers to knowledge integration:

- *Knowledge diversity*: knowledge from multiple domains has to be assembled, melded, and integrated. Different domains (such as technical disciplines, industries, or professional practice areas) have different rules and heuristics for creating, expressing, and sharing knowledge, and these are not always easy to reconcile.

The authors are grateful to Dr Mark Hunter, Senior Research Fellow at INSEAD, for his help in gathering, organizing, and interpreting data from HP and ST, and to Peter Smith Ring and Jose Santos for their helpful comments.

- *Knowledge dispersion*: as a corollary to diversity, but also just because knowledge is increasingly dispersed around the world in most fields, geographical distance, and the differences in cultural, institutional, and semantic contexts across locations have to be overcome.

- *Knowledge complexity*: contextual differences across domains and between locations matter all the more that the requisite knowledge may not be fully explicit and articulated – hence, sharing via the mere communication of information is at best partial, and probably misguiding and ineffective. Tacit, collectively held knowledge, and knowledge the interpretation of which is context-dependent are particular challenges.

- *Knowledge ownership*: lastly, knowledge useful to an innovation is likely to be the property of various companies, individuals, government centers, or other organizations who have to each contribute some knowledge and work on its melding and further development in close collaboration. Sometimes there are also agreements with third parties that may enable or kill collaborations.

Yet, innovating by tapping into the world for dispersed and differentiated sources of knowledge can be particularly effective, enable innovation and new industry creation that would not otherwise take place,[1] and allow global innovators to enjoy an advantage more localized competitors find hard to match. This has been called the "metanational"[2] advantage, in comparison to more traditional approaches to the internationalization of firms and innovations (Doz Y., Santos J. and Williamson P. 2001).

---

[1] One of the striking findings in research on the creation of the flat screen-display industry, for instance, has been the observation that all of the successful early entries in that emerging industry were alliances and joint ventures involving firms from at least two continents: no single region of the world had all the required technologies and skills available (Murtha, Lenway, and Hart, 2001).

[2] The prefix "meta," borrowed from the Greek, signals an entity that goes beyond, but not above, or across the borders of nations. We chose this term to convey the sense that metanational companies no longer use nation states as definition of territories and national subsidiaries as their organizational building blocks, but locally rooted sites and global links between those sites. Metanational companies do not draw their competitive advantage from their home country, nor even from a set of national subsidiaries. Metanationals view the world as a global canvas dotted with pockets of technology, market intelligence, and capabilities. They see untapped potential in these pockets of specialist knowledge scattered around the world. By sensing and mobilizing this scattered knowledge, they are able to innovate more effectively than their rivals.

In fact, such dispersion and differentiation, as well as the complexity of emergent knowledge, are conditions for time knowledge creation, and innovation. While specific knowledge creation is a situated localized act, innovation results from the melding of knowledge from different origins.

In this chapter we combine the real life experience of two of the co-authors (Andrea Cuomo and Julie Wrazel) and its conceptualization as a process of self-reflection on one's own experience (Schön, 1983) with that of a third co-author who brings an academic researcher's perspective to the observation of their collaboration.

We first explore and analyze the evolution of a co-innovation which faced all four challenges of knowledge diversity, knowledge dispersion, knowledge complexity, and knowledge ownership. Although the innovation was ostensibly technological, combining microelectronics and micro fluidics knowledge to design and build inkjet printers' cartridges between HP and ST, the most innovative aspect of their co-innovation was how the collaboration process itself was designed and managed over time. Our purpose, though, is not to just document and report a "case study," but rather to use the analysis of this innovation as a springboard to develop a conceptual framework for leading innovation, and show how building a true alliance relationship – rather than the more traditional narrowly contractual customer-supplier relationship – allowed a more effective and flexible collaboration to flourish. The technical innovation is but an artifact for the organizational innovation it triggered.

Our approach is both Schumpeterian and relational: relational contracting and an evolving relationship between HP and ST allowed both companies to fully benefit from the growth and value creation potential offered by the inkjet printer innovation.

Thus the research questions addressed in this paper are two-fold:

(1) How did the two companies transform what could have been but "the" usual quasi arm's-length outsourcing customer – supplier relationship into a true alliance?
(2) Why did this alliance design and structure, as well as its leadership over time, prove better than typical outsourcing contracts? (and how much and in which ways did they really differ from them?)

Of course, some of the specific circumstances that allowed the collaboration between HP and ST to flourish were of a fortuitous and idiosyncratic nature. However, we hope more generalizable implications can be drawn from the specific example on the governance and management of collaborative innovation between firms.

In keeping with a spirit of inductive theory development from managerial testimonies, we have refrained from writing an academic and theoretical paper, and privileged the analysis of the actual experience of HP and ST, and of their managers, in the collaboration process.

10.1   The collaboration and its interpretation

10.1.1   Initial Conditions

In the summer of 1992, when HP an ST started their negotiations (ST had been approached by HP as a potential supplier for the cartridge print head circuits HP needed to control ink firing chambers for higher definition thermal inkjet printers) HP was facing a capacity problem. With sales of HP inkjet printers growing rapidly, the need for cartridges to serve the expanding installed base would soon outstrip HP's own limited integrated circuit production capacity. Unavailability of replacement cartridges would have disastrous consequences on HP's printer business and overall credibility. Yet, the very capital-intensive nature of integrated circuit manufacturing and HP's already large investments in printers' manufacturing capacity made it reluctant to further invest in integrated circuits, in particular as this would have to be specific dedicated capacity. Further, HP's inkjet business unit lacked the technologies needed for integrating logic functions into the print head controllers (with higher resolution needs for printing, "intelligence" had to be embedded into the cartridges) and welcomed complementary expertise in these areas. Although HP was concerned with the risks of technology leakage, the need for significant capital investments and the awareness that complementary semiconductor technologies would be useful led them to seek external suppliers.

Coincidently, ST had an older, largely amortized plant in the United States it was about to close, but could be refurbished at reasonable cost to produce for HP. ST, then a $1.6 billion company, the result of the 1987 merger of two floundering state-owned semiconductor companies, Thomson Semiconducteurs in France and SGS MicroElectronicca in Italy, was keen to become a strategic supplier to one of the world's leading electronics companies.

Both ST and HP were trying to build and sustain leadership with limited resources, both financial and technological, and both were facing opportunities that would stretch these resources considerably. HP lacked both the willingness to devote their funds to a non core activity and the competencies to invest in "next-generation" cartridge circuits, in particular as logic functions were to be incorporated into the print heads. ST was concerned with using partly depreciated capacity, finding new applications for its bundle of technologies (combining mixed signal and power/control functions), and escaping the accelerating race to higher and higher level of integration and smaller and smaller scale, imposed by the leaders of the semiconductor industry.

In more abstract terms, the two companies were seeking to

(1) overcome resource limitations through shared risks and investments, and the combination of complementary skills;
(2) tap into global learning opportunities by combining widely dispersed expertise, from Italy, Ireland, various North American locations, and Singapore, from both firms;
(3) foster a joint entrepreneurial process that would allow them to exploit the opportunity and, for ST at least, leverage the learning from that exploitation toward new applications.

Although HP had started with an outsourcing supplier contract mindset, it quickly became clear that a true alliance was needed. The point was brought home in part by the refusal of most of the twenty or so potential suppliers they had approached to entertain serious discussions with them, and in part through the realization, on the part of HP's team, that a potential supplier would consider the opportunity differently from HP.

In any alliance, both HP and its partner would be putting substantial dependencies at risk. Cartridges were a critical component for HP, and one which generated a significant profit stream. Proprietary HP technology was also embedded in cartridge print heads. This would lead initially to a logistically awkward split of tasks, with ST in charge of wafer production and a few initial processing steps – or "moves" – as the process steps are known in the semiconductor industry, and semi-processed wafers being shipped to HP, in Oregon, for further processing. HP also had to convince itself that ST not only would be able to deliver current technology at a high enough yield, and low enough cost, but also would be able to develop for HP next-generation technologies, and would not divert resources, both manufacturing capacity and engineering talent, to more profitable opportunities at each upturn of a highly cyclical industry.[3]

---

[3] In essence, HP was committing to steadily growing volumes, given that cartridges are consumables for an installed base of printers, but at prices much lower than usual in the semiconductor industry, particularly in periods of high demand.

ST would be required to refurbish an existing plant and dedicate it to the HP contract. However, it was less confident than HP that inkjet printers would grow quite as successfully as HP planned.

Both companies were thus making high relation-specific asset commitments and incurring significant risks. The scope of the initial collaboration was defined by a small group, essentially Andrea Cuomo and Giuseppe Mariani on ST's side and Julie Wrazel and Ken Rick on HP's.

Although at the time, their convergence on a collaboration structure was an iterative and intuitive process, the outcome can be interpreted in more conceptual terms.

10.2   The alliance design

First, they came to separate clearly joint value creation from each partner's own value capture concerns. Trying to create through contractual arrangements what amounted (as they later came to call it) a "virtual joint venture," they set up a whole series of specialized task teams (the interlocking pieces of a puzzle in Figure 10.1). In their operations these teams favored joint problem-solving and developed a strong emotional commitment to avoiding escalation. The priority of these teams was value creation. Separate, but complementary, were the business, strategy, and financial teams, essentially dealing with the economic aspects of the alliance, and more concerned with the value-capture priorities of each partner.

Second, the partners were able to separate their own internal performance incentives from the joint work between them. Although transfer prices between ST and HP were set on a "cost plus" basis, ST used as a basis a "worldwide standard," setting an average cost for ST for cost factors its management did not control, and allowing plant managers to be incentivized to lower the costs they could control, without the resulting savings being automatically shared with the partner.[4] Although the alliance was, in principle, an "open-book" one, even in the absence of visibility of its costs to HP, ST had no incentive to over price its contribution or to be lax in its cost-saving efforts.

Third, the two companies created an expectation of future collaboration relatively early in their discussions. Although HP was initially reluctant to share some of its technological expertise with ST, and hence retained most

---

[4] The contracts between the partners, though, made provisions for price decrease as a function of volume and experience, but these were contractually set irrespective of the actual cost decrease achieved by ST. Obviously, though, the attractiveness of the HP contract to ST, given that dedicated capacity was used, hinged on the actual versus contractual rates of cost reduction over time and as a function of volume.

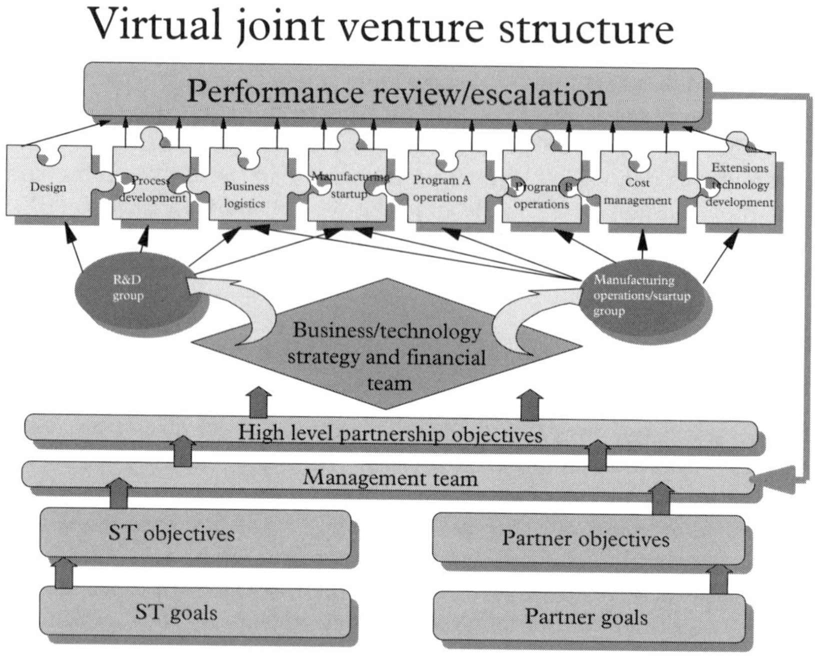

Virtual joint venture structure

Figure 10.1   Virtual joint venture structure

of the circuit-processing steps in house, the complexity of splitting the production process between locations – in particular the ones in Italy and HP's offshore manufacturing sites in Singapore and Ireland were involved – led HP to rely increasingly on ST for more and more of the process steps. Further, the partners engaged in coordinated R&D on a next-generation product concept relatively early in the relationship, creating the expectation that they would work together over the longer term. Finally, the size of the joint business grew rapidly, given the success of HP's inkjet printers in the consumer and small office markets.

Fourth, although they separated internal performance measurement from alliance measurement, the fact that ST needed to invest additional dedicated capital, and achieve a target return on it, led ST to share their actual cost data with HP, and get HP's help in managing cost reduction jointly. This avoided conflicts over price, essentially by making ST's actual performance visible to HP. Over time the relationship developed into a true "open book" one. Mutual dependency and high co-specialization created a strong pressure for partners to help each other and, hence, to be open about problems and performance.

The pace of the project allowed time for learning, for moving from early commitments which were still far from technical and operational clarity, and where the precise agreement between the partners could not be pinned down, because too little was known – and shared – between them.

HP chose ST, as a supplier, in early 1993, following initial contacts in August 1992. Through 1993 various task teams were put in place, starting with manufacturing, between HP's Corvallis plant, and ST's Carrolton and Rancho Bernardo plants. In parallel, known to ST, HP had also partnered with Sanyo in Japan as a second source. HP saw this clearly as a way to mitigate risks and to keep ST exposed to changes in the supply balance, should ST not deliver up to HP's expectations. Production started in early 1994. HP agreed that, if quality and productivity levels were satisfactory, capacity utilization allocation would remain constant between HP (20%), Sanyo (30%), and ST (50%). Over time, as more capacity would be required, HP's share would decline, and the Corvallis operation would be less and less of a benchmark, given technological obsolescence and smaller scale production than at either of HP's partners.

In sum, from the beginning, the two partners created conditions favoring cooperation: complementarity and co-specialization, sharing of significant risks, both calling for mutual dependence, and the basis of a collaboration structure that would avoid the usual haggling on transfer prices so common in customer – supplier partnerships.

The fact that both partners negotiated from a position of relative weakness with a large upside potential to the collaboration versus alternatives was also critical in establishing early cooperation. In the Fall of 1992, HP was negotiating against a time bomb. In order not to run out of capacity it had to commit fast to a supplier partnership, or to an internal investment. It was known to Julie Wrazel and Ken Rick, who were middle managers in a large division, that committing HP to invest on its own would be difficult, given the mandate to look for an external supplier and the significant amounts needed to build a new plant. Going far down the road in negotiations with any supplier, to have them then fail was a huge risk, of which the reluctance of most potential suppliers to even entertain HP's proposal was a constant reminder. ST had an old fab it was about to close, which was available for this; hence its upside potential was considerable.[5] Extra investments were nevertheless large.

---

[5] An additional benefit was that print head processing used gold, which was a dangerous pollutant to usual integrated circuit production. A dedicated plant avoided this pollution risk. Had capacity not been dedicated, an "open book" policy might have been more difficult to adopt, undermining one of the key enablers of the relationship.

Any commitment which offered a decent return on the extra investment needed to refurbish the plant was superior to alternatives. Further, as ST's engineers and manufacturing process experts started to understand what HP needed, the learning potential from the alliance became clearer. ST hoped to pick up insights on microfluidics and thermodynamics that could potentially be relevant to other applications.[6]

ST's upside potential, and HP's growing need to find a successful collaborative outcome as time went by, as well as, in particular, Ken Rick's cleverness in overcoming some of HP's reluctance to share its technology and to become dependent on third parties for a key element of its business system, provided the impetus for the development of a more collaborative approach than usual.

In retrospect, both ST and HP teams saw the process they engaged in over the years, between 1993 and 1998 as composed of several distinct phases, the first of which they labeled "architecting."

10.3   Evolution

The small team (essentially four people) who started negotiations in the summer of 1992 developed the architecture of a dual innovation: the technological innovation was the concept of a "next-generation" print head, using MOS process technologies, and other competencies that ST could bring to HP to develop and manufacture affordable, high-reliability cartridges in very high volume. Perhaps more important, in retrospect, and certainly of greater interest from an innovation management standpoint, was the architecting of the virtual joint venture as a governance form for collaboration that would allow the partners to combine complex knowledge from multiple locations and overcome the various barriers to knowledge melding we outlined in the introduction: domain, distance, context-dependency, and inter-organizational differences.

---

[6] In retrospect, ST would ascribe another learning, of a higher order, to the HP experience. In the early 1990s, ST was starting to formulate a "partnership model" of business, where, by focusing on needs and various customer applications, that required the hard–to–achieve combination of multiple technologies and functions on a single chip that was ST's emerging core competence, but that did not require the most advanced and most densely integrated chips, it could escape the relentless race to miniaturization imposed by Intel in processors, or Korean companies in memories, and define for itself a new market space between standard and fully custom chips. The HP relationship contributed in at least two ways to this learning: (1) It helped ST articulate the model, and (2) it provided Andrea Cuomo, and others in the management team of ST an opportunity to reflect on the enabling conditions and design and process features of a successful collaboration.

The focus of the architecting phase is matching needs and cap-abilities. This is both a cognitive and a political exercise, when done across partners. In the early steps of architecting, in the Fall of 1992, HP was willing to depart from a simple supplier – customer model to acknowledge that success, given that the nature of the products, and its markets and technologies required a more creative approach toward partnering. ST, perhaps, because the decision was simpler from its standpoint (against the alternative of closing down an old plant), fol-lowed suit easily.

A combination of familiarity with the underlying technologies – and what they enabled, or not – of business savvy and of creative flexibility in designing and structuring agreements allowed a collaborative and innovative approach to alliance governance.

As they got into the actual collaboration process, a joint identity started to develop around the core team. Rather than be barriers, cultural dif-ferences became the basis for shared "insider" jokes between ST and HP, for instance, on the agenda and flow of meetings. As more people got involved, that shared identity got stronger, in particular because new people had to appreciate the relationship and because sometimes mem-bers of the joint teams would have to confront the doubts, or hostility, of their own organization. For instance, HP Singapore tended to view ST more conventionally as a supplier and found it difficult, from a distant cultural context, to appreciate the closer collaboration that had developed in the United States and between the United States and Italy.

The two organizations developed a "culture of guilt," that is, of self-criticism rather than mutual blame, in particular because at the outset HP's reluctance to fully share its technology made it difficult for ST to meet HP's expectations.[7]

This was very helpful in the sense that when things went wrong, they each would first blame themselves rather than harp on the partner.

Over time, as the level of confidence and trust increased between the two companies, and as the cost of having ST to provide the wafer-based circuits and HP to provide the actual print head became increasingly clear, more of the manufacturing steps – including the print head – moved to ST.

R&D proceeded successfully, but with less of a business impact than initially hoped for reasons beyond the control of the virtual joint venture.

---

[7] It also became apparent that a lot of HP's know-how was tacit and collective, residing in Corvallis where HP's pilot plant was located. Transferring such know-how to ST required either its explicitation and codification in Corvallis or an extent of collaboration and co-practice HP was not willing to engage into. The process of collaboration was thus more arduous and difficult to start than anticipated.

In essence, the new cartridge concept, with a "semipermanent" head and replaceable ink tanks, might have undermined HP's cartridge business, and HP took the decision to confine its use only to high-end printers. Very different approaches to R&D, with HP being more thorough and systemic and ST trying more creative flair with smaller teams, also made collaboration in R&D challenging since executives at HP also regarded ST as a manufacturing subcontractor – or "toll manufacturer"- and remained reluctant to fully collaborate with ST in R&D.

On the manufacturing side the relationship continued to deepen and extend, leading HP to close some of its own capacity to rely on ST, and ST dedicating an additional plant to the HP relationship.

Reflecting back on the experience, we can identify a second phase, following alliance innovation architecting. The focus of the first phase was identifying relevant resources, mobilizing them, and bringing them together in a way that would allow the innovative potential of their combination to be released. In the second phase, the focus shifts to organizing the effort and providing for an ongoing operational collaboration.

This phase essentially corresponds to the implementation of the initial architecture: products, processes, locations, interfaces, the ramping up of production, yield improvement, and scale-up. In our experience, this phase started in earnest after the negotiation – or even in parallel to negotiations – and continued through 1996–1997, when the production system was expanded to a network of plants, worldwide. This "organizing" phase is characterized by its emphasis on joint value creation, effectiveness of the collaboration process, growth (involving more and more people in and around the relationship), more explicit leadership, and gaining full legitimacy within each partner's organization, with time to market providing urgency, rhythm, and flow to the joint effort.

The collaboration process then moves to a third phase, in which the effort – particularly on the manufacturing side – moves into a more organizationally embedded and operationally routinized activity. The emphasis shifts progressively toward efficiency, day-to day execution, and the fit with ongoing control systems and operating procedures.

Figure 10.2 provides a graphic summary of the evolution across the three phases. In fact, as ST applied the cooperation heuristics it pioneered in the collaboration with HP (and even earlier with disk-drive manufacturers such as Seagate or Western Digital), the specifics of each phase became more clearly defined. What also emerged from ST's experience with a widening portfolio of alliance relationships for co-innovating with customers was the key role played – often repeatedly – by a small core group of individuals in shepherding the process through its various phases. We called them "impresarii."

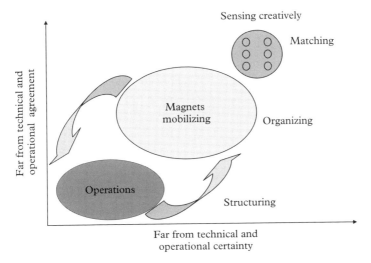

Figure 10.2   The innovation process

Beyond the priority and activity mix differences from phase to phase noted earlier, we observed that the required leadership skills and critical leadership priorities differ significantly across phases.

Phase 1 is built on the creative matching between what is likely to be – or become soon enough – technologically feasible, based on a sense of how much of a stretch – or of a risk – can be imposed, and what would create real customer value. Customer knowledge of a visionary nature – that is, not just what they do, but what they may find beneficial should it become feasible – is required here. In some cases the challenge was obvious. In one of the early partnerships a disk-drive manufacturer threw a gauntlet to ST: "If you can do this on a single chip, we will buy it from you." This gave birth to the first system on a chip. Sometimes the need is obvious, in particular when the nature of the need shifts from simple products to complex systemic innovations. A good impresario will play that visionary integration role.[8] Figure 10.3 summarizes the impresario skills through the various phases.

A way to consider this phase is to see it as an act of collective creative entrepreneurship, of co-creation of a product and of organizational concept between a customer and a supplier (Prahalad and Ramaswamy 2004). This requires a quality of openness and an adaptive process that

---

[8] Many successful serial entrepreneurs play such a visionary integration role repeatedly in areas they know well. Alex Zaffaroni, with his stream of pharmaceutical industry-related ventures is an outstanding example of a phase 1 impresario. Steve Jobs in the IT industry can be considered in a similar way. In the HP–ST venture this role was not so salient since HP had already specified the product.

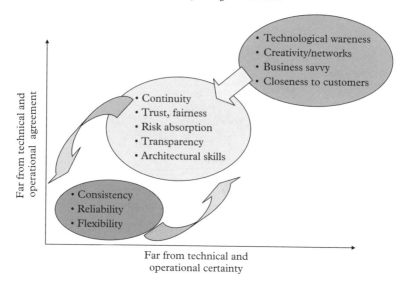

Figure 10.3   Summary of impresario skills through various phases

the small teams, on both sides, that initiated the collaboration between HP and ST displayed. The "collective entrepreneur" exercises a form of collective sense-making about the opportunity (something seen even more clearly in other ST partnership, with customers who had a less clear idea of their needs than HP) and the "framing" flexibility to understand and share the strategic nature of the joint opportunity, and make such framing evolve through several iterations.(Weick 1995; Szulanski Doz, and Ovestky, 2004).

Phase 2 is a real test of the will to lead, in the face of uncertainty ambiguity, and perhaps, skepticism or hostility from one's own organization. The ability to act personally as a risk absorber, to maintain trust and fairness at difficult phases in building the relationships, are key. For instance, when tensions crept up in the relationship, the original creators of the relationship repeatedly came to the rescue and kept signaling the need for collaborative behavior.

Impresario skills in phase 3 are mostly around providing the level of reliability and consistency that will allow a routine organization – such as ST's fabs – to embrace and serve the alliance and respond to the partner.

10.4   Conclusion

We started the chapter with an assertion and an interrogation. The assertion this chapter tested – on the instance of the HP–ST collaboration in inkjet print head – is that a Schumpeterian and evolutionary perspective on collaboration, when implicitly adopted by managers, may allow value-creating opportunities to be jointly pursued when such opportunities might be squashed by a more traditional arm's length contracting approach rooted – implicitly or explicitly – in transaction cost economics. Of course, the value of an example is limited, but we hope the "story" summarized sheds some light on the debate and shows in this one case, the value of a relational and evolutionary approach.

We also started with a question on how collaboration can overcome knowledge diversity, dispersion, complexity, and ownership issues, when the combination of complex knowledge is required for a collaborative effort to succeed. Here too, the answer lies in the quality of the collaboration process set up in the United States, and between the United States and Italy, among the small teams that negotiated, started, and scaled up the project. The closeness, the cultures of guilt rather than blame, the fact that in the negotiations the upside had been considerable, and the United States. downside – of failing to negotiate successfully – daunting, as well as the embracing and expansive nature of the Italian culture, all contributed to the informality of collaboration that, in retrospect, we see underpinned success.

Difficult problems of manufacturing process (for instance "adhesion" difficulties between the wafer-processing "moves" performed at ST and those performed at HP could be solved only because the high quality of collaboration between the partners allowed IPR and other concerns for proprietary knowledge that might have made solving these problems impossible were solved thanks to the closeness and openness of collaboration.

References

Doz Y., S.J. and Williamson, P. 2001. *From Global to Metanational: How Companies Win in the Knowledge Economy.* Boston, MA: Harvard Business School Pres.

Murtha, T.P., Lenway, S.A. and Hart, J. 2001. *Managing New Industry Creation: Global Knowledge Formation and Entrepreneurship in High Technology,* Stanford, CA.: Stanford University Press

Prahalad, C.K. and Ramaswamy, V. 2004. *The Future of Competition: Co-creating Unique Value with Customers.* Boston, MA: Harvard Business School Press

Schön, D.A. 1983.*The Reflective Practitioner: How Professionals Think in Action.* Newyork: Basic Books

Szulanski, G., Doz, Y. and Ovestky, Y. 2004. "Incumbents' Framing: How Three Established Firms Responded to the Emergence of the Internet", *Advances in Strategic Management* 21: 77–106

Weick, K.E. 1995. *Sensemaking in Organizations.* Thousands Oaks, CA: Sage

Comments to Chapters 9 and 10:

Innovation and organizations: comments and perspectives*

Sidney G. Winter

We have here two valuable chapters on innovation,[1] one deriving general insights from careful study of a significant empirical case and the other directing attention to some important gaps in the broad theoretical structure that many of us employ in our thinking about innovation processes in organizations. Given that the two chapters differ so substantially in method and viewpoint, it is not surprising that they do not add up to a rich dialogue. While it is not exactly the case that they pass each other like ships in the night, it is certainly the case that they do not *collide* like ships in the night, and neither is it true that they are pursuing parallel courses. In this note, I comment on the two contributions individually, and then discuss briefly the area of overlap between the two Chapters.

Doz, Cuomo and Wrazel: managing an alliance for innovation

In commenting on this chapter, it is important first of all to commend it as representing an uncommon and clearly valuable art form. Professor Doz here offers a case study that is at the same time a sort of memoir for two key players in the case, who appear as his co-authors. In general, it is not unusual or inappropriate to wonder whether interpretations of a case offered by an academic researcher would pass muster with the actual participants; in this example we have substantial reasons to lay such

---

*Extension of remarks at Schumpeter 2004 Conference, Milan, Italy.

[1] Yves Doz, Andrea Cuomo and Julie Wrazel, "From Leadership to Management: Mobilizing Knowledge for Innovation in Strategic Alliances," and Daniel Levinthal, "Bringing Selection Back into Our Evolutionary Theories of Organization," Chapters 9 and 10, this volume.

concerns to rest. Whether this trio of co-authors actually "have it right" is a question that inevitably remains open to some extent, but at least we have good reason to think that this account has objective grounding in the reality "out there" – the one that lies beyond the powerful reality of academic research agendas, paradigms, and reward systems.

The story offered by Doz, Cuomo and Wrazel (DCW) is a story of innovative success. It details how Hewlett Packard and STMicroelectronics (HP and ST) collaborated in the 1990s to produce an innovative type of inkjet printer cartridge. They seized a very attractive joint opportunity framed by the conjunction of HP's need for certain technological competencies and manufacturing capacity and ST's ability to meet those specific requirements. The details put forward in this story include a wealth of interesting guidance on how to manage a high-stakes, high-tech alliance. In my view, this guidance has great credibility because it is nicely positioned at the margins of the familiar (too familiar implies recognition without respect, too unfamiliar implies unthinking rejection). However, from a broader perspective (such as that represented by Levinthal's paper), this success story triggers the generic success story response: (1) Granting that this is a success, is it a success of such an unusual kind that it challenges, all by itself, existing understanding of what is possible? (2) Taking a more statistical view, can we judge whether observing one success like this would be unexpected, given the population of tries and recognizing the random elements in outcomes – what is the count of failures that, in other respects, closely resembles this success?[2]

Much of the theoretical interest in the chapter lies in the fact that the answer to the first question is not entirely apparent. Some, I think, would react by saying that this account presents an extreme case, and it is therefore either an overly sunny view of the reality or else it implies that the possibilities for genuine cooperation between independent partners in an alliance are substantially broader than we had imagined. In my view, the case is not *that* extreme. This alliance had a lot of things going for it – above all, big incentives on both sides to achieve a success. The authors make clear that it would have taken a lot of discouragement to tempt the parties (especially HP) to write the whole thing off, and that level of discouragement never arose. That fact, plus the fact that the attempted innovation was clearly incremental at the technical level and focused on a clear market objective, tipped the scales toward ultimate success. This conclusion does not diminish the interest in the question

---

[2] In calling these questions "generic" I do not mean to imply that asking them is common – but it should be.

of *how* success was achieved, or (certainly) suggest that the successful outcome was in any sense foreordained. The partners might easily have botched the opportunity, but instead they rose to the occasion and made a success of it. From my perspective at least, this observed outcome does not seem so extreme as to require a major rearrangement of the theoretical furniture.

On the second question, there is at least enough systematic evidence to provide some perspective. It appears that the overall failure rate for technical alliances is high and that the task of managing them effectively is one that firms find quite challenging (Kale, Dyer and Singh 2001) Indeed, the premise of the preceding discussion is that the HP–ST case is clearly somewhere off in the right tail of the success distribution, and the issues are "how far?" (question 1) and "why – luck or leadership?" (question 2). The details offered by DCW make a credible case that distinctive aspects of the management of the alliance were a major factor.[3] Most notably, the authors describe a multilayered system of managerial steps taken in the attempt to anticipate, constrain, and deal with the threat to cooperation posed by the underlying divergence of interests between the parties, as well as the more mundane lower-level conflicts that frequently arise as teams address challenging tasks. The authors also mention some other possible success factors that seem to be closer to the borderline between policy and luck. For example, ST's contribution to the alliance in the early stages included a dedicated plant with no profitable alternative use – a situation which helped to make cost issues relatively transparent and thus facilitated a relationship in which ST shared its cost data with HP and the parties worked together to reduce ST's costs.

The authors make passing reference to transaction cost economics at the end of the chapter, and that reference offers a contrast between the successful collaborative approach of ST and HP and "a more traditional arm's length contracting approach rooted – implicitly or explicitly – in transaction cost economics" (p. 293). Yet the story of the alliance can easily be read as an informative exercise in applied transaction cost economics, one that could plausibly find a place on an advanced reading list of the subject. After all, the basic claim of TCE is that the parties to a transaction are capable of anticipating the principal hazards to which it is subject and, when they choose to go forward, devise appropriate governance arrangements to control those hazards and keep the

---

[3] As for the luck part, DCW do acknowledge that "Of course, some of the specific circumstances that allowed the collaboration between HP and ST to flourish were of a fortuitous and idiosyncratic nature."

transaction on track.[4] Transactions differ in their attributes, and "governance structures are aligned to the needs of transactions in a discriminating way" (Williamson 1985). The story of the HP–ST alliance seems to be a success story of exactly that kind, in which the hazards foreseen are very much the ones that a TCE theorist would identify, and the governance response to them is notably sophisticated, pre-emptive, and well-aligned. Certainly the DCW account makes clear that these companies and their key managers are not naïve enthusiasts of mutual trust; they are, rather, sophisticated *engineers* of mutual trust, both cognizant and respectful of the parties' need to protect themselves from each other. To the extent that there is some tension between these facts and typical TCE perspectives, it relates to the question of whether the governance arrangements adopted turn out to work "unbelievably well" (in a literal sense) – as suggested in my question 1. One might reasonably speculate that Professor Williamson would be among those who take the affirmative view on that. Be that as it may, the question is certainly an interesting one – and novel in proportion as the governance measures are themselves novel – that is appropriately addressed in the TCE framework. The case for putting the DCW paper on the TCE reading list is strong.[5]

Levinthal: the role of selection in innovation

A vigorous and persistent debate in the theory of biological evolution concerns the question of the "unit of selection" (see, e.g., Lloyd 2000). What precisely are the entities that are "selected" as evolution proceeds? Genes? Entire organisms? Populations? Species? Ecologies? Other? All of the above? Clearly, survival tests of different kinds are encountered by entities at different levels of organization, so the answer "all of the above" has a lot of face appeal. Such an (indecisive?) answer, however, only opens the door to the obviously challenging question of how the levels relate to each other, and whether the most important evolutionary "action" occurs at some particular level. Levinthal's discussion of selection in economic evolution confronts many of the same abstract issues addressed in the units of selection controversy. In both contexts, a particularly interesting and challenging problem is understanding

---

[4] TCE is certainly *not* a brief for arm's length contracting, or against "relational" contracting, as the authors seem to imply. To the extent that it is ever right to caricature, the right caricature of TCE is that it is a brief for vertical integration – which does not seem like a realistic possibility in this case.

[5] While I am revising that reading list, I suggest that the DCW paper could well be placed near (Mayer and Argyres 2004; Mayer 2006).

how relatively blunt environmental feedback (survival or not, profit or not) can provide discriminating guidance for the evolution of a complex entity (mammals, multibusiness corporations). How can there be enough information in that feedback signal to guide anything so complex?

Although "Feedback, not Foresight!" could serve as a shared battle cry for the entire evolutionary camp, there are big differences between the evolutionary mechanisms of biological systems and those of human economic activity. Skeptics and critics of the evolutionary viewpoint in economics often point to the sharp contrast between the manifestly goal-oriented character of human activity and the emphatic rejection of teleological elements in the theory of biological evolution. They say that it is not credible to attribute observed economic progress to the mere accumulation of random mutations. On this narrow point the critics are quite right (though they would be hard pressed to find anybody who actually supports the incredible claim). The mechanisms are different – and it's a good thing too, for if they were not we should expect economic change to occur on roughly the same timescale as biological evolution.

I am increasingly persuaded that it is crucial to appreciate, and emphasize, the distinction between (locally) instrumental behavior and behavior guided by (accurate, long-run) foresight.[6] As individuals and in organizations, we often undertake elaborate investment programs to enhance skills and capabilities, to create new artifacts, and to design and build elaborate organizational structures for particular purposes. To do this, we typically exploit a high degree of local rationality and control – an ability to assess what would constitute progress and to take actual steps in the right direction, or to fail in that relatively promptly, reassess, and try again. The typical motive for this elaborate activity is to generate a complex and coherent behavioral alternative that can be a promising contender in some future competitive arena – shaped by forces coming from well beyond the local context – where wealth or other good things may be won.[7] Thus the skills acquired in school may ultimately attract a job offer, the early R&D results an extension of project funding, the marketing campaign a wave of new buyers, and so forth. Accurate foresight about how the anticipated competitive encounter will turn out would be extremely valuable in the alternative generation stage, especially if available conditionally for all alternatives that might be put forward. For a variety of reasons, such foresight is typically very hard to

---

[6] Expansive versions of rational choice modeling, featuring actor expectations that are fundamentally accurate if not single-valued, tend to blur this distinction.

[7] Such an alternative is indeed "intelligently designed," not a random program. But it is not designed with objectively foreseeable ultimate consequences in view.

come by: first because it is *competitive* arena that is ultimately addressed, second because of the diffuse etiology and unfamiliar character of the forces operating there, third because even an effective oracle couldn't answer questions that cannot be identified or posed, and then several others. Foresight is inevitably weak, and it is (above all) this highly imperfect anticipation of the environmental verdict that lends the quasi-random element to economic evolution, making its outcomes hard to predict and elevating "variety" to the status of a key concept. It is not a matter of random mutation at some "genetic" level.

To an extent, the combination of human intentionality with local rationality and control offer a "solution" to the problem of how, in economic evolution, complex things come to be proposed for environmental testing, a solution that is not available in the biological sphere. Levinthal, however, makes it clear that this solution is actually little more than a sketch; it still needs a lot of working out and some attendant qualification. Although his discussion does not explicitly emphasize intentionality and local rationality in the manner of the preceding paragraph, his analysis of selection is fully applicable to the sort of process described there. Note first that my account of alternative generation and *subsequent* environmental testing tumbled right into the very trap Levinthal identifies as common in formal learning models, the too-easy acceptance of the assumption that the testing is somehow deferred until after the learning is complete. In many cases, learning or design is taking place in a house built of straw, and the Big Bad Wolf of the external environment may blow through that door in no time. In such cases, Levinthal observes, viable patterns of innovative behavior must mature quickly to a stage where the external testing can likely be survived.

Even when the external threat is effectively forestalled for a substantial period, selection mechanisms of a different kind are operating on a nascent innovation. While the unknown objective future plays no causal role, subjective images and aspirations toward the future play a key shaping role in the evaluation mechanisms that define "a step in the right direction" and can sometimes deliver the ominous message "this isn't working out." On the other hand, those evaluation mechanisms can also be powerfully shaped by considerations that are *not* much grounded in images of the future – for example, by habitual ways of thinking about what is "sensible," or by institutionalized demands for short-run accountability that implicitly deny the relevance of the long-run payoff function. How these sorts of process work out is not well understood, and Levinthal is persuasive that we would be well advised to pay more attention to them.

A challenging aspect of such a line of inquiry is that the same key issues recur at a sequence of "meta levels" – in a manner analogous to, and in fact related to, the case of dynamic capabilities (Winter 2003). Thus, for example, the evaluation processes that characterize the internal selection environment for a new product design are themselves designed – and designed partly with an eye to how these higher-level processes will affect survivability in the face of external tests over an extended future. As in the case of dynamic capabilities, the trail going up the hierarchy of logically possible levels is one that is likely to disappear into a fog of observational ambiguities, long time horizons and dominance by chance events. At some point, there are certainly diminishing returns to trying to assess and understand patterns at the higher levels, but at this stage we have little idea of where those limits are.

Internal selection in the HP–ST alliance

The DCW case study offers some points of illustration and support, but also some contrasts, with Levinthal's discussion of selection. The case certainly illustrates the point that the internal selection environment matters, and that the internal environment is derivative of the external environment only in a rather weak sense. True, the initiation of the project was motivated by perceived threats and opportunities in the external environment – but beyond that starting point it seems that it was the internal hazards that posed the big threats and demanded managerial effort and ingenuity. The DCW account contains a couple of references to strictly technological/ operational challenges, and one to a revision of the image of the projects fit to the market environment (HP restricted the product to the high end of its line). But the challenging environment featured in the study is not the external environment, nor is it an internal attempt to anticipate or synthesize that external environment – it is rather the internal dynamics of the project itself, together with the organizational context established by the alliance partners. Analysis in the Levinthal style would suggest that an organization is not well served when its internal selection criteria are at best orthogonal to the external criteria. That appraisal seems consistent with the view taken by the managers of the project, who were more concerned with buffering it against the internal hazards of myopia and parochialism than with worries about the degree to which its purpose and progress were aligned with the anticipated external tests.

Of course, the fact that the case is about an alliance is a key principal factor that qualifies its direct relevance to Levinthal's theoretical analysis. The DCW paper is concerned with how the project managers were able to establish a "true alliance relationship" – that is, to buffer the promising effort at joint value creation from the stresses intrinsic to alliance itself. It is a foregone conclusion, at least in the case study, that those stresses tend to create a form of intermediate selection that is not really in the interest of the parties.

References

Kale, P., Dyer, J. and Singh, H. 2001. "Value Creation and Success in Strategic Alliances: Alliancing Skills and the Role of Alliance Structure and Systems," *European Management Journal* 19: 463–71

Lloyd, E. A. 2000. "Groups on Groups: Some Dynamics and Possible Resolution of the Units of Selection Debates in Evolutionary Biology," *Biology and Philosophy* 15: 389–401

Mayer, K. J. 2006 "Managing Positive and Negative Spillovers: Evidence from Information Technology," *Academy of Management Journal*, 49:69–84

Mayer, K. J. and Argyres N. 2004. "Learning to Contract: Evidence from the Personal Computer Industry." *Organization Science* 15: 394–410

Williamson, O. 1985. *The Economic Institutions of Capitalism.* New York, NY: Free Press

Winter, S. G. 2003. "Understanding Dynamic Capabilities," *Strategic Management Journal* 24: 991–95

Part 6

Innovation and entrepreneurship

11 Schumpeterian legacies for entrepreneurship and networks: the social dimensions of entrepreneurial action

Piera Morlacchi

Introduction

Schumpeter was a rich thinker and many aspects of his work have yet to be explored

(Swedberg, 1991)

Schumpeter was a scholar of awesome breadth and depth (Galambos, 1996). His legacy is a gold mine of ideas, insights and intuitions on a number of topics such as economic change and growth, entrepreneurship, innovation, capitalism and democracy. Some of his early ideas, insights and intuitions have yet to be fully explored.

Several scholars have built upon Schumpeter's ideas and tested them empirically. Specifically, the last twenty years has seen a revival of interest in his work from evolutionary economics[1]. The result of these efforts is an improvement in our understanding in a number of areas such as the dynamics and evolution of specific industries (Breschi et al., 2000); the dynamic relations among technology, growth and trade (Fagerberg, 2003); systems of innovation (Freeman, 1994); and evolutionary modelling (Nelson and Winter, 1982).

However, in some areas that are the core of Schumpeter's work our knowledge has not progressed evenly; one of these areas is entrepreneurship. Schumpeter made several major contributions to this area of knowledge and provided powerful insights on a number of issues, such as

I am indebted to Stefano Brusoni, Nick von Tunzelmann and Ed Steinmueller for their comments on this chapter.

[1]See Fagerberg (2003) for an appraisal of the recent contributions in evolutionary economics inspired by Schumpeter's work.

theorising about entrepreneurship being a crucial factor in economic development, highlighting the creative character of entrepreneurial action and arguing the complementary role of economics, sociology and history in analysing economic change in general, and entrepreneurship in particular. Moreover, along with Weber and few other intellectual giants, Schumpeter understood 'how particular agents shaped history and how social structures shaped the actions of those agents' (Galambos, 1996: 930); this is a very sophisticated and challenging way of viewing the phenomenon of social interaction that we call 'networks'. We could argue that were Schumpeter able to observe the current shift towards a decentralised entrepreneurial economy he would certainly be adding 'networks' to his set of ideas and tools. Many of the ideas and lines of inquiry put forward by scholars such as Schumpeter have either been forgotten (Dosi and Teece, 1996) or over time we have lost touch with the insights and intuitions in the original contributions of the founding fathers of our disciplines. 'Standing on the shoulders of giants' is a common approach in the social sciences. One consequence of the cumulative character of knowledge creation is that some ideas and concepts become institutionalised: they get taken for granted and black-boxed. The meanings attached to the concepts of the founding scholars are re-interpreted, and adapted by later scholars to create new knowledge. However, somewhere along this process the connection with the original meanings of these concepts is lost.

Furthermore, various disciplines within social sciences use different frameworks and tools to examine the same phenomena. For instance, economic exchange, communication and power are examined by economics, sociology, psychology and anthropology. Knowledge about these phenomena is codified in ideas and concepts, which take on particular meanings depending on the context of each discipline. The interaction between two characters from Lewis Carroll's (1971: 190) 'Through the Looking Glass – Alice's Adventures in Wonderland' can illustrate this point. In the story Humpty Dumpty says to Alice that 'When I use a word ... it means just what *I* choose it to mean – neither more nor less.' And Alice's answer to this remark is: 'The question is whether you *can* make words mean so many different things', Alice's point being where does the power to establish the meaning of a word, and of a concept, lie. For Humpty Dumpty the question is, who is the master of what gets said. The Humpty Dumpty perspective on words dominates much of our understanding and use of words, and also of the ideas, concepts and tools in different disciplines of social sciences. Each discipline has, at least, one master and every concept has many masters.

Occasionally, it can be useful to let our 'Alice side' dominate. Meaning that, on a fairly regular basis, we should question our use of different ideas, concepts and tools taking into account that different disciplines, and perspectives within them, use the same concepts, but may attach different meanings to them. Problematising ideas, concepts and tools that we might have begun to take for granted enables us to investigate whether new insights can be gained from different perspectives, and whether perhaps different perspectives can be integrated into a more powerful synthesis. Going back to the original work of the intellectual giants or founding fathers of our disciplines is a process, which, if practised cyclically, enables us to stay in touch with their legacies, and exploit their insights and intuitions. The new and enriched understanding that we can derive from questioning, problematising and returning to the original sources of ideas, concepts and tools can lead to a deeper knowledge of some of the phenomena relevant in our society, such as entrepreneurship and networks.

In imitation of Alice's challenging and somewhat impudent reply to Humpy Dumpty, this chapter examines the legacies of Schumpeter within our understanding of entrepreneurship and networks. The main argument in this chapter is that we have explored some of Schumpeter's ideas about entrepreneurship, but have rather forgotten his views on entrepreneurial action. Analysing Schumpeter's views on entrepreneurial action can improve our understanding of entrepreneurship and networks.

This chapter is structured as follows. In the next section, current understanding of the phenomena of entrepreneurship and networks is briefly discussed, starting with the basic questions of 'what is entrepreneurship?' and 'what are networks?' and re-examining the historical development of the academic discourse on these topics and their nexus. Next, Schumpeter's work on entrepreneurship (and networks) is examined with the objective of highlighting his view of 'entrepreneurial action'. The chapter concludes with some implications for the role of entrepreneurship and networks in society derived from a close examination of the work of Schumpeter on these topics and their overlap, and its implications for future research agendas.

11.1   What do we know about entrepreneurship and networks?

This section is not intended to be a review of the bodies of knowledge on entrepreneurship and networks. It aims to provide an overview of the main ideas about entrepreneurship and networks in the social sciences; to use Humpty Dumpty's words, to describe the main 'masters' that dominate what gets said, or to summarise our knowledge on the two topics. Selected empirical studies are used to exemplify the main points. The overview highlights that some clarification of both concepts and their associated areas of knowledge is required.

A starting point for our discussion of entrepreneurship is to ask the questions 'what is entrepreneurship?' and 'what do we know about entrepreneurship?'. Depending on which stream of literature we consider, we find different answers to these questions. For instance: (1) entrepreneurship is an economic function; or (2) entrepreneurship is the creation of a new business venture; or (3) entrepreneurship is a form of behaviour; or (4) entrepreneurship is a set of individual characteristics. This diversity of answers shows that not only is there no agreement about the concept of entrepreneurship, but there is also disagreement about the questions that should be posed in relation to this phenomenon. For instance, much scholarly effort has been expended on establishing why some people become entrepreneurs while others do not, and why some recognise entrepreneurial opportunities and act upon them while others do not. Until recently, the question of how people become entrepreneurs has received little attention (Sarasvathy, 2003).

This diversity of views on the phenomenon is in part due to the development of the field of entrepreneurship, to which many social science disciplines have contributed. After an initial ownership by psychology (McClelland, 1961) and economics (Casson, 1982), the field of entrepreneurship is now dominated by management and business studies: entrepreneurship education is delivered mainly by business schools, and the sets of concepts and tools adopted are primarily those of business. The business approach to entrepreneurship is rarely complemented with insights from disciplines such as sociology, history, political science or anthropology[2] Furthermore, entrepreneurship-related topics have been explored by different sub-disciplines within management and business studies, each of them focusing on some aspects of the phenomenon. For instance, social psychology investigates the characteristics of the entrepreneur to understand the determinants of entrepreneurial behaviour (e.g. McClelland, 1961); industrial economics and evolutionary economics focus on the consequences of entrepreneurial activities in terms of the creation or establishment of new firms, and industry evolution.

The focus of the research on entrepreneurship in management and business studies is on the process of discovery, evaluation and

---

[2] For a notable exception see *Entrepreneurship Theory and Practice*, special issue, Winter, 1991, and recent efforts in sociology (e.g. Thorton, 1999; Swedberg, 2004).

exploitation of opportunities and their effects on the creation of new firms (Shane and Venkataraman, 2000). The study of entrepreneurial outcomes – the firm or the venture – and some aspects of the entrepreneurial process have received more attention than others (Low and MacMillan, 1988). For instance, the entrepreneurial process can be analysed at the individual, organisational and inter-organisational levels. At the individual level, we have some knowledge about the characteristics of the entrepreneur. We have developed a partial understanding about the process of creating a new firm and how the transformation of an entrepreneurial idea into an organisation requires the acquisition or coordination of resources. At the inter-organisational or industry level we know that entrepreneurship leads to the creation of new firms and new industries and thus the creation of new jobs (Audretsch, 1995), but its connection with competitiveness, which is important in terms of policy, has not been clarified.

Moreover, different disciplines have brought to the analysis of the entrepreneurship phenomenon their own theoretical ideas and tools. For example, economics brought to entrepreneurship concepts such as market structure, and tools such as mathematics and statistics.[3] The new evolutionary economists went even further than the neo-classical economists in the sophistication of their mathematical tools and techniques, bringing in dynamic modelling. Different dynamic modelling efforts, derived from Schumpeter's theories of innovation, which were built around entrepreneurship, the core of creative destruction, have been proposed (Kirchoff, 1991).

Gartner (2001) summarised the state of the debate on entrepreneurship and the division of labour in entrepreneurship research in a curious and intriguing question 'Is there an elephant in entrepreneurship?' He suggested that different disciplines' attitudes to entrepreneurship are like the six blind men and the elephant. The blind men touch different parts of the elephant and on this basis offer different descriptions of its characteristics. Gartner (2001: 28) claimed the story "offers a syllogism for thinking about the problems of integrating different views of a large and complex phenomenon" such as entrepreneurship. His own opinion is that there is no elephant in entrepreneurship, that is "the various topics in the entrepreneurship field do not constitute a congruous whole" (2001: 34). Other scholars,

---

[3] Based on biographical material, we know that from the 1930s Schumpeter was concerned with mathematicising economics. From 1950s mathematics became the favourite tool of economists.

however, do believe in the elephant in entrepreneurship (Shane and Venkatraman, 2000).

When we ask the same basic questions about networks – 'what are networks?' and 'what do we know about networks?' we also encounter many masters and Humpty Dumpty perspectives in this body of knowledge. There is an interplay of action and structure in the network concept: networks are both structures and processes (Mitchell, 1969) because "Networks are constructed when individuals, whether organizations or humans, interact" (Salancik, 1995: 345). Ronald Burt argued that "network theory builds its explanations from patterns of relations" (1982: 106). Network analysis is based on a set of implicit assumptions that can be summarised in a rejection of the statement that human behaviour or social processes can only be explained in terms of the attributes of the actors, whether individual or collective. However, in practice, the process side of networks and the role of agency were quickly ignored in network studies and the focus narrowed to some types of structures connected with economic exchange and communication.

Network ideas, originally developed in anthropology in the 1960s, have become one of the most developed sets of theoretical ideas and tools of modern sociology. However, since the 1980s the idea of networks started to be used as a tool in a variety of disciplines outside sociology, such as management and business studies, psychology and economics. Different disciplines adopted to various degrees both the theories and/or the tools of network analysis. The power and flexibility of network analysis determined the diffusion of its concepts and tools, and generated insights in a number of domains (e.g. the labour market). Various studies have empirically tested the usefulness of different network strategies, in some cases extending the ideas of social or personal networks to inter-organisational networks (e.g. Uzzi, 1997; Greve and Salaff, 2003). Among others Gulati and Gargiulo (1999) illustrated that formal inter-organisational networks develop out of existing inter-personal or social relations.

The spread across different fields of inquiry has some positive aspects, but also some drawbacks. The positive aspects are that more people are applying networks to answer research questions, thus creating new knowledge and methodologically advancing the tools. The drawbacks are related to the progressive detachment of the network tools or social network analysis ('weak network analysis') from network theories ('strong network analysis'). Network analysis has become a tool, that is used without any basic understanding of the concepts and theories that generate and support the tool, thereby undermining the value of findings from studies based on unclear and contradictory implicit assumptions.

Some scholars have argued for the use of networks to study entrepreneurship, and from the beginning of the 1990s there was an increase in the number of studies adopting this approach. Networks are a very powerful set of theories and tools that can be used to study the phenomenon of entrepreneurship. Network analysis might allow the study of processes, outcomes and contexts of entrepreneurship to be integrated in the research design. However, currently, the structure of entrepreneur networks has received more attention in empirical studies than the dynamics and processes of networks.

A piece of exemplary work in the study of the entrepreneur's networks is Burt's (1992) study of structural holes. He studied the social structure of competition, emphasising the instrumental action of the individual actor or the 'strategic action of the entrepreneurs'. He proposed an 'imaginative' network theory of entrepreneurship: in his view, entrepreneurial action is about brokering information in disconnected parts of the network, that is, structural holes. Networks are the mechanisms used by entrepreneurs in the process of creating a new organisation, to acquire such resources as information and knowledge.

Overall, the network studies of entrepreneurship assume that entrepreneurial action is both constrained and enabled by networks. Where these networks and their structures come from is rarely investigated. Furthermore, agency and creativity are not included in the actions of entrepreneurs in their own networks.

This brief and schematic overview of the main ideas in the bodies of knowledge on entrepreneurship, networks and their crossroad presented in this section indicates that to benefit from combining entrepreneurship and networks in our analysis, we need to develop a better understanding of both concepts, and how they are linked. The objective here is to show how Schumpeter's work can help to generate this understanding. A key insight, which has been forgotten but could prove very fruitful for our analysis, is Schumpeter's view of entrepreneurial action. This is elaborated on in the next section.

11.2   Schumpeter's legacy for entrepreneurship and networks

In this section I discuss Schumpeter's ideas and insights on entrepreneurship, networks and their nexus. The reasoning proposed is circular and relational. First, Schumpeter's ideas about the entrepreneur and the entrepreneur's function in economic change are outlined to highlight his conception of entrepreneurial action. Next, Schumpeter's conception of entrepreneurial action is discussed in reference to the literature that addressed and investigated the topic after him. Finally, building on the

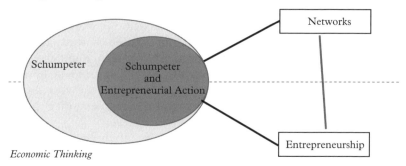

Sociological Thinking

Economic Thinking

Figure 11.1 A relational approach to Schumpeter's legacies on entrepreneurship and networks

observation that the idea of networks was implicit in Schumpeter's theorising about the 'entrepreneurial function' and the 'view of the economy as a whole' (Becker and Knudsen, 2002) the discussion in the last part of this section illustrates the work of scholars who have elaborated and theorised Schumpeter's insight by using networks to model the embeddedness of all economic actions, including the entrepreneurial one, in the social context.

Figure 11.1 depicts the structure of this section, illustrating its relational nature by showing how the three key concepts – entrepreneurship, entrepreneurial action and networks – are discussed through their connections. Schumpeter's thinking, and recent developments in economics and sociology, provide the context for the discussion.

11.2.1 *Schumpeter and entrepreneurship*

The ideas of entrepreneur and the entrepreneurial function in Schumpeter's work highlight the contextual nature of the embodiment of the entrepreneurial action. Schumpeter viewed economic change as an evolutionary process, characterised by variation, selection and retention. The entrepreneurial function is what generates variation in the economic realm. The principle of selection was based on the idea of 'creative destruction', whereas 'New combinations were retained by the social structure, the underlying mechanism being the interaction among social agents, and social agents are nested within social sectors' (Becker and Knudsen, 2002: 396).

To explain the dynamics of economic change Schumpeter introduced the role of entrepreneur to fulfil a function, that of combining new things or innovating, that is, introducing a new product or new quality in a product, a new method of production, a new market, a new organisation within an industry. The notions of entrepreneurship and innovation are clearly interrelated for Schumpeter, but recently his ideas on innovation have been receiving more attention at the expense of his work on entrepreneurship. The so-called 'Schumpeter hypotheses' on market structure, firm size and innovation have been investigated empirically to analyse the different roles of small and large firms in innovation. Schumpeter's central concerns about the role of the entrepreneur in economic change lost ground and were rather overlooked until recently.

What Schumpeter was trying to convey through the concepts of entrepreneur and entrepreneurial function, and how their meanings changed, or not, in his thinking, have been the subject of intense debate and provoked fierce discussion and many pages in the literature. One view is that in his early work he presented the entrepreneur as a powerful leader who changed the economy through new combinations of things; in his later work he played down the heroic characteristics of the entrepreneur, portraying him as much less individualistic. Another view is that there is no dichotomy between the 'old' and the 'young' Schumpeter positions. Analysing Schumpeter's work on the entrepreneur and the entrepreneurial function raises some core questions: 'Who is/are the entrepreneur/s?', 'What do entrepreneurs do?', and 'What is entrepreneurship?' which in their turn initiate some interesting lines of inquiry. Starting from Schumpeter's insights and contributions, these questions and their answers should contribute to shaping our theory of entrepreneurship and our understanding of the entrepreneurship phenomenon.

Schumpeter's first key contribution was focusing on and theorising about the role of the entrepreneur in economic change. He did so by building on the ideas from a well-established academic tradition on the entrepreneurial function in economic development, which includes the work of scholars such as Cantillon, Say, Smith, Walras and Knight (Hebert and Link, 1982), but also differentiating his views from other scholars on some topics. For example, for both Schumpeter and Say the entrepreneur is an industrial leader and is pivotal in the economy, but, differently from Knight, Schumpeter gave great importance to the entrepreneur's assumption of risk. By selectively choosing among the ideas that existed at the time about the entrepreneur, Schumpeter developed an original theory of the entrepreneur, which comprised the entrepreneur and innovation, and their connection. A key contribution of this theory is in stressing the presence of an agent in economic change, that is, changes occur through the purposeful actions of entrepreneurs.

Schumpeter argued that the entrepreneurial function is not necessarily embodied in a physical person and particularly not necessarily in a '*single* physical person':

> The entrepreneurial function need not be embodied in a physical person and in particular in a single physical person. *Every social environment has its own ways of filling the entrepreneurial function.* For instance, the practice of farmers in this country has been revolutionised again and again by the introduction of methods worked out by the Department of Agriculture and by the Department of Agriculture's success in teaching these methods. In this case then it was the Department of Agriculture that acted as an entrepreneur (Schumpeter, 1949: 260, emphasis added).

The entrepreneurial function thus can be embodied in an organization other than companies, such as the Department of Agriculture – even if Schumpeter's focus was mainly on large corporations.

Again the entrepreneurial function may be and often is filled *co-operatively*. With the development of the largest-scale corporations this has evidently become of major importance: aptitudes that no single individual combines can thus be built into a corporate personality; on the other hand, the constituent physical personalities must inevitably, to some extent, and very often, interfere, with each other. In many cases, therefore, it is difficult or even impossible to name an individual who acts as 'the entrepreneur' in a concern (Schumpeter, 1949: 260, emphasis added).

Schumpeter recognised the possibility of a collective embodiment of the entrepreneurial agent of change, but in general he theorised about the contextual nature and embodiment of the entrepreneurial function. A limitation to his theorising may lie in his not elaborating further the different forms of organised and cooperative entrepreneurial action, compared with actions of large companies, but in his work he was driven by observation of the context in which he lived. He did not predict the embodiment of the entrepreneurial function in networks, which is what characterises our current society, but it is not difficult to imagine that he would have added networks as a possible embodiment of the entrepreneurial function if he were observing society today.

A final, but important, contribution of Schumpeter's theory of the entrepreneur is highlighting and analysing the 'creative component' of entrepreneurial action. The entrepreneur exemplifies creative or dynamic energetic action in the economic sphere, which Schumpeter counterpoised with the static-hedonistic action of the manager (Santarelli and Pesciarelli, 1990; Dahms, 1995). So, it seems necessary and important to analyse Schumpeter's view on entrepreneurial action, and compare it with our current thinking in the context of entrepreneurship.

11.2.2 The forgotten legacy of Schumpeter for entrepreneurship and networks: entrepreneurial action as rational, social and creative

Schumpeter's view of entrepreneurial action is the forgotten legacy that we need to reintroduce in our research agenda. Entrepreneurial action is a form of economic action, and for Schumpeter entrepreneurial action, unlike other forms of economic action such as managerial action, has a creative component. Schumpeter's entrepreneur is no 'homo economicus'; the drivers of his action are the joy of creating and the will to conquer (Schumpeter, 1934). He adopted Nietzsche's view on the 'will to power' and on 'creative activity' as the creation of the new, but he also moderated his position on entrepreneurial action by inserting also a 'rational component', which was more in line with the economic thinking of his time (Santarelli and Pesciarelli, 1990). Rationality of action was a topic that interested Schumpeter for many years and he was very critical of the economists' treatment of rationality.

Moreover, Schumpeter considered as outmoded the attempt to explain certain phenomena, such as institutions, through the actions of the individual as opposed to those of the group. In his view, economic sociology or Sozialökonomik should study institutions such as 'private property' and 'the family', which are 'partly economic' and 'partly non-economic', and methodological individualism should not be used. He argued that the actions of entrepreneurs structurally affected the system of stratification in capitalist society (Swedberg, 1991). This observation and many others, such as that innovations come in 'swarms', and that entrepreneurship is related to social classes, point to the sociological way that Schumpeter reasoned, that social structures shape the actions of agents, and agents shape history (Galambos, 1996).

Schumpeter's view of entrepreneurial action, constituted by both creativity and rationality and influenced by institutions, is very contemporary. It reminds us of the contributions of scholars such as Herbert Simon and James March on economic action and more on human action more generally.[4]

---

[4] Building on these contributions, Sarasvathy (2001a) proposed a preliminary, but exciting, theory of entrepreneurial action that she called 'effectuation', which is described in the discussion section of this chapter.

However, in the social sciences, entrepreneurial action, similar to any other type of economic action, is mainly characterised as rational. This is because our understanding of entrepreneurial action is influenced by the approaches to economic action that prevail in the two major disciplines of economics and sociology. Economic and sociological approaches adopt stylised interpretations of economic action, which are consistent with their assumptions, to elaborate their theories at the cost of oversimplifying reality. Mainstream economics assumes that all economic actions are rational and constrained by tastes and by scarcity of resources, and that actors are uninfluenced by other actors. Modern theories of organizational economics introduce limits on human rationality with the assumption of 'bounded' rationality incorporated in their theorising about human action (Simon, 1955), retaining the methodological individualism of mainstream economics. Sociological approaches to economic action assume that many different types of economic action exist, including rational ones, and they are constrained by the scarcity of resources, by the social structure and by meaning structures. It therefore follows that actors' actions are connected and influenced by other actors' actions (Smelser and Swedberg, 1994).

Empirical observations of actual human behaviour indicate that economic action is both creative and 'boundedly' rational, and also social, that is, it is influenced by other actors' actions or interactions and meaning structures. As already mentioned, the social aspect of economic action or its embeddedness in the social context was implicit in Schumpeter's thinking related to the 'entrepreneurial function' and the 'view of the economy as a whole' (Becker and Knudsen, 2002), but he did not theorise about his insights. The social aspect of economic action has been analysed by other scholars and more recently has been associated with the idea of networks (Granovetter, 1985).

11.2.3 Schumpeter and networks

Since the mid-1980s the field of economic sociology has been developing some of Schumpeter's insights into sociological analysis of the 'whole economy' and in particular the influence of social interaction on the economic actions of the individual actor.

Swedberg (2004) gave primacy to Schumpeter's contribution to the genealogy of economic sociology, in strongly advocating the need for certain economic problems – and economic change in particular – to be tackled by different branches of the social sciences. He viewed 'economic theory', 'economic sociology', 'history' and 'statistics' as fundamental components for a recasting of conventional economic theory in a more dynamic direction (Swedberg, 1991). Schumpeter was especially acquainted with the sociological theories of his time. He deeply admired, and also collaborated with, Weber and Parsons, the founding fathers of economic sociology.[5] Based on recent biographical work by Swedberg, which re-establishes the proper intellectual context of Schumpeter's work, 'in practice ... he often let his economic theory slip into historical and sociological ways of reasoning' (ibid: 41) and 'Schumpeter's theory of innovation leads in a natural way into a sociological analysis' (ibid: 59).

Economic sociology goes some way towards an elaboration and theorisation of Schumpeter's insights in relation to the 'social character of economic action' through the concepts of 'embeddedness" and networks. Mark Granovetter underscored the importance of embeddedness; he contended that economic actions are 'embedded in concrete systems of relations' and 'social relations are preferably, if not necessarily, to be understood in network terms' (Granovetter, 1985 and 1990, quoted in Swedberg, 1997: 165). Zukin and DiMaggio (1990) added a constructive criticism and extended the concept of embeddedness proposed by Granovetter in maintaining that structural embeddedness, or embeddedness in networks and social structures, is very important, but there are also other types of embeddedness, such as political, cognitive or cultural. However, the empirical challenges of these other types of embeddedness, and the influence of Granovetter's views, acted to legitimise one of the traditions in contemporary sociology, network analysis, as the preferred way to operationalise embeddedness.

Since Granovetter's (1985) landmark article, several important theoretical and empirical works have built a strong connection between economic actions and networks. However, theory-building inevitably oversimplifies the reality. Current network analysis cannot be applied to investigate social structure as embedded in time or in process, or to conceptualise human agency, culture and institutions (Emirbayer and Goodwin, 1994).

Returning to Schumpeter's legacy, using network analysis we can investigate the embeddedness of economic action in social structure, but the human agency and the creativity involved in economic action is still out of our empirical reach. Padgett and Ansell's (1993) article on the 'robust action' of Cosimo de Medici is a notable exception: the two scholars use network analysis to investigate empirical manifestations and features of human agency. Their historical study illustrates a type of

---

[5] Schumpeter was one of the original reviewers of Parsons's (1937) *Structure of Social Action* (Swedberg, 1991).

human action – which they defined as 'robust' – through the dynamics of networks. In my opinion, the study of Padgett and Ansell comes close to the investigation of the rational, social and creative entrepreneurial action that Schumpeter advocated and theorised.

11.3   Discussion

Ideally, our studies on entrepreneurship and networks should be based on a theory of entrepreneurial action that includes both rationality and creativity, and in which institutions, social relations and culture play a part. But, the first step in our research strategy to achieve this final objective could be to evaluate and to exploit some of the research opportunities offered by Schumpeter's legacies for entrepreneurship, networks and their overlap. We can attempt to operationalise some of Schumpeter's ideas on entrepreneurship, such as the determinants of entrepreneurial behaviour, the entrepreneurial function and process, and the entrepreneurial context, and determine which of Schumpeter's legacies for entrepreneurship benefits from adopting a 'network perspective' and which do not.

The first of Schumpeter's insights, which certainly benefits from a network perspective, is related to the 'sources or determinants of entrepreneurial behaviour'. Schumpeter offered two explanations: they can be unborn, or related to families and other social groups. Work on the entrepreneur's ego-network has examined entrepreneurial behaviour as a dependent variable in the structure of the network (e.g. Burt, 1992). Other network studies have analysed the relationship between ethnicity and entrepreneurship (e.g. Aldrich and Zimmer, 1986). Furthermore, a network perspective might be useful to operationalise the embodiment of the 'entrepreneurial function', that is, carrying out new combinations of things. Schumpeter observed that in his time and context a single physical person or an organisation could embody this function. Becker and Knudsen (2002 and 2004) argued that in his late work Schumpeter emphasised the organisation of the entrepreneurial function, over the individual entrepreneur, with a 'de-personification of the entrepreneurial function' (2002: 394). These two authors, examining newly translated material, pointed out that an important but overlooked concept in Schumpeter's work is that of social interaction. They claimed that he was 'very optimistic regarding the possibility of constructing an explanation of development by understanding social interaction' (Becker and Knudsen, 2004: 17) and that 'new combinations were retained by social structure, the underlying mechanism being the interaction of social agents' (Becker and Knudsen, 2002: 396). Would it be impossible that the entrepreneurial function is embodied in a network of people? The network design necessary to answer this question differs from the structural network approach that characterises the majority of current studies. Network processes need to be reintroduced into the picture besides network structures (Salancik, 1995). One example of this line of inquiry is the work of Powell et al. (1996), which focuses on the locus of innovation, that is, the embodiment of the entrepreneurial function in networks of relations.

Networks might also enable modelling the 'entrepreneurial context' or the environment surrounding the entrepreneurial behaviour and its function. Although Low and Macmillan (1988) argued that networks are an important aspect of the context of entrepreneurship, subsequent studies focused mainly on the role of networks for the acquisition of resources available in the environment to exploit the entrepreneurial opportunity.

Overall, the research of many aspects of the entrepreneurial phenomenon – such as the sources and outcomes of entrepreneurial action, processes, structures and context – can benefit from using a network lens and from integration of these aspects in a single research design. But integrative studies of this kind are missing. The challenge for social sciences is to build theory 'where contexts, structures and individual actions interact and change together' (Granovetter, 2001: 26). The same applies to our theory of entrepreneurship.

After identifying and discussing the research opportunities that can be derived from examining Schumpeter's legacies for entrepreneurship, networks and their overlap, I would like to conclude this section by pointing out three areas that deserve more space in the entrepreneurship-research agenda. Each area could be closely related to Schumpeter's work and offers interesting implications for our understanding of the entrepreneurial phenomenon. The first area is 'entrepreneurial action'. This issue was central to Schumpeter's thinking, because it represents a nexus of several of his ideas such as rationality in human action, the entrepreneur and the entrepreneurial function. Recently, Sarasvathy (2001a and b) has taken a fresh look at entrepreneurial action. She investigated entrepreneurial expertise empirically and proposed a 'theory of effectuation' (Sarasvathy, 2001a). She argued that 'effectuation' – choosing many possible effects, using a particular set of means, and 'causation' – choosing means to create a particular effect – are integral parts of human reasoning that can occur simultaneously. In particular, effectuation fits with the way entrepreneurs act. The consequences of her theory of effectuation for entrepreneurship studies still need to be fully derived, but her work deserves attention. Her promise to frame entrepreneurship as 'economics with imagination' is particularly appealing (Sarasvathy, 2001b).

The second topic that warrants further research is 'entrepreneurship policy'. The key research questions in this line of inquiry could be 'why is entrepreneurship worth fostering?' and 'what are the barriers to entrepreneurship?' (Sarasvathy, 2003). But also 'what is the connection between entrepreneurship and competitiveness?' is a question that must be examined. David Hart (2003) in his edited book on the emergence of the entrepreneurship policy field in the United States rightfully argued that entrepreneurship is too important to be ignored by public policy and left in the hands of business management practitioners and academics.

Finally, 'entrepreneurial history' deserves a more prominent place on the research agenda. Schumpeter was personally involved in the creation of the Research Center for Entrepreneurial History, which operated in Harvard between 1948 and 1957. The acquaintance of these scholars with sociological theory produced high-quality work that retained the narrative mode of writing history, but was informed by theory. Entrepreneurial history no longer exists as a line of inquiry within a broader field, and since the 1960s the focus of research in economic history has shifted to business history. However, during the short existence of entrepreneurial history as separated area of inquiry its scholars produced a series of studies that addressed important issues in entrepreneurship such as the diversity of its forms across countries and historical periods, creativity and national entrepreneurial styles.

To conclude this discussion, there is clear need and opportunity to build a richer and more integrated body of knowledge, based on clear concepts and powerful tools, which will influence and reshape the nature and role that entrepreneurship and networks play in our society. Therefore, re-examining our research agenda on entrepreneurship and networks by building on Schumpeter's legacies and rediscovering his insights on topics like entrepreneurial action is a challenging but necessary exercise.

11.4   Conclusions

The objective of this chapter was to identify and discuss Schumpeter's legacies for the study of entrepreneurship and networks. To achieve a better understanding of these phenomena, we must problematise our tools and ways of thinking. By examining and discussing the concepts of entrepreneurship, entrepreneurial action and networks I have tried to demonstrate the legacies of Schumpeter's work, including his often-forgotten insight on entrepreneurial action. The existing work on entrepreneurship, networks and their overlap was illustrated with the purpose to derive some research opportunities that we should evaluate and may be exploit.

I conclude by entering a plea for the aspects of Schumpeter's work that have not been explored to be analysed. His vision and contribution are uncommonly broad and rich, and he provided a synthesis of knowledge that affects our ideas and ways of thinking about certain phenomena, such as economic change and innovation. Schumpeter continues to shape our theorising by providing the intellectual maps 'we need if we are going to determine where we have been, where we are and where we might be going' (Galambos, 1996: 928). He may even approve of the creative destruction of some of the maps that he proposed.

References

Aldrich, H. and Zimmer, C. 1986. 'Entrepreneurship through Social Networks', in Sexton, D. and Smilor, R. W. (eds.) *Art and Science of Entrepreneurship*. Cambridge, MA: Ballinger

Audretsch, D. B. 1995. *Innovation and Industry Evolution*. Cambridge, MA: MIT Press

Becker, M. C. and Knudsen, T. 2002. 'Schumpeter 1911. Farsighted Visions on Economic Development', *American Journal of Economics and Sociology* 61 (2): 387–403

Becker, M. C. and Knudsen, T. 2004. 'The Role of Entrepreneurship in Economic and Technological Development: The Contribution of Schumpeter to understanding of Entrepreneurship', Schumpeterian Conference 2004 Proceedings

Breschi, S., Malerba, F. and Orsenigo, L. 2000. 'Technological Regimes and Schumpeterian Patterns of Innovation', *The Economic Journal*, 110: 388–410

Burt, R. 1982. *Toward a Structural Theory of Action*. New York: Academic Press

Burt, R. 1992. *Structural Holes: The Social Structure of Competition*. Cambridge, MA: Harvard University Press

Carroll, L. and Dodgson, C. L. 1971. *Alice's Adventures in Wonderland and Through the Looking-Glass*. Oxford, UK: Oxford University Press

Casson, M. 1982. *The Entrepreneur – An Economic Theory*. Cheltenham, UK: Edward Elgar

Dahms, H. F. 1995. 'From Creative Action to Social Rationalization of the Economy: Joseph A. Schumpeter's Social Theory', *Sociological Theory* 13 (1): 1–13

Dosi, G. and Teece, D. 1996. 'Weber/Schumpeter: Some Preliminaries', *Industrial and Corporate Change* 5(3): 905

Emirbayer, M. and Goodwin, J. 1994. 'Network Analysis, Culture and the Problem of Agency', *American Journal of Sociology* 99(106): 1411–54

Fagerberg, J. 2003. 'Schumpeter and the Revival of Evolutionary Economics: An Appraisal of the Literature', *Journal of Evolutionary Economics* 13: 125–59

Freeman, C. 1994. 'The Economics of Technical Change: A Critical Survey,' *Cambridge Journal of Economics* 18(5): 463–514

Galambos, L. 1996. 'End of the Century Reflections on Weber and Schumpeter – With Karl Marx Lurking in the Background', *Industrial and Corporate Change* 5(3): 925–31

Gartner, W. B. 2001. 'Is There an Elephant in Entrepreneurship? Blind Assumptions in Theory Development,' *Entrepreneurship Theory and Practice* 25(4): 28–39

Granovetter, M. 1985. 'Economic Action and Social Structure: The Problem of Embeddedness,' *American Journal of Sociology* 91(3): 481–510

Granovetter, M. 2001. 'A Theoretical Agenda for Economic Sociology', in M. F. Guillem, R. Collins, P. England, and M. Meyer(eds.), *Economic Sociology at the Millenium*. 2001. New York: Russell Sage Foundation

Greve, A. and Salaff, J. W. 2003. 'Social Networks and Entrepreneurship,' *Entrepreneurship Theory and Practice* 28(1): 1–22

Gulati, R. and Gargiulo, M. 1999. 'Where Do Interorganizational Networks Come From?', *American Journal of Sociology* 104(5): 1439–93

Hart, D. 2003. (ed.) *The Emergence of Entrepreneurship Policy. Governance, Start-ups, and Growth in the US Knowledge Economy*. Cambridge, UK: Cambridge University Press

Hebert, R. F. and Link, A. N. 1982. *The Entrepreneur*, New York: Praeger

Kirchoff, B. A. 1991. 'Entrepreneurship's Contribution to Economics', *Entrepreneurship Theory and Practice* 16(2): 93–112

Low, M. B. and MacMillan, I. C. 1988. 'Entrepreneurship: Past Research and Future Challenges', *Journal of Management* 35: 139–61

McClelland, D. C. 1961. *The Achieving Society*, Princeton, NJ: Van Nostrand

Mitchell, J. C. 1969. 'The Concept and Use of Social Networks' in J. C. Mitchell (ed.), *Social Networks in Urban Situations*. Manchester, UK: Manchester University Press, pp. 1–50

Nelson, R. and Winter, S. 1982. *An Evolutionary Model of Economic Change*. Cambridge, MA: Harvard University Press

Padgett, J. and Ansell, C. 1993. 'Robust Action and the Rise of the Medici, 1400–34', *American Journal of Sociology* 98: 1258–1319

Parsons, T. 1937. *The Structure of Social Action*. New York: Free Press

Powell, W. W., Koput, K. and Smith-Doerr, L. 1996. 'Interorganizational Collaboration and the Locus of Innovation: Networks of Learning in Biotechnology,' *Administrative Science Quarterly* 41(1): 116–45

Salancik, G. R. 1995. 'Wanted: A Good Network Theory of Organization' *Administrative Science Quarterly* 40: 345–49

Santarelli, E. and Pesciarelli, E. 1990. 'The Emergence of a Vision: The Development of Schumpeter's Theory of Entrepreneurship', *History of Political Economy* 22(4): 677–96

Sarasvathy, S. D. 2001a. 'Causation and Effectuation: Towards a Theoretical Shift from Economic Inevitability to Entrepreneurial Contingency', *Academy of Management Review* 26(2): 243–88

Sarasvathy, S. D. 2001b. 'Entrepreneurship as Economics with Imagination', *Business Ethics Quarterly. The Ruffin Series*, 3: 95–112

Sarasvathy, S. D. 2003. The Questions We Ask and the Questions We Care About: Re-Formulating Some Problems in Entrepreneurship Research, Working Paper, University of Maryland

Schumpeter, J. A. 1934. *The Theory of Economic Development*. Cambridge, MA: Harvard University Press

Schumpeter, J. A. 1949. 'Economic Theory and Entrepreneurial History', in Clemence, Richard, V. (ed.), *Essays on Entrepreneurs, Innovations, Business Cycles, and the Evolution of Capitalism*. New Jersey, NJ: Transaction Publishers, 1991

Shane, S. and Venkataraman, S. 2000. 'The Promise of Entrepreneurship as a Field of Research,' *Academy of Management Review* 25(1): 217–26

Simon, H. A. 1955. 'A Behavioral Model of Rational Choice', *Quarterly Journal of Economics* 69: 99–118

Smelser, N. J. and Swedberg, R. (eds.) 1994. *Handbook of Economic Sociology*. New York: Princeton University Press and Russell Sage Foundation

Swedberg, R. 1991. *Joseph Schumpeter: His Life and Work*. Cambridge: Polity Press

Swedberg, R. 1997. 'New Economic Sociology: What Has Been Accomplished, What Is Ahead?', *Acta Sociologica* 40(2): 161–82

Swedberg, R. 2004. *Principles of Economic Sociology*. Princeton, NJ: Princeton University Press

Thornton, P. H. 1999. 'The Sociology of Entrepreneurship', *American Review of Sociology* 25: 19–46

Uzzi, B. 1997. 'Social Structure and Competition in Interfirm Networks: The Paradox of Embeddedness', *Administrative Science Quarterly* 42: 35–67

Zukin, S. and Di Maggio, P. (eds.) 1990. 'Introduction', *Structures of Capital. The Social Organization of the Economy*. Cambridge, UK: Cambridge University Press

12 Knowledge-based entrepreneurship: the organizational side of technology commercialization

Ulrich Witt and Christian Zellner

Introduction

New knowledge which shapes and supports technological advance continually emerges in the academic institutions. It is a result of publicly financed, scientific problem-solving. As such, its generation is not (primarily) guided by application interests. However, such knowledge usually carries some commercial business potential. National economies differ substantially both in their capacity to exploit the opportunities and in the pace of doing so. These differences have been found to be a major source of competitive advantages in global markets (Porter 1990). New production technologies and products drive the process of economic growth and allow innovation rents to be appropriated. In recent years, one question has therefore attracted increasing interest both in economic research and in politics (Nelson 1993, Edquist and McKelvey 2000, Salter and Martin 2001). How does new knowledge from scientific research find its way into the commercial part of the innovation system? How does it support technological advance?

In this chapter it is argued that the transfer is essentially an entrepreneurial process. On the one hand, to understand that process, it is necessary to recognize the kind of actions and services involved in the entrepreneurial reshaping of the division of labor. In general, entrepreneurship requires command over suitable resources. In the case of knowledge-based entrepreneurship, these are, in particular, resources enabling the access to, and the exploitation of, new technological knowledge. Therefore, an essential part of entrepreneurial activity here is the organization of the knowledge transfer from academic research to commercial production and marketing activities.

On the other hand, the entrepreneurial process cannot fully be grasped without recognizing the constraints under which it operates. Entrepreneurship always faces obstacles such as barriers to entrepreneurial entry, lack of qualified resources, and/or organizational rigidities. Depending on entrepreneurial skills, some of these can be overcome and some cannot. Difficulties like these vary with the institutional and political conditions in the different national economies. They can impede the entrepreneurial commercialization of new technologies (Henrekson and Rosenberg 2001) just as much as they can impede any other entrepreneurial activity. In the case of knowledge-based entrepreneurship, there are, however, additional obstacles. They result from the peculiarities of scientific knowledge and its mode of transfer. It will be claimed here that the attempt to overcome them shapes the manner in which the commercialization of new technologies is organized (Zellner 2003).

To explain this, it is useful to introduce the common distinction between tacit and encoded (or implicit and explicit) knowledge which goes back to Polanyi (1967). Encoded knowledge exists in the form of written information. It is accessible to commercial users as long as their training allows them to understand the context and content. Tacit knowledge, in contrast, can only be acquired by experience on the job – in the case of scientific tacit knowledge, by conducting scientific research. This knowledge is hard, or even impossible, to encode and therefore has to be carried in embodied form. However, in a rapidly progressing research environment, state-of-the-art tacit knowledge is constantly changing. Privately held tacit knowledge is therefore subject to relatively rapid decay unless it is quasi-automatically updated on the job in a continued involvement in scientific research.

Technological and disciplinary fields differ with respect to what form of knowledge is relevant for commercial applications, and so, too, does the actual organization of the knowledge transfer. Yet, in any case, the service of scientifically trained personnel is necessary to achieve the transfer. This very special resource requirement implies substantial overhead costs for the venture, unless the entrepreneur is able to provide that service in person (as in the case of start-up firms run by former researchers as entrepreneurs). This fact may explain why the transfer and commercial exploitation of technological knowledge indeed tends to be a matter of either small entrepreneurial start-up firms or large, incumbent firms (Cohen, Nelson, and Walsh 2002).

In the first case, the knowledge transfer is accomplished by the entrepreneur who embodies the necessary technological knowledge in person when setting up the business. In the case of the large, incumbent firm, by contrast, technological expertise that is complementary to the organizational capabilities already existing has to selectively be acquired on the labor market. Usually this is done by expanding the existing R&D staff – a specialized organizational unit a firm can afford to support only if its operations are on a large scale (cf. Keck 1993 and Reinhardt 1997 for historical examples). The two contrasting forms of firm organization represent very different stages of organizational development (Witt 2000, Rathe and Witt 2001). They therefore have different strengths, face different problems, and pursue different entrepreneurial strategies in exploiting knowledge-based business opportunities. However, both organizational forms compete for the same resource: the human capabilities needed to realize the knowledge transfer.

The competition for human capabilities takes place in a self-sorting process among the scientists and engineers who are about to migrate from academic research into the private sector. They can realize the commercial value of the knowledge that have acquired in academic research either by setting up an own entrepreneurial start-up venture or by becoming employees in someone else's entrepreneurial business. Accordingly, they either become the subject or the object of knowledge-based entrepreneurship. Yet, in either case, the peculiarities of scientific knowledge and its transfer mode imply constraints for exploiting knowledge which are changing systematically over time. As will turn out, the – differing – responses to these constraints result in organizational arrangements which seem to be characteristic concomitants of the commercialization of new technologies.

The paper proceeds as follows. To set the stage, Section 12.1 focuses on the different forms of scientific knowledge and, correspondingly, of knowledge transfers. These represent the constraints for the entrepreneurial choices about how to carry knowledge across the institutional boundaries between science and commerce. Section 12.2 gives a brief review of the actions and services involved in the entrepreneurial reshaping of the division of labor. It then turns to the logic of the competitive process by which people sort themselves into entrepreneurs and employees. The understanding of this sorting also helps to explain certain features of the knowledge-transfer process. Section 12.3 looks into the case of start-up firms. Special attention is given to the problems they encounter with the manner in which they organize the transfer of tacit knowledge – problems that may, in the longer run, undermine their capacity to exploit new knowledge-based business opportunities. Section 12.4 explores the problems which large, incumbent firms face with their way of organizing the knowledge transfer under pressure from the competitive sorting process. A connection with human resource development plans is made which – in the form of career path options offered to scientifically trained staff – turns out to be a major organizational provision taken into account for these problems. Section 12.5 offers some conclusions.

12.1   The nature of scientific knowledge and its transfer conditions

Scientific research creates new knowledge in basically two different forms. These forms also differ with respect to the manner in which knowledge is transferred across the institutional boundary between scientific research and commercial application. One form is propositional knowledge. It is the object of most empirical studies of knowledge transfer (cf. Grupp 1998). Propositional knowledge can be encoded and stored by means of some information medium like, for example, a written document. Transmitted by the information medium, it can, in principle, be acquired by all potential recipients in parallel when they gain access to the medium and command sufficient interpretative knowledge. Once published, encoded technological knowledge can be used indiscriminatingly for commercial purposes by any interested party. No personal contact is required with those who originally created the knowledge and disclosed information about it.

Indeed, the innovation performance and long-term competitiveness of firms in high-technology environments seems to hinge critically on their capacity to monitor and tap, in an anonymous form, scientific and technological developments that have originated elsewhere (Cohen and Levinthal 1989, Rosenberg 1990). This means that new propositional knowledge emerging in academic research and more generally in the international innovation systems needs to be traced. Its usefulness, in terms of complementarities to the firm's own capabilities, needs to be assessed. Prerequisite for doing so is that the firm has sufficient cognitive "absorptive capacity," that is, sufficient own knowledge to understand the context and meaning of the information transmitted.[1] The transmission of knowledge in encoded form is, however, not possible for all aspects of scientific research. Besides theoretical and empirical insights

---

[1] Cf. Cohen and Levinthal (1989). The entrepreneur in a knowledge-based business venture may, but does not have to, possess that cognitive absorptive capacity in person. To the extent to which new technological developments are documented and are accessible in the form of encoded knowledge, the entrepreneur can link up with them by hiring suitably trained human resources (as part of the overall resources) who have acquired the necessary cognitive absorptive capabilities in academic education and research.

(whose publication in scientific journals constitutes the main objective of academic research), a variety of other forms of knowledge are essential to the scientific process. They are mainly differentiated from the former by their procedural characteristics and imply a substantial degree of tacitness.

These additional, tacit elements of scientific knowledge can be classified according to their specificity and the role they play in the generation of new scientific knowledge (Zellner 2003). Scientific skills presuppose the command of methodological knowledge about experimental procedures and research strategies applied in the respective (sub-)disciplines. Furthermore, scientific skills include the analytical skills necessary for the recognition, formulation, and solution of complex problems. (Even though these analytical skills are substantially less specific to the discipline and are difficult to observe, they represent one of the most important ingredients in the expansion of scientific knowledge.) In addition, the operation of the physical infrastructure of the research process requires a substantial amount of practical, procedural knowledge. This includes knowledge about, and experience with, physical instrumentation and laboratory equipment. (In fact, progress in, e.g., experimental physics or chemistry is often based on the development of a specific experimental setup which sometimes also absorbs the bulk of the resources invested.) Finally, an ever more important prerequisite of scientific research today is knowledge of how to use information technologies and how to develop own applications in the form of the design of simulations and software programming. The substantial requirements for data analysis and processing have made modern scientific research one of the high-end users of hardware technology.

All these forms of knowledge are hard, or even impossible, to encode and, hence, cannot be transmitted to potential recipients by information media.[2] Their acquisition by individual scientists is based on repeated practice and continuing interaction with senior scientists, usually in the context of noncommercial research activities. For many technologies like, for example, in molecular biology and chemistry, these tacit knowledge components represent an important part of what is needed to

---

[2] There is yet another reason playing a role here. Besides the limits to codification that are inherent to the elements of knowledge discussed, there are institutional factors that can inhibit the codification of knowledge that is instrumental to the production of propositional knowledge. Academic science is based on an incentive system that rewards priority in disclosure (Dasgupta and David 1994, David 1998). To the extent that scientific skills and technicalities confer a competitive advantage to individual scientists or their laboratories, the incentives to encode and disclose such knowledge are probably limited or even negative. This observation may be particularly relevant to skills and technicalities that are specific to the (sub-)discipline or line of research.

accommodate – and ultimately to commercially exploit – the state-of-the-art technological knowledge.

Since transmission in encoded form is not possible here, a different transfer mode is relevant. Knowledge of this form, often containing substantial procedural elements, can only be carried from scientific research to the commercial sphere by scientists and engineers who migrate from academia to business. They embody the forms of tacit knowledge mentioned after acquiring them "on the job" in noncommercial research where these skills and technicalities are developed and refined in a usually costly and time-intensive way not subject to profitability considerations. Thus, the attraction of scientifically trained staff to business firms does not necessarily only serve the purpose of creating cognitive absorptive capacity in these firms. It also enables the firms to get a hold of tacit state-of-the-art knowledge from scientific research which cannot be transferred in other ways.

However, by the very mode of transfer – the physical migration of the knowledge carriers into the commercial sphere – the knowledge carriers are cut off from the further development of that knowledge in the academic sphere. Even with considerable communication effort, it is, in most cases, not possible for a scientist who has left academic research to keep up with the rapid development of the tacit knowledge components there. Therefore, the problem of transferring tacit state-of-the-art knowledge can always only temporarily be solved. To retain the tacit knowledge transfer, ever new cohorts of scientists and engineers from academic research need to be attracted into the commercial sphere with a frequency that depends on the decay time of the tacit knowledge they embody.

This organizational feature of the tacit knowledge transfer causes a problem. In many national innovation systems the boundary between the academic and the commercial sphere is not equally permeable in both directions. Moreover, it tends to be the less so, the less close a scientist's field of specialization is to applied research. This means that the return option for scientists who cross the institutional boundary between the academic and the commercial sphere is more or less uncertain. The question therefore arises of whether they can be induced to take such a potentially irreversible step in their professional development. If their employment prospects with the hiring firm were limited to that period of time during which their tacit knowledge is state of the art, these people might be reluctant to accept an offer from a business firm. This is especially true for the most talented researchers and/or if that period of time is only a relatively short episode in a professional life span.

The problem for the commercial employer thus is how to create additional incentives for nonacademic employment without raising to a prohibitive level the costs of acquiring state-of-the-art tacit knowledge. The problem is particularly difficult, because the contribution a scientifically trained employee will actually make to the firm's future profitability cannot be judged safely at the time of hiring the employee. Moreover, the future employee's effort in exploiting complementarities between her or his embodied knowledge and the knowledge that exists in the firm organization cannot easily be observed and, hence, is difficult to contract. Before discussing in more detail what provisions are taken to account for this problem, it is useful, however, to consider in the next section who the potential migrants to the corporate R&D departments are. As mentioned in the introduction, this is decided in a competitive self-sorting process that determines how people can exploit the knowledge they have acquired in academia.

12.2   Competition between business conceptions: the entrepreneurial sorting process

The mode of transferring new knowledge from noncommercial research to commercial R&D, production, and marketing activities depends, it has been argued, on whether the knowledge to be transferred is tacit or encoded. In both cases, however, for the transfer people are required who have previously acquired the corresponding knowledge and capabilities in academic training and research. These people play a central role as interpreters of encoded information and/or as carriers of the state-of-the-art tacit knowledge. Making them available for the commercial sphere and, thus, organizing the knowledge transfer, is a core element of knowledge-based entrepreneurship. In fact, the service provided by knowledge-based entrepreneurship has yet other, equally important aspects.[3]

In order to trigger the transfer it is necessary, first, for someone to see business opportunities in new scientific knowledge. This is an often overlooked, genuinely entrepreneurial, achievement (Witt 1998). Newly discovered scientific insights and techniques do not by themselves suggest new products or production processes that may be commercially successful. Such opportunities have to be imaged by combining at least some ideas about the new technology with conjectures about market conditions (Shane 2000). Once a more or less concrete business

---

[3] Like in Penrose's classical definition, entrepreneurial services are contrasted here "... with managerial services which relate to the execution of entrepreneurial ideas and proposals and to the supervision of existing operations" (Penrose 1959, p. 32).

conception has been created it must, second, be realized by attracting resources and coordinating their interactions.

In the case of knowledge-based entrepreneurship these are, in particular, knowledge resources. Sometimes they may even be embodied by entrepreneurs (who have been scientists). Yet this is not necessary. The entrepreneur can always try to access technological knowledge by hiring human resources properly trained to accomplish the knowledge transfer across the institutional boundaries. Where firms already exist, a third entrepreneurial task is to integrate the newly mobilized technological knowledge into the established organization and, where necessary, to adjust the business conception. Only if new knowledge can be made complementary to already existing organizational capabilities can it successfully be exploited by the firm. This may sometimes require major organizational restructuring.

A question that arises, and that is particularly important for understanding the conditions of the knowledge-transfer process, is who will be, or will become, the entrepreneurs trying to accomplish these tasks? In principle, everyone who has thought up a business conception for exploiting new technological knowledge commercially could make an attempt to hire employees and other resources. And large numbers of business conceptions can be imagined. What determines which of these are indeed turned into concrete ventures? As has been claimed elsewhere (Witt 1987), the answer to these questions is a competitive self-sorting process that precedes the founding of a new firm or the signing of an employment contract.

In the context of knowledge-based entrepreneurship, the key role in the competitive self-sorting process is played by those scientifically trained researchers who are about to migrate from academia to the commercial sphere. As one alternative, they are offered (nonentrepreneurial) employee positions, preferably in R&D, as catalysts of the knowledge transfer discussed. The offer is, of course, based on the expectation that the future employee is ready to acquiesce to the entrepreneurial business conception pursued by the entrepreneur(s) of the hiring firm. As another alternative, these migrating researchers can, by using their technological knowledge and capabilities, develop an own business conception and consider realizing it as entrepreneurs. In this case, they would found an own technology-oriented start-up enterprise and try to attract the complementary resources.

The decision about who sorts, and how, has many facets. Economic reasons are not necessarily the most important ones. However, they do have an influence. As an entrepreneur, a migrating scientist would have to tackle the two tasks just mentioned: attracting resources and coordinating their interactions in a way that is conducive to realizing the own business conception. Accordingly, the own coordination skills have to be assessed, as would the profitability of the own business conception. The income stream that is imagined to be retained after paying the hired resources has to be compared with the income stream to be earned by accepting an employee's position in someone else's firm, and so too have the expected workloads and the nonpecuniary features of the two alternatives. For the subjective opportunity cost assessment the question of how much – and how long – the own, embodied knowledge will be valued by the future employer is obviously important. Indeed, it may be decisive in the case of a potentially decaying state-of-the-art tacit knowledge. A hiring firm may therefore be compelled to provide a long-term professional perspective.

Thus, even though there is no market for entrepreneurial business conceptions, something like an anticipated (and necessarily very subjective) market contest is actually taking place when scientists and engineers who leave academic institutions sort themselves into entrepreneurs and employees in knowledge-based business ventures. The subjective opportunity cost assessments, and thus the outcome of the contest, may, of course, turn out to be wrong. For this reason, the results of the sorting process are not always stable. A startup firm may not succeed and will then go out of business, bringing its founder back on the labor market. Conversely, someone who has become an employee in a firm in a first sorting may later reconsider that decision. Based, for example, on a better assessment of the value of the own embodied knowledge and/or changed market conjectures, the chances of own business conception may be rated better than before. If this happens, a former scientist may want to leave the firm and set up something of her or his own.

12.3   Start-up firms and the decay of founder knowledge

To become an entrepreneur, and to found a technology-oriented start-up firm, is one way in which a migrating scientist can contribute to the knowledge transfer from academic science to the commercial sphere. This organizational solution of the transfer problem, which has attracted considerable political interest (cf., e.g., BMBF 2001), is of course, confined to knowledge and capabilities acquired by, and embodied in, the entrepreneur in person.[4] Initially, the fate of a startup firm is

---

[4] For the transfer to succeed, the innovation – new knowledge-based products and/or production processes – introduced by the start-up venture do not necessarily have to be

independent of which of the two forms of knowledge, tacit or encoded, are transferred. In both cases, the founding entrepreneurs exploit the knowledge they embody in order to realize innovative production processes or products. If successful, start-up firms find their niche in the markets – perhaps even a growing niche or market.

However, depending on the pace of progress in noncommercial scientific and technological research, improvements and diversifications in the firm's processes and/or product may sooner or later become feasible. A continued transfer of new technological knowledge would then be useful or even necessary to keep up with competitors. In the case of technological knowledge that is accessible in encoded form, the founding entrepreneurs can support the transfer themselves, provided they continue to use their existing cognitive absorptive capacity to adopt the new knowledge and, where necessary, to update that capacity. Thus, if run by former scientists, even comparatively small business ventures do not necessarily have a disadvantage in keeping up with technological developments and in exploiting new commercial opportunities in this case.

A different situation arises where new technological knowledge is largely tacit and where the state of the art needs to be transferred in embodied form. Former scientists, who founded their start-up firm with the current state-of-the-art capabilities they acquired previously may then face problems. Owing to the peculiar conditions of the transfer of tacit knowledge discussed in Section 12.1, they are cut off from further development of the technology in the sciences and may sooner or later experience decaying relevance or up-to-dateness in their tacit knowledge. When the products and process with which their startup succeeded age and improvements and/or diversification are needed, they will lack the required updated knowledge base unless they find ways of attracting migrating scientists or engineers who embody what has then become the state-of-the-art knowledge.[5]

---

profitable. Ouring to the public good features which new technologies at least partially have it is sufficient for them to be introduced into the markets and to allow profitable business later, perhaps by firms other than the start-up venture that introduced the innovation. For the question of how scientists, who migrate at a future point in time, perceive their alternatives in the entrepreneurial sorting process – and, hence, their willingness to also try a start-up – the fate of earlier knowledge-based founding activities may, of course, be decisive, cf. Fornahl (2005).

[5] Small firms, particularly those in science-based industries, often try to nurture and maintain links with academic institutions in the form of consulting and/or collaboration projects. These efforts underscore the firms' continuing dependence on tacit knowledge. However, as discussed, they presuppose that the firms have a sufficient absorptive capacity. Precisely for this reason, maintaining such organizational links is not a permanent substitute for a repeated physical migration of scientists. To ensure a

The attempt to attract such resources means participating as a potential employer in the contest in which a new generation of researchers migrating from academic institutions sort themselves into entrepreneurs and employees. As mentioned, the problem for all commercial employers is to create incentives for these people, first, to indeed leave academic research and, second, to become employees (rather than self-employed). At the same time, these incentives should not drive the costs of acquiring state-of-the-art tacit knowledge to a prohibitive level. From the point of view of the migrating scientists and their subjective opportunity cost assessment, the question of how much – and how long– the own, potentially decaying state-of-the-art tacit knowledge they embody will be valued by the future employer is a crucial variable. Hence, an important incentive for them to sign on with a firm is a professional long-term perspective offered by an employer that is independent of a possible future knowledge decay.

Such a long-term perspective is provided by a career path option for scientifically trained R&D staff which allows employees to proceed at a certain stage from technical to managerial tasks and to maintain or even increase income.[6] An entrepreneur who can offer professional career options therefore has a competitive advantage over other entrepreneurs who cannot. However, in order to be able to make a credible personal promotion promise, the entrepreneurial startup must have already grown to an organizational size where it can create profitable employment opportunities for scientifically trained managers. In many industries a pace of growth of the firm sufficient to meet this requirement may be hard to achieve.

If so, other ways of attracting the latest brand of scientific expertise into the firm to strengthen its capabilities have to be found. One of these, though, in a sense, an expensive one, is the following. In order to be able to compete in the contest, entrepreneurs who have founded a startup can offer partnerships to migrating former scientists, particularly if the firm organization is still small. In view of the uncertainties about the actual value of the tacit knowledge to be acquired, such a step is not without risks. It can occasionally be observed when the future partners

---

sustained knowledge transfer, the firms' absorptive capacity needs to be kept up-to-date by occasionally acquiring new state-of-the-art embodied knowledge.

[6] In large firms, the transition from technical to managerial tasks is a distinct career feature (Biddle and Roberts 1993). It amounts to a change from the manipulation of objects or processes to the coordination of other employees. While the demarcation between corporate functions may not always be sharp in these terms (e.g., when someone takes on project responsibility in mainly technical functions like R&D or production), it is *the* major qualitative change in a career.

already know each other from their work in the scientific institution from which they migrate.

If even this option is not feasible while the competitive pressure to adopt state-of-the-art tacit knowledge is high and is increasingly reducing the business prospects of the entrepreneurial knowledge-based firm, there is always a default strategy. The founder entrepreneur can escape from further decline by trying to sell the startup venture and to "cash in". Scientists who embody the tacit knowledge that would be needed and who are about to migrate usually lack the financial resources to purchase a firm. However, other (sometimes rival) companies may acquire the business and merge it with their existing activities. If the acquiring firms do not already have the organization size necessary to be able to offer attractive career terms in the competition for the needed tacit knowledge, mergers and acquisitions may be the way to reach that size.

12.4 Incumbent firms and the role of intra organizational career paths

As a matter of fact – notwithstanding the recent attention paid to entrepreneurial start-up firms as an effective mechanism of knowledge transfer – a significant, or perhaps the largest, share of new technological knowledge is transferred to, and exploited by, large, incumbent firms. These firms have grown out of entrepreneurial founding activities in the past. As many examples from business history show, the origins have often been classical cases of what would now be called knowledge-based startups, for example in the electrical, pharmaceutical, chemical, and many other industries (cf., e.g., Heuss 1946, Galambos and Sewell 1995, McKelvey 1996, Buenstorf and Murmann 2003, Murmann 2003). The survival and growth of these firms was based on entrepreneurial business conceptions which informed and motivated the application of a technology in the form of new products and processes. Those business conceptions were the basis on which knowledge resources and other resources could be acquired, coordinated, and used to create commercial value.

However, over time, what were initially innovative products and processes are always in danger of becoming easily and cheaply copied standard practice. Even worse, if not improved or replaced these products and processes are prone to being outdated by further technological progress elsewhere. The continued involvement of large, incumbent firms in the further commercialization of new technology (often outside the technological field in which they started) therefore follows an own logic. It is different from, and more complex than, the immediate grasping of a market opportunity in startup ventures. While startups usually aim at gaining a foothold in the market and expanding the business, large incumbent firms usually try to maintain or improve a competitive advantage they have been able to attain in the past. In dynamic markets, this means that they have to keep track of the latest technological developments relevant to their industry and to spot commercial application opportunities that may be emerging from these. This is no trivial task. In fact, large firms face two kinds of problems in doing this.

First, unlike a migrating scientist founding a startup venture as an entrepreneur who can exploit the latest brand of technology by her or his own embodied knowledge and capabilities, large firms have to find ways to acquire and absorb the most recent technological knowledge. The second problem which large, differentiated firm organizations face is that spotting technological opportunities requires considerable coordination efforts. (This is obviously different from the case of small startup firms in which the entrepreneur integrates all necessary functions into one person.) The various people in charge of conceiving new business opportunities, of deciding about them, and of actually (re-)organizing the internal resources, all need to be coordinated. New technological knowledge has to be aligned with the firm-specific knowledge, capabilities, and routines that have been accumulated in the past.[7] Moreover, as a consequence, the adjustment of (up to now successful) business conceptions may be necessary and has to be mastered.

The second kind of problems all relate to the services of knowledge-based entrepreneurship as discussed above. After all, entrepreneurship is no less important for large, hierarchical firms than for the small start-up venture. However, in large, hierarchical firm organizations, the entrepreneurial function is divided and distributed over various people who have the role of subentrepreneurs. These are usually paid managers at different layers of the hierarchy who provide managerial and some entrepreneurial services in one person. They need to be coordinated with an overarching, socially shared business conception as the cognitive frame for their various, distinct operations in order to ensure that the entire organization will not lose corporate coherence and will not risk

---

[7] Occasionally, the latter may be more of a burden than an asset for pursuing new business opportunities inside the organization. Taking a strategic management perspective, Leonard-Barton (1995) observes that the firm's "core capabilities" can sometimes turn into "core rigidities," as they become strongly embedded in the people who deploy their competencies in specific ways. Leonard-Barton not only identifies cognitive biases and "signature skills" as obstacles to change, but even asserts that there are rigidities arising from "knowledge-base-specific values."

frictions resulting from rivaling orientations (Witt 2000). This is easier to accomplish when those performing an entrepreneurial function have been "socialized" inside the firm (cf. Penrose 1959, Chapter 4, who claims that the availability of such personnel *inside* the firm is a limiting factor in the expansion and diversification of the firm).

Under these conditions, an ingenious option for large firms is to try to solve the two problems – that of securing a continued knowledgetransfer and that of accounting for its organizational coordination – with one and the same measure. This measure is the intraorganizational career paths for scientifically trained employees which lead from technical to management functions.[8] In offering such options, large firms can, on the one hand, strengthen their attractiveness as employers for migrating researchers who are about to sort themselves into entrepreneurs and employees. As mentioned in the section, the prospect of professional in-house promotion is an important source of motivation for getting migrating scientists to sign on with firms.[9] The offer of intraorganizational career plans can thus bolster the firm's effort to acquire the latest technological knowledge.

On the other hand, this measure at the same time allows large firms to build a resource base in house in which technical expertise and entrepreneurial talent can be combined. Employees differ in their abilities, their aspirations, their task responsiveness, their working attitudes, and, not least, their entrepreneurial talents. Since personal promotion is usually supposed to be contingent on the employee's demonstrated performance, there is time for an attempt to verify the employees'

---

[8] There may, of course, be still other motives for encouraging turnover within the firm organization through which employees with technical functions in corporate R&D are promoted to managerial ones. One of these may be the expectation that the individual creativity and productivity profiles of the former scientists in R&D decline after a certain age as they have been shown to do for academic researchers, cf. Levin and Stephan (1991). This motive is independent of the form of knowledge to be acquired by the firm. It suggests also offering career path options where scientifically trained staff is hired so as to create an absorptive cognitive capacity for the transmission of encoded knowledge.

[9] From the point of view of the hiring firm, there is yet another potential advantage to relying on this measure. Compared with an attempt to completely compensate the potential decay of embodied knowledge by monetary incentives, the career path option is less costly. In addition, the offer of personal promotion, emphasizing the stability of the employment relationship and possible assignment of increasing responsibility depending on performance tends to foster the employees' intrinsic motivation. Intrinsic motivation, in turn, is a precondition for the transfer of tacit knowledge among employees, and hence for coordination, cf. Osterloh and Frey (2000). In knowledge-*producing* functions, such as R&D, outputs are often difficult to attribute to individuals, rendering direct monetary compensation inappropriate or even detrimental. This is particularly true when tasks are ambiguous or multiple.

personal skills and abilities.[10] In this way, a sorting process like the one discussed in the section can be emulated within the firm's internal labor market. The firm can identify and seek to deploy the entrepreneurial talent of employees who are scientifically trained and have gained a background in the technical capabilities and procedures of the firm. (The importance of technological expertise in this combination is underscored by the fact that large firms often try to confer business competencies to employees with a background in the sciences and technology by providing the relevant courses and training sessions.)

During their individual careers, such employees would typically move along the spectrum from technical problem-solving tasks toward problems that require entrepreneurial knowledge coordination across internal functional boundaries. The sequential exposure to functional areas in the organization means that knowledge and competencies can be accumulated which span many of the firm's internal tasks, projects, and functional areas. When the employee eventually ends up in an entrepreneurial position, the accumulated firm-specific experience allows the views from several subdivisions to be integrated and diverging interests to be arbitrated. Particularly with respect to the integration of corporate R&D with other functional divisions in the firm, career trajectories starting in R&D can serve an important bridging function. Career trajectories create the personal basis for making formal and informal contacts. Moreover, people familiar with the differing subcultures of the divisions can mediate a better mutual understanding of needs and constraints and of goodwill in the informal interactions between those divisions.[11] With these exceptional capabilities, employees selected by

---

[10] However, to be credible and, within limits, predictable for the employee, promotion plans have to a certain extent to be standardized and supported by examples of career trajectories that can be observed by those in lower hierarchical ranks. How the human resources should be managed and promoted within the firm organization to attain the multiple goals is a question to which justice cannot be done in this paper. There is evidence that scientists in corporate R&D rate nonscientific managerial work very differently, and not overwhelmingly positively (Bowden 2000). However, the relatively flat hierarchies in corporate R&D rule out the possibility of "pure" R&D careers for a significant number of people employed there. The rule therefore is that career success is inevitably associated with the transition to management (Biddle and Roberts 1993, Bowden 2000). Yet it is not certain that all scientifically trained employees in R&D can indeed be offered managerial functions at a later point in time. If the number of promotions is too small, this may strain the credibility of the career path option and, in turn, the firm's knowledge-sourcing strategy based on recruitment.

[11] In the context of his analysis of Japanese telecommunications firms, Fransman (1999, p. 128) reports that engineers are regularly transferred between R&D and the operating divisions in either direction to facilitate the development and transfer of new technologies. The effects on knowledge coordination of that kind of job rotation seem to be similar. However, the temporary nature of job rotation excludes the motivational effects related with personal promotion plans as discussed before.

personal promotion plans should best be able to conceive of, and assess, new business opportunities. They may be given the opportunity to develop an own business conception in a subdivision of the firm organization and to participate in the possible profits.

Indeed, it may be necessary for the incumbent firm to offer entrepreneurially minded employees such an opportunity in order to account for those cases where the initial sorting of migrating scientists into entrepreneurs and employees tends to be unstable. Some of the former scientists with an entrepreneurial talent may have initially underrated their capacity to create an own firm and have therefore become employees. In the meantime, they may have learned enough about their firm's knowledge base, about technological opportunities, including those not pursued by their employing firm, and about running a commercial venture. They may therefore have arrived at an improved own business conception. If so, they may be inclined to give it a try in a "second" sorting and, thus, to quit and to start what appears from outside as a spin-off (cf. Klepper 2001).

Even worse for the incumbent firm, with their improved business conception, the former employees could try to attract and entice away other members of their former firm whose productivity they may have learned to appreciate. This would amount to a "fissioning" of a whole group of employees as it can occasionally be observed particularly in new, highly innovative, industries (Ziegler 1985). If the firm wants to prevent this, those employees who may be inclined to re-sort themselves have to be transferred to a more highly valued position in the organization. To ensure that they can still be profitably employed, this must usually be a higher position in the firm hierarchy with some entrepreneurial responsibility – precisely what the notion of a career commonly stands for.[12] Sometimes the site of the higher positions may, of course, be in affiliated companies or joint ventures.

12.5 Conclusions

The knowledge transfer from academic research into commercial production and marketing activities is based, it has been claimed in this chapter, on several entrepreneurial services. The transfer does not

---

[12] When Stephan (1996, p 1211) asks "why do companies adopt compensation strategies that impair the productivity of scientists by tying salary increases to the assumption of managerial responsibilities?" an answer may therefore be given as follows. This may not be a strategy discriminating against scientists in corporate R&D. Rather, it may be a strategy that raises the opportunity costs for those who could be tempted to opt out into an own (and possibly competing) spin off firm as just described.

happen unless someone conceives of business opportunities in new scientific knowledge in the first place. Then these (often only vague) imaginings have to be transformed into conceptions of how to run a business firm. On that basis, resources – foremost the necessary knowledge resources – have to be attracted and coordinated. Where business organizations already exist, the new knowledge resources must, furthermore, be integrated with the organization's expertise and capabilities to yield a coherent business conception. It has been argued that each of these services is a core element of *knowledge-based* entrepreneurship.

Like all forms of entrepreneurship, knowledge-based entrepreneurship is confronted with certain constraints. However, there are some specific constraints that result from the peculiarities of scientific knowledge and its mode of transfer. These specific constraints – which only become apparent once the simplistic reduction of academic science to a knowledge output in encoded form is abandoned – have been center stage in this chapter. It has been shown that they shape the way in which the commercialization of new technologies is organized. The procedural nature of the knowledge that typically underlies new scientific insights implies that it can only be acquired by being actively engaged in scientific research. In other words, such tacit knowledge is hard, or even impossible, to encode and therefore needs to be carried in embodied form. However, in a rapidly progressing research environment, state-of-the-art tacit knowledge is changing constantly. Without the automatic update on the job in a continued involvement in scientific research, privately held tacit knowledge is therefore subject to a relatively rapid decay.

Depending on the form of knowledge, the proper entrepreneurial organization of the knowledge transfer varies. Two ways of organizing the transfer have been discussed here: start-up firms run by former scientists as entrepreneurs and large, incumbent firm organizations with specialized R&D staff. Both compete for the human resources needed to realize the knowledge transfer. From the point of view of the scientists and engineers who are about to migrate from academic research into the commercial sector that competition results in a self-sorting process. These former scientists can realize the commercial value of the technological knowledge they have acquired in academic research either by setting up an own entrepreneurial start-up firm or by becoming employees in the large, incumbent firms. As was explained, the constraints implied by the tacit, embodied form of knowledge force both organizational solutions to adopt particular knowledge-transfer strategies. Start-up firms have to find measures to cope with the decay of the founder's knowledge as time elapses. The large, incumbent firms have to find ways to make migrating scientists decide in favor of becoming employees and to stabilize this initial outcome of the sorting process in order to prevent spin-offs and fissoning at later stages.

References

Biddle, J. and Roberts, K. 1993. "Private Sector Scientists and Engineers and the Transition to Management," *Journal of Human Resources* 29: 82–107

BMBF – Bundesministerium für Bildung und Forschung 2001. *Zur technologischen Leistungsfähigkeit Deutschlands*. Bonn: BMBF, Referat Öffentlichkeitsarbeit

Bowden, V. 2000. *Managing to Make a Difference – Making An Impact On the Careers of Men and Women Scientists*. Aldershot: Ashgate

Buenstorf, G. and Murmann, P. 2003. "Ernst Abbe's Scientific Management: Theoretical Insights from a 19th Century Dyanmic Capabilities Approach," *Papers on Economics and Evolution*, Max Planck Institute Jena, n. 0312

Cohen, W. M. and Levinthal, D. A. 1989. "Innovation and Learning: The Two Faces of R&D," *The Economic Journal* 99: 569–96

Cohen, W. M., Nelson, R. R., and Walsh, J. P. 2002. "Links and Impacts: The Influence of Public Research on Industrial R&D," *Management Science* 48: 1–23

David, P. A. 1998. "Common Agency Contracting and the Emergence of "Open Science" Institutions," *American Economic Review* 88 (2): 15–21

Dasgupta, P. and David, P. A. 1994. "Toward a New Economics of Science," *Research Policy* 23: 487–521

Edquist, C. and McKelvey, M. (eds.) 2000. *Systems of Innovation: Growth, Competitiveness and Employment* Vol. I, Cheltenham: Edward Elgar Publishing

Fornahl, D. 2005. "The Impact of Regional Social Networks on the Entrepreneurial Development Process," in Fornahl, D., Zellner, C., and Audretsch, D. B. (eds.), *The Role of Labour Mobility and Informal Networks for Knowledge Transfer*. Dordrecht: Springer, 53–78

Fransman 1999. *Visions of Innovation – The Firm and Japan*. Oxford: Oxford University Press.

Galambos, L. and Sewell, J. E. 1995. *Networks of Innovation: Vaccine Development at Merck, Sharp & Dohme, and Mulford 1895–1995*. Cambridge: Cambridge University Press

Grupp, H. 1998. *Foundations of the Economics of Innovation*. Cheltenham: Edward Elgar

Henrekson, M. and Rosenberg, N. 2001. "Designing Efficient Institutions for Science-based Entrepreneurship: Lesson from the US and Sweden," *Journal of Technology Transfer* 26: 207–31

Heuss, T. 1946. *Robert Bosch – Leben und Leistung*. Tuebingen: Wunderlich und Leins

Keck, O. 1993. "The National System for Technical Innovation in Germany," in Nelson, R. R. (ed.), *National Systems of Innovation: A Comparative Analysis*. Oxford: Oxford University Press, 115–57

Klepper, S. 2001. "Employee Startups in High-Tech Industries," *Industrial and Corporate Change* 10: 639–74

Leonard-Barton, D. 1995. *Wellsprings of Knowledge – Building and Sustaining the Sources of Innovation*. Boston: Harvard Business School Press

Levin, S. G. and Stephan, P. E. 1991. "Research Productivity Over the Life Cycle: Evidence for Academic Scientists," *American Economic Review* 81 (1): 114–32

McKelvey, M. 1996. *Evolutionary Innovations: The Business of Biotechnology*. Oxford: Oxford University Press

Murmann, J. P. 2003. *Knowledge and Competitive Advantage*. Cambridge: Cambridge University Press

Nelson, R. R., (ed.) 1993. *National Innovation Systems*. Oxford: Oxford University Press

Osterloh, M. and Frey, B. S. 2000. "Motivation, Knowledge Transfer, and Organizational Forms," *Organization Science* 11 (5): 538–50

Penrose, E. 1959. *The Theory of the Growth of the Firm*. Oxford: Oxford University Press

Polanyi, M. 1967. *The Tacit Dimension*. Garden City: Doubleday

Porter, M. E. 1990. *The Competitive Advantage of Nations*. New York, N Y: Free Press

Rathe, K. and Witt, U. 2001. "The Nature of the Firm – Static versus Developmental Interpretations," *Journal of Management and Governance* 5: 331–51

Reinhardt, C. 1997. *Forschung in der chemischen Industrie – Die Entwicklung synthetischer Farbstoffe bei BASF und Hoechst, 1863 bis 1914*. Technische Universität Bergakademie Freiberg, Freiberg: TU, Bergakademie

Rosenberg, N. 1990. "Why Do Firms Do Research (With Their Own Money)?," *Research Policy* 19: 165–74

Salter, A. J. and Martin, B. R. 2001. "The Economic Benefits of Publicly Funded Basic Research: A Critical Review," *Research Policy* 30: 509–32

Shane, S. 2000. "Prior Knowledge and the Discovery of Entrepreneurial Opportunities," *Organization Science* 11: 448–69

Stephan, P. E. 1996. "The Economics of Science," *Journal of Economic Literature* 34: 1199–235

Witt, U. 1987. "How Transaction Rights Are Shaped to Channel Innovativeness," *Journal of Institutional and Theoretical Economics* 143: 180–95

Witt, U. 1998. "Imagination and Leadership – The Neglected Dimension of an Evolutionary Theory of the Firm," *Journal of Economic Behavior and Organization* 35: 161–77

Witt, U. 2000. "Changing Cognitive Frames – Changing Organizational Forms: An Entrepreneurial Theory of Organizational Development," *Industrial and Corporate Change* 9 (4): 733–55

Zellner, C. 2003. "The Economic Effects of Basic Research: Evidence for Embodied Knowledge Transfer via Scientists' Migration," *Research Policy* 32 (10): 1881–95

Ziegler, C. A. 1985. "Innovation and the Imitative Entrepreneur," *Journal of Economic Behavior and Organization* 6: 103–21

Comments to Chapters 11 and 12

Maureen McKelvey

"Innovation and Entrepreneurship" are key issues within economic analysis inspired by Joseph A. Schumpeter. Such issues are addressed in the two chapters, "Schumpeterian Legacies on Entrepreneurship and Networks: Notes from a Distal Point of View" by Piera Morlacchi and "Knowledge-based Entrepreneurship: The Organizational Side of Technology Commercialization" by Ulrich Witt and Christian Zellner.

Schumpeter's key insight was that change is endogenous to economic systems: it is not imposed from without, but rather is generated within, as explored in these two chapters. Schumpeter argued that fundamental change in existing activities as well as the introduction of entirely novel activities would keep providing the 'fuel' to the capitalist engine (Schumpeter 1947: 82–83). Through innovations and creative destruction, the existing and old activities are modified but keep existing – as well as help give rise to new ones. Nelson (1996: 87) argues that this is "Schumpeter's most consistent and elaborated argument about innovation and economic transformation, that it fundamentally involves disequilibrium and that standard equilibrium theory in economics cannot cope with it and its economic consequences." In other words, innovation and entrepreneurship continue to disrupt the economy, thereby fundamentally changing activities and moving the economy into new directions. This is a forward-looking view of the economy. This is an economy where learning, heterogenous expectations, and actions among a range of different organisations and individuals keep shifting and reorienting economic activities along new trajectories of development. As such, the nature of industrial dynamics, structural change and transformation depend upon flexibility and inertia, upon transformation and stability (McKelvey and Holmén 2006: 11–15). Despite innovation and entrepreneurship being key issues in this type of economic analysis, many research questions remain about how to analyze and explain the relationship between them and different aspects of economic transformation.

The two chapters, respectively, by Morlacchi and by Witt and Zellner take up this challenge in different ways, thereby making separate contributions to the field.

"Schumpeterian Legacies on Entrepreneurship and Networks: Notes from a Distal Point of View," the chapter by Piera Morlacchi, argues that network concepts and tools can be applied to analyzing entrepreneurship. The chapter first reviews Schumpeter's legacies, in terms of how the idea of entrepreneurial action represents entrepreneurs as individuals as having a function in the economic system, for example, that economic changes occur through the purposeful actions of individuals. By stressing economic action as embedded in the social context, Morlacchi argues that contemporary analysis of networks, as found especially in economic sociology, provides concepts and tools in order to model the environmental context of entrepreneurship. Morlacchi concludes by discussing the new research opportunities afforded. He explores existing and future lines of research which could use network theory, especially the sources and determinants of entrepreneurial behaviour, entrepreneurial action, entrepreneurial policy, and entrepreneurial history. Thus, Morlacchi is arguing that network analysis is a very useful way to explore how and why entrepreneurship as action affects, and is affected by, the broader context.

As quoted in this chapter, Swedberg (1991: 3–77) has explored related issues within Schumpeter's own writings, albeit without dealing with networks. Swedberg provides extensive arguments for when, and where, Schumpeter distinguished the value of "economic theory" as a separate realm of understanding but also as a complement to understanding acquired through history and statistics. In relation to this exposé, Morlacchi's argument seems plausible that "the idea of social interaction of heterogeneous actors, what we would call networks, was implicit" in Schumpeter's work, and thereby an avenue for further research. Swedberg (1991: 56), however, also makes a different point. He stresses that Schumpeter consistently excluded sociology from his analysis in early work and also stressed the importance of economic history, but not as a substitute for theory.

From this interpretation, we can reassess Morlacchi's main claim. To follow in a Schumpeterian legacy, it is vital that network analysis will be used not only as a descriptive tool. Instead, the value of these network concepts and tools will lie in further explaining empirical observations in terms of theory, and in more systematically linking empirical observations and theoretical explanations. Theoretically, the main issues that networks analysis must therefore address is how and why these entrepreneurial actions and function affect further changes in the economic system. For example, Powell, Koput, and Smith-Doerr (1996) argue that the structure of the network and the position of agents within it fundamentally determine agents' access to relevant sources of scientific and technological knowledge and therefore innovative activities and performances (see also Walker, Kogut, and Shan, 1997). This type of network analysis has been used to explain the trajectories and performance of different types of organizations which may express the entrepreneurial function – with implications for understanding the competitiveness of individuals, firms, and regions. Hence, clearly network analysis can be used not only to engage in empirical analysis of nodes and relationships but also to help explain theoretically the relationships between actors, networks, and different trajectories of economic development.

The chapter "Knowledge-based Entrepreneurship: The Organizational Side of Technology Commercialization" by Ulrich Witt and Christian Zellner takes up the challenge of understanding innovation and entrepreneurship from the perspective of knowledge and organizations, developing over time. This chapter argues that knowledge transfer, from academic institutions to firms that commercialize it, is essentially an entrepreneurial process. They begin by arguing that entrepreneurship requires command over suitable resources. Hence "in the case of knowledge-based entrepreneurship these are, in particular, resources enabling the access to, and the exploitation of, new technological knowledge." This chapter first examines the institutional boundaries between science and commerce, in order to understand the entrepreneurial choices of how to conduct knowledge transfer. With respect to state-of-the-art tacit knowledge components, the main mode of transfer, they argue, is the hiring of scientists from academic institutions to firms.

One of the main issues explored in this chapter, in relation to both start-up firms and large incumbent firms, is how to create incentives for employment, without prohibitive costs, and how to stimulate and renew the flow of state-of-the-art technological knowledge into the firm. On the one hand, knowledge-based entrepreneurship depends upon several more general elements, such as conceiving business opportunities, conceiving how to run the business firm, coordinating resources, and integrating knowledge into existing organizational strengths. On the other hand, this chapter argues that there are particular opportunities and constraints, due to the pecularities of scientific knowledge and its mode of transfer. The authors stress that start-up firms run by former scientists as well as large, incumbent firms with specialized R&D departments represent different organizational forms, which must solve the competitive problem of realizing the commercial value of scientists' technological knowledge.

This chapter by Witt and Zellner make one key assumption, namely that once the scientist leaves an academic institution, the value of his or her tacit knowledge decays over time. First, they follow much of the existing literature by conceptualizing scientific and technological knowledge in relation to codified and tacit aspects, where the latter includes skills, experimental procedure, and procedural knowledge (Foray 2004). Witt and Zellner then go on to argue that tacit knowledge components are the most valuable in state-of-the-art technological fields. Second, the authors discuss the opportunities and constraints to the firm, of the entrepreneurial action of knowledge transfer as the hiring, firing, and loss of academic scientists. They can focus on this mode of transfer, owing to the assumption that scientists gain their main skills while doing science at the *university, research institute, or other academic institution*. This assumption is crucial to the authors' argumentation – but also raises a number of interesting questions, with implications for understanding knowledge-based entrepreneurship.

Without detracting from the authors' elegant expose, a simple question arises, namely what happens if the employment of (new) scientists is not the main mode of transfer of knowledge. What if the scientist leaves the academic institution, but instead of the value of their knowledge decaying over time, the scientist retains skills because he or she continues to engage in the production of new scientific and technological knowledge in the firm? Studies using histories of science and technology make such claims. They have presented many case studies of technologies and of industries, where firms are very actively involved in further development not only of the commercial product but also of the scientific and technological knowledge (Constant 1980, Vincenti 1990, Houndshell 1996, McKelvey 1996). These empirical studies strongly suggest that firms need to solve different domains of problem-solving in technological knowledge found in universities, not least in order to integrate and coordinate diverse knowledge into a commercial product. As such, this suggests that scientists as well as engineers do not necessarily 'lose' their tacit skills – instead, they retain and develop them in the firm, thereby forming new technologies and new areas of knowledge (Rosenberg 2004).

This would thereby imply that scientists employed in start-ups and in incumbent firms can renew and even develop new technological knowledge within the firm's boundaries. The problem of knowledge-based entrepreneurship thereby becomes somewhat more complex, because of the need for organizational forms to facilitate the renewal and creation of new knowledge competencies. The firm may need other mechanisms, such as making sure that the engineers and scientists continue reading literature, translating new research results into the goals of the firm, and engaging in scientific activities such as publishing and conferences. Hence, some parts of the literature on science, technology, and innovation would suggest that the organizational challenges of science and technology lie in developing and renewing this type of knowledge resource, by working on problems of interest to the firm.

In concluding, these two chapters provide insights into how networks and the organization of scientific and technological knowledge resources enable competition through innovation and thereby "fundamentally involve disequilibrium". Morlacchi's emphasis on entrepreneurial action – set in relation to individuals and context – suggests links between the novelty introduced through entrepreneurship and the stability (or structure) represented by the network. Witt and Zellner stress the value of new scientific and technological knowledge, as an entrepreneurial problem, as set within the perception of opportunities and their realization in start-ups and incumbent firms. In different ways, both chapters point to interesting areas of new research, within Schumpeter-inspired economics.

References

Constant, E. 1980. *The Origins of the Turbojet Revolution*. Baltimore: Johns Hopkins University Press

Foray, D. 2004. *The Economics of Knowledge*. Cambridge, MA: The MIT Press

Houndshell, D. 1996. "The Evolution of Industrial Research in the United States" in Rosenbloom and Spencer (eds.), *Engines of Innovation: U.S. Industrial Research at the End of an Era*. Harvard Business School Press

McKelvey, M. (1996). *Evolutionary Innovations: The Business of Biotechnology*. Oxford: Oxford University Press

McKelvey, M. and Holmén, M. 2006. *Flexibility and Stability in the Innovating Economy*. Oxford: Oxford University Press

Nelson, R. 1996. *The Sources of Economic Growth*. Cambridge, MA: Harvard University Press

Powell, W., Koput, K.W., and Smith-Doerr, L. 1996. "Inter-Organizational Collaboration and the Locus of Innovation: Networks of Learning in Biotechnology," *Administrative Science Quarterly* 41: 116–45

Rosenberg, N. 2004. "Science and Technology: Which Way Does the Causation Run?," Paper presented at the opening of the Stanford University, "Center for Interdisciplinary Studies on Science and Technology", http://www.econ.upf.edu/crei/activities/sc_conferences/23/papers/rosenberg.pdf, accessed on November 1, 2004

Schumpeter, J. A. [1947] 1975 second edition. *Capitalism, Socialism and Democracy*. New York, NY: Harper

Swedberg, R. 1991. *Joseph Schumpeter: The Economics and Sociology of Capitalism*. Princeton, NJ: Princeton University Press

Vincenti, W. G. 1990. *What Engineers Know and How They Know It: Analytic Studies from Aeronautical History*. Baltimore: John Hopkins University Press

Walker, G., Kogut B., and Shan, W. 1997. "Social Capital, Structural Holes and Formation of an Industry Network," *Organization Science* 8 (2): 109–12

Part 7

Innovation and evolution of the
university system

13 Academic entrepreneurs and technology transfer: who participates and why?

Janet Bercovitz and Maryann Feldman

Introduction

According to Schumpeter, innovation is about entrepreneurship: the implementation of new ideas that change established procedures and alter organizational practices (1934). While the idea of creative destruction is compelling, there are few opportunities to observe the characteristics of change agents in situations where new practices emerge. University technology transfer – the realization of commercial value from university research – presents such an opportunity. While universities are an important source of invention and new knowledge, there is great variation across universities in the commercial realization of academic discoveries (Nelson, 2001). This result is understandable when we consider that university technology transfer has only become a formal activity for most universities in the United States in the last twenty-five years. In this regard, a series of changes marked by the passage of the 1980 Patent and Trademark Law Amendment Act (Pl 96–517), commonly known as the Bayh–Dole Act, represent gales of change as universities embrace new objectives that value active technology commercialization over older routines that promoted passive knowledge diffusion. However, the commercial realization of academic discoveries is ultimately dependent on the personal decisions and actions of the faculty, as faculty invention disclosures form the basis for university patents and subsequent licenses. Though the Bayh–Dole Act specifies that faculty members are to disclose their inventions to the university technology-transfer office, enforcement of this requirement

We wish to acknowledge support from the Andrew W. Mellon Foundation for this Chapter as part of a larger project on evolving university–industry relationships. We are indebted to the technology-transfer personnel, research administrators, and medical school faculty at Duke University and Johns Hopkins University for generously sharing their time and expertise in identifying salient issues. This paper has benefited from discussions with Irwin Feller and Rich Burton.

has proven difficult. When individual faculty members choose to disclose their discoveries to the university's technology-transfer office, they signal that they are entrepreneurial in adopting the new initiative that aids in the transfer of knowledge out of the university to established companies or to use in the formation of new companies.

This chapter examines the process of university technology transfer and delves into the question of which faculty members will act entrepreneurially by selecting to participate in the technology-transfer process. We examine the disclosure behavior of individual faculty members at the medical schools of two prominent research universities, Johns Hopkins and Duke University. Both universities are late entrants to technology transfer, as defined by Mowery and Ziedonis (2002). Until the mid-1980s, neither institution had significant technology-transfer activity or a dedicated office of technology transfer and licensing. This was well behind similarly ranked universities with medical schools (AUTM, 2002). However, by 1990, the administrations at both these universities embraced technology transfer as an institutional objective and created incentives to encourage faculty to participate. During the decade of the 1990s, they made substantial, yet varying, progress in carrying out this new mission.

This chapter begins by examining university technology transfer as a new routine for universities. We consider the factors that influence an individual faculty member's decision to disclose inventions. The next section draws upon the literature and the results of interviews with technology transfer managers and faculty members to develop a set of propositions about the individual faculty member's decision to disclose new inventions. The third section of the chapter introduces the data and methodology. To investigate the characteristics of faculty members who engage in this entrepreneurial initiative, we examine the faculty in the medical schools at Johns Hopkins and Duke university. The majority of technology-transfer activity is focused on biomedical applications, and medical schools have been a focus of this activity. We find evidence that a faculty member's participation in technology-transfer activities is a function of that individual's inventive capacity and their entrepreneurial propensity. The fourth section provides empirical results.

13.1   The technology-transfer process

Research commercialization has emerged as a new mission for American universities that differs from the older norms favoring the open dissemination of research discoveries (see Nelson, 2001). Faced with budgetary difficulties, universities now attempt to actively market their discoveries to industry, use their inventions to form new companies, and engage in commercial activity related to economic development. These initiatives are described by Slaughter and Leslie (1997) as marking a new era of academic capitalism while Etzkowitz (1983) uses the term entrepreneurial universities.

The process of change occurred in three distinct phases. First, while there was limited early institutional experimentation with technology transfer, the passage of the 1980 Bayh–Dole Act provided a broad legislative mandate and legitimized these activities. The Bayh–Dole Act allows for the transfer of exclusive control over government-funded inventions to universities for the purpose of further development and commercialization. Universities are then permitted to license the inventions to other parties and retain any licensing fees that may result. This was followed by a second stage of implementation as institutions set up dedicated technology-transfer offices, adopted the policies and procedures to minimize conflicts of interest, and established royalty-sharing incentives, among other supportive activities. By 1998, every Carnegie I and II American Research University (Feldman et al. 2002) had a dedicated technology-transfer office.

Still, even with these institutional changes, there was a noted performance gap in the realization of these new initiatives (Siegel, Waldman, and Link, 2003). Becoming an entrepreneurial university active in technology transfer requires the participation and commitment of the faculty. The entire technology-transfer process is predicated on individual faculty members disclosing their inventions to the university's technology-transfer office. After all, if individual faculty members do not disclose their inventions, then there can be no patenting and no subsequent downstream licensing and licensing revenue. Without disclosures, regardless of the amount of research dollars, quality of the faculty at the institution, or any other asset measure, the institution will not be productive in technology transfer. Thus the individual faculty member's decision to disclose inventions is critically important. These invention disclosures are evaluated as to their commercial potential and the next stage is protecting the ideas in the invention disclosure through formal intellectual property protection. An invention disclosure may result in one or several patents. Once patents have been granted, the intellectual property may be licensed to firms.

On face value, it seems that disclosing research results should be straightforward. First, increased technology-transfer activity has become an articulated goal of the university administration and is espoused as a strategic initiative at most academic institutions. Royalty-sharing incentives have been adopted and technology-transfer offices actively encourage faculty participation. Second, disclosing research results to the technology-transfer office is a stipulation of federal research grants, which constitute the largest source of US university research funding. Third, the costs associated with disclosing an invention are low with the required forms available online. Fourth, there are limited quality barriers, as there are no objective standards that faculty discoveries are required to meet to warrant filing an invention disclosure with the technology-transfer office. Finally, technology-transfer managers are trying to encourage faculty to disclose since the number of faculty disclosures is one criterion used to evaluate the performance of the technology-transfer office. To the extent that technology-transfer managers are trying to encourage greater disclosure by the faculty, there does not appear to be any screening device that would discourage faculty. Indeed, Mowery, Sampat, and Ziedonis (2000) note that only about twenty percent of disclosures were patented after six years. While the cost associated with filing a disclosure is negligible, the cost of filing a patent is approximately $100,000, indicating that greater scrutiny accompanies this later stage of the technology-transfer process.

Thursby, Jensen, and Thursby (2001) argue that invention disclosures represent a subset of university research with commercial potential, and later suggest three reasons why faculty would choose not to disclose research results (Thursby and Kemp, 2002). First, faculty who specialize in basic research may not disclose because they are unwilling to spend time on the applied R&D required to interest businesses in licensing the invention. This is perhaps countered by the trend toward patenting basic scientific results from the human genome. Second, faculty may not disclose inventions because they are unwilling to risk publication delays that may be required to interest industrial partners in licensing the technology. Interviews with both faculty members and technology-transfer and licensing officials, however, indicate that this is more a perceptual problem than a reality. There are strategic ways to accommodate both academic and commercial interests, but this requires a sophisticated understanding of the technology-transfer process. Trusted peers who are familiar with the process can communicate strategies to accommodate both academic and commercial interests. Third, faculty members may not disclose because they believe that commercial activity is not appropriate for an academic scientist. This view certainly represents the older norms of academic science. However, when faculty members disclose inventions, they appear to have adopted the newer norms, choosing to become academic entrepreneurs.

Few studies have examined the internal process of disclosing academic inventions and the factors that underlie the decision to disclose (cf. Bercovitz and Feldman, 2006). The question examined in this chapter is what are the characteristics of faculty members who disclose their inventions, choosing to become entrepreneurial academics? To develop hypotheses, we rely on interviews with technology-transfer officials and faculty members. Given that the profitability of disclosing is uncertain and also that every individual in the faculty is engaged in research, we assume that every individual in the medical school has the potential to disclose inventions. We expect that faculty would be responsive to financial incentives and that there would be a direct relationship between licensing royalty distribution rates and the amount of technology-transfer activity across universities. While we may expect that certain fields of research would be more amenable to disclosure, the lack of objective standards indicates that it is an individual decision. Both universities we examine have a similar distribution rate with one-third of future revenue going to the individual faculty member, one-third going to the central administration, and one-third to the department. Departments "sweeten the deal" by distributing a share of their third of the royalties directly to the inventing faculty member's lab. This practice was first used to encourage technology transfer; however, by 1991, the first year that we consider it was well established across departments in both universities.

Our interviews suggest that two key attributes – inventive capacity and entrepreneurial propensity – are believed to differentiate academic entrepreneurs from the traditional academic preoccupation with publish or perish.

13.1.1   Inventive capacity

The breadth of an individual's training is expected to influence inventive capacity. Koestler (1990) argues that bisociation – the ability to relate two seemingly unrelated concepts – is at the root of creativity and invention. Considering breadth of knowledge as a necessary condition for bisociation, entrepreneurial research has shown that individuals with inter-disciplinary educational backgrounds and expansive prior knowledge are better positioned to recognize, and then act upon, innovation opportunities (Venkataraman, 1997; Shane, 2000). Individuals who commonly encounter multiple theoretical perspectives in their professional role – boundary-spanners who occupy positions in multiple functional or technological departments, for example – are also more likely to be skilled in evaluating, integrating, and responding to diverse information (Cohen and Levinthal, 1990). That is, these boundary-spanning individuals are more apt to have the unique talent of being able to "think outside the box" and take novel approaches to solving existing scientific puzzles. Further, these boundary-spanning individuals are likely to be "central" players in a network interacting with a greater range of individuals and having access to a broader and more diverse set of information. This too will enhance the likelihood of discovery (Burt, 1992; Aldrich, 1999). Within universities some faculty members hold positions in multiple departments and can be identified as boundary-spanners. As such, we hypothesize:

Hypothesis 1 (H_1): Individuals that hold boundary-spanning positions will be more likely to disclose innovations.

Numerous studies of product development and innovation show that using cross-functional teams – to bring together individuals with differing but complementary skills sets – is associated with improved performance (Bantel and Jackson, 1989; Clark and Fujimoto, 1991; Dougherty, 1992). The ability to integrate diverse knowledge pertaining to both upstream (research) and downstream (application/market) issues is held to be a key source of these performance advantages (Brown and Eisenhardt, 1995). Similar to "cross-functional" teams, dual-degree individuals can leverage the diverse skills acquired through their doctoral (research-oriented) and medical (application-oriented) training to generate innovations with commercial potential. Individuals who hold both a Ph.D. degree and an M.D. are expected to have training that encompasses research and practical application. In this respect, they may be in an advantageous position to develop new science with an eye to the commercial potential for such innovations. This in turn may lead these individuals to embrace technology-transfer activity, which leads us to:

Hypothesis 2 (H_2): Individuals that hold dual degrees will be more likely to disclose innovations.

13.1.2   *Entrepreneurial propensity*

Inventive capability alone is not enough, however. This capacity needs to be linked with an entrepreneurial propensity if new ideas are to be commercially exploited. There are several factors that may contribute to an individual's tendency to be entrepreneurial. These factors may be innate or learned.

Two often cited traits associated with entrepreneurs are comfort with risk-taking and a strong internal locus of control – the belief that individual actions can and will influence outcomes (Thornton, 1999). Though these personality attributes are associated with entrepreneurial behavior, they are also associated with achievement in organizational contexts. As such, the presence of these traits appears to be necessary but not sufficient for the emergence of entrepreneurial activity.

These same traits appear to play a role in immigration decisions of highly skilled individuals such as scientists, engineers, and doctors. Our interviews suggest that professionally trained individuals who self-select to leave their home country and seek opportunities abroad exhibit both a willingness to take risks as well as confidence in their ability to succeed in new environments. Further, evidence is growing that such skilled immigrants do show an increased likelihood to become involved in entrepreneurial activity. For example, Saxenian (2002) finds a strong presence of new immigrant entrepreneurs in Silicon Valley, noting that close to one-quarter of the high-technology firms started in this region during the 1980s and 1990s are run by either Chinese or Indian immigrants.

It is relatively common for American universities to hire faculty members who trained outside of the United States. This is particularly true for technology-based fields since epistemic communities are international, credentials are easily transferred, and diversity is valued. According to the National Science Foundation (NSF) (2000), close to seventy percent of the total 1997 doctorate degrees awarded in natural and biological sciences globally were earned in Western Europe and Asia. If the willingness to immigrate signals entrepreneurial propensity, we would expect the following relationship to hold:

Hypothesis 3 (H₃): Individuals who trained outside of the United States will be more likely to disclose inventions.

Entrepreneurial propensity can also be a learned behavior. There is a diverse body of literature on social imprinting that gives background on how norms associated with training influence subsequent behavior and drive the adoption and diffusion of new practices. Many authors have argued that social institutions, with educational institutions being a key subset of this group, mold individual perspective by promoting, both implicitly and explicitly, a particular set of norms and/or values of "how things ought to be done" (Schein, 1985; Locke, 1985; Haas, 1992; Calori et al., 1997; Biglaiser, 2002). DiMaggio and Powell (1983: 153) emphasize the role universities play in this socialization process, stating that those "drawn from the same universities and filtered on a common set of attributes, . . . will tend to view problems in a similar fashion, see the same policies, procedures and structures as normatively sanctioned and legitimated, and approach decisions in much the same way." Support for these arguments may be found across academic domains. For example, in a series of studies, Frank, Gilovich, and Regan (1993) find evidence that economics students, particularly those trained by an instructor with research interests in game theory, are more likely to adopt self-interested behavior than their peers. Similarly, recent work in political economy shows that the presence of US-trained economists is a key predictor of the adoption of various types of neoliberal reform (e.g. tariff rate reduction, capital account liberalization) in emerging markets (Biglaiser and Brown, 2003; Chwieroth and Fellow, 2003). In a similar vein, Finnemore and Sikkink (1998: 905) argue: "Professional training does more than simply transfer technical knowledge; it actively socializes people to value certain things above others." In sum, professional training can instill a particular set of norms and ideas and in acting according to these norms, students serve as a critical conduit for the diffusion of new ideas and practices.

In our context, this logic implies that individuals who trained at institutions where participation in technology transfer was accepted and actively practiced will be more likely to adopt these practices in their own careers. Interviews and anecdotal evidence are supportive of this conjecture. For example, one professor active in technology transfer indicated that his graduate school mentors had disclosed and licensed their technology. He learned about disclosing by observing their experiences and this dictated his expectations for a professional career. While he recognized that when he joined the Hopkins faculty the culture did not support technology transfer, he believed that disclosing would provide a vehicle for implementing his ideas. Similarly, William Brody, current president of Johns Hopkins, started as Assistant Professor of Radiology in 1972. Brody learned about technology transfer during his graduate study at Stanford University's Medical School and Department of Electrical Engineering, a very active department in terms of involvement with industry. Once at Hopkins, he continued to actively disclose inventions and subsequently started a company. His expectation was that technology transfer would be part of his career. In contrast, faculty who received their medical school training at institutions where technology transfer was not perceived as a legitimate activity often questioned the long-term impacts of this activity both on their careers and on the broader pursuit of science. Several, including the chair of a Duke department who trained at Cornell, had no intention of disclosing and expressing strong sentiments against technology-transfer-pursuits even though such activity was now strongly supported by the university administration. This foundation of theoretical logic and anecdotal evidence leads to the following testable hypothesis:

Hypothesis 4 (H_4): Individuals whose graduate training incorporated technology-transfer objectives will be more likely to disclose innovations.

Selection bias may be a potential concern if individuals' familiarity with technology transfer enters into the hiring decision. However, our interviews reveal that this should not be a problem. The reputation of the two medical schools and the resources offered will be more important to an academic scientist. Academic culture relative to technology transfer would not be a primary criterion for accepting a position. From the medical schools perspective, individual excellence as measured by academic publications and standing in the field are the dominant hiring criterion. The ability or propensity to engage in commercial activity is not part of the criteria used in the individual's hiring decision.

13.2   Data, variables, and methods

Our empirical analysis is based on an original database compiled from the technology-transfer office records and other administrative data at Duke University and Johns Hopkins University. Our point of departure is the individual faculty member. We have data for 1779 faculty members across fifteen departments in two medical schools for the years 1991–1999. We elected to examine medical school departments because most technology-transfer activity originates within medical schools. We chose departments for which there was variation in disclosing rates across the universities. Our selection was constrained by the degree to which departments were present in both universities. Under the advice of medical school faculty we selected matching departments – that is places where similar work was being done although the titles of the academic departments are slightly different.

Our unit of observation is the individual faculty member. The dependent variable is equal to zero if the individual did not file an invention disclosure in the three-year window for 1996–1998. The dependent variable is equal to one if the individual filed one or more disclosures. We use a three-year window to track disclosures. This was chosen to capture a larger number of disclosures and also to capture a reasonable time period during which a faculty member might have results to disclose. Thus, we examine faculty who were at the university consistently from academic year 1996–1997 to academic year 1998–1999. Personnel records, university course catalogues, and archival data were used to build records for faculty members. Data on the disclosures and licenses are from the records of the technology-transfer offices at the two universities. The probability of disclosing is estimated using a PROBIT model.

The Table 13.1 summarizes the variables used to test our propositions and their predicted signs. Table 13.2 presents descriptive statistics.

Table 13.1 *Predictions for measures for the two effects*

Greater propensity to disclose if		Variable description	Expected sign
Inventive capacity	H1: Individuals that hold boundary-spanning positions will be more likely to disclose innovations.	Individual has joint appointments in multiple academic departments (1 = if yes; 0 = no)	+
	H2: Individuals that hold dual degrees will be more likely to disclose innovations.	Individual holds both a Ph.D. and M.D. (1 = if yes; 0 = no)	+
Entrepreneurial propensity	H3: Individuals that trained outside of the U.S. will be more likely to disclose inventions.	Individual holds degree from a non-U.S. institution (1 = if yes; 0 = no)	+
	H4: Individuals whose graduate training incorporated technology-transfer objectives will be more likely to disclose innovations.	Count of the number of patents applied for at the individual's graduate institutions during the time of their training	+

The independent variables are measured for the prior time period. Thus, our model is that disclosing in the current time period is a function of activity during the prior time period. We include several control variables in the estimation. Career-stage may be important. Thus we control for experience. Experience is measured as the number of years since the last advanced degree. We expect that disclosure behavior will be influenced by the amount of resources available for scientific investigation. To control for any such influence, we include a dollar measure of the National Institute of Health (NIH) awards received by each faculty member in the previous five-year period. NIH funding is the most prominent source of medical school funding and carries the provision that invention disclosures be filed on the resulting

Table 13.2 *Descriptive statistics correlation matrix*

		Mean	Standard deviation	Minimum	Maximum	1	2	3	4	5	6	7	8
1	Current disclosures	0.169	0.375	0.00	1.00	1.00							
2	Boundary-spanning individuals	0.338	0.473	0.00	1.00	0.17	1.00						
3	Dual degree, holds PhD and MD	0.076	0.266	0.00	1.00	0.16	0.16	1.00					
4	Non-U.S. degree	0.114	0.316	0.00	1.00	0.05	0.00	-0.01	1.00				
5	Graduate institution patenting	5.308	9.937	0.00	114.0	0.15	0.04	0.12	-0.18	1.00			
6	Individual NIH awards	620.1	1869	0.00	19600	0.16	0.15	0.06	-0.02	-0.04	1.00		
7	University (0 = Hopkins; 1 = Duke)	0.494	0.500	0.00	1.00	-0.04	-0.23	-0.07	-0.05	-0.14	-0.03	1.00	
8	Years since last graduate degree	23.22	9.40	4.00	60.00	-0.15	-0.01	-0.10	0.01	-0.49	0.10	0.08	1.00

discoveries. Because grants are received infrequently we use a five-year time window. We also include departmental fixed effects to control for technological opportunity. Finally, we add in a control variable for University to account for variation in the policies and procedures at the technology-transfer offices at the two universities.

13.3   Results

Table 13.3 provides results from the Probit analyses of faculty disclosure behavior. Model (1) provides the baseline model. Several of the control variables are significant. First, we find that resources matter. The likelihood of a faculty member disclosing an invention increases significantly ($P < 0.001$) with the amount of NIH funding that individual received in the prior five-year period. Second, there is a clear vintage effect. Experience years, calculated as the number of years since the last graduate degree, is negatively and significantly ($P < 0.001$) related to participation in technology-transfer activities. The probability that a faculty member will disclose an invention decreases by one percent for each year since the completion of graduate study. Somewhat counter to human capital arguments, we find that the newer, rather than the more established faculty members seek to leverage their research activities for commercial gain. We also ran the model with experience years squared to investigate potential non linearities. The coefficient on the squared variable was not significant.

Finally, we find substantial departmental effects. Faculty in the basic science departments – cell biology, genetics, immunology, microbiology, and neurobiology – have a greater propensity to disclose inventions as compared with faculty in clinical departments – obstetrics/gynecology, pediatrics, and psychiatry – that are more focused on providing patient care. Interestingly, we find no significant differences in disclosure activity across the two universities.

Model (2) builds on the baseline model by adding the inventive capacity and entrepreneurial propensity measures. The explanatory power of the model increases significantly with the addition of the independent variables of theoretical interest. A likelihood ratio test comparing Model (2) to Model (1) is significant with a P-value less than 0.01. The coefficient on the first inventive capacity variable – boundary – is positive and significant ($P < 0.001$). Boundary-spanning individuals, those with appointments in more than one academic department, are seven percent more likely to disclose than peers associated with only one department. In support of H1, we find that having expertise across multiple technological areas is predictive of inventive activity. Similarly, the coefficient on the dual degree dummy variable is positive and

Table 13.3 *Empirical results: PROBIT model: faculty members dependent variable = disclosure filed(0,1)*

Independent variables	Model 1		Model 2	
Boundary-spanning			0.306	***
			(0.085)	
Dual degree			0.426	***
			(0.126)	
Non-U.S. degree			0.362	**
			(0.118)	
Graduate institution patenting			0.009	*
			(0.004)	
Years since last graduate degree	−0.031	***	−0.025	***
	(0.005)		(0.005)	
Individual NIH awards	0.092	***	0.087	***
($1,000)	(0.017)		(0.018)	
University	−0.113		−0.003	
	(0.078)		(0.082)	
Anesthesiology	0.095		0.090	
	(0.162)		(0.164)	
Cardiology	0.438	*	0.531	**
	(0.180)		(0.185)	
Cell biology	0.370	+	0.347	+
	(0.202)		(0.205)	
Genetics	0.723	**	0.613	*
	(0.268)		(0.276)	
Immunology	1.011	***	0.917	***
	(0.232)		(0.236)	
Microbiology	0.780	***	0.739	**
	(0.240)		(0.246)	
Opthalmology	0.298	+	0.351	*
	(0.173)		(0.176)	
Pathology	0.361	*	0.319	*
	(0.158)		(0.160)	
Pharmacology	0.867	***	0.737	***
	(0.200)		(0.203)	
Radiology	0.110		0.131	
	(0.179)		(0.182)	
Obstetrics	−0.443	*	−0.361	
	(0.217)		(0.221)	
Pediatrics	−0.112		−0.101	
	(0.141)		(0.145)	
Psychiatry	−0.583	***	−0.513	
	(0.172)		(0.176)	
Neurology	0.715	***	0.645	***
	(0.167)		(0.169)	
Constant	−0.453	***	−0.888	***
	(0.138)		(0.174)	
N	1779		1779	
Log likelihood	−705.717		−685.655	
LL ratio improvement			(20.062)**	

Notes +P < 0.10; *P < 0.05; **P < 0.01; ***P < 0.001.

significant ($P<0.001$). Individuals having earned both an MD and a PhD show greater propensity to disclose inventions than colleagues with a single degree. Holding both degrees increases the probability of disclosing by eleven percent. This result, which supports H2, indicates the inventive benefits of having both upstream and downstream knowledge.

We also find support for H3 and H4. Individuals trained outside the United States are significantly more likely to disclose inventions than those trained in the United States. Holding a degree from a foreign institution increases the likelihood of disclosure by twelve percent. Thus, it appears that individuals willing to uproot themselves in pursuit of opportunity are also likely to take the leap and pursue academic entrepreneurship. Finally, we find evidence that entrepreneurial propensity can be learned. Faculty members that trained at institutions that were heavily involved in technology-transfer activities internalize this norm showing a heightened propensity to adopt such commercialization behavior in their own career. Specifically, we find that a one percent increase in patenting activity at the institution where an individual received their graduate training is associated with a one percent increase in the probability that the individual will disclose.

13.4   Reflective conclusions and further research

If we are going to think creatively about public policies toward increasing university technology transfer, we need to reflect on the process of disclosing and to understand who discloses and why. The faculty members at prominent universities face a variety of demands on their time from teaching to research to patient care and publishing. The option to act entrepreneurially and participate in the commercialization of academic research has only recently been endorsed at the institutional level, and to date, faculty members have not universally embraced this activity. Given the fundamental need for faculty involvement in the technology-transfer process, understanding the factors that differentiate individual faculty members in terms of their inclination to become academic entrepreneurs is clearly important.

In this study, we pursue the question of which faculty members will act entrepreneurially by selecting to participate in the technology-transfer process by investigating the disclosure activity of medical school faculty at Duke University and Johns Hopkins University. We find evidence that a faculty member's participation in technology-transfer activities is a function of that individual's inventive capacity and their entrepreneurial propensity. With respect to the inventive capacity, our analysis shows that breadth of knowledge – whether technological or functional – positively contributes to disclosure activity. Our results suggest that faculty members who have earned both an MD and a PhD show greater propensity to disclose inventions than colleagues with a single degree. Moreover, boundary-spanning individuals, with appointments in more than one academic department, are more likely to disclose inventions than peers associated with only one department. Indicators of entrepreneurial propensity – past willingness to grapple with uncharted territory and/or exposure to pro-commercialization norms during training – are also predictive of academic entrepreneurship. Individuals who trained outside the United States were more likely to pursue academic entrepreneurship. Our results also suggest that faculty members who trained at institutions that were heavily involved in technology-transfer activities also demonstrated a heightened propensity to continue this entrepreneurial behavior in their own career.

We find that faculty in the basic science departments – cell biology, genetics, immunology, microbiology, and neurobiology – have a greater propensity to disclose inventions as compared with faculty in other departments, specifically clinical departments, such as obstetrics/gynecology, pediatrics, and psychiatry. These departments are more focused on providing patient care and may have more limited technical opportunity.

Though, in this piece, we have focused on individual characteristics that catalyze entrepreneurial behavior, future research in this stream needs to explore the influence of social context on academic entrepreneurship. Even though both universities have renowned medical schools the variation that exists among academic departments suggests that social interactions and peer expectations may influence participation in technology transfer (Bercovitz and Feldman, 2006). Individuals bring with them certain entrepreneurial proclivities when they join an organization. Our results in this chapter suggest a training effect as individuals form expectations about technology transfer from the institutions where they trained. However, it is also likely that their actions are shaped, to some degree, by the environment they encounter within the organization they join. Academic institutions, in particular, are known to possess strong and enduring cultures. These cultural norms can be communicated and encouraged through the actions of leaders. Simultaneously, the conduct of peers can provide information regarding accepted and supported behavior. The question remains, however, as to the extent social context, when supportive, can prompt a nonentrepreneurial individual to change and join the ranks of academic entrepreneurs, or the extent when a nonsupportive, social context can curb the entrepreneurial behavior of individuals who are so inclined.

Schumpeter (1934) suggests that entrepreneurs have a greater ability to perceive opportunity, accept challenges and organize resources. In this chapter, we build on the Schumpeterian tradition by exploring the entrepreneurial decision as a complex mix of individual characteristics, propensities, and perceptions of opportunities.

References

Aldrich, H.E. 1999. *Organizations Evolving*. London: Sage
AUTM. 2003. *The AUTM Licensing Survey: FY 2002*. Northbrook, IL: Association of University Technology Managers
Banerjee, A.V. 1992. "A Simple Model of Herd Behavior," *Quarterly Journal of Economics* 107(3): 797–817
Bantel, K.A. and Jackson, S.E. 1989. "Top Management and Innovations in Banking: Does the Composition of the Top Team Make a Difference?," *Strategic Management Journal* 10: 107–24
Bercovitz, J.E.L. and Feldman, M.P. 2006. "Academic Entrepreneurs: Organizational Change at the Individual Level.' Working paper
Biglaiser, G. 2002. *Guardians of the Nation? Economists, Generals, and Economic Reform in Latin America*. Notre Dame: University of Notre Dame Press
Biglaiser, G. and Brown, D. 2003. "What's the Big Idea? An Ideational Explanation for Tariff Reform in Latin America, 1981–1995," paper presented at the American Political Science Association, Philadelphia, PA
Brown, S.L. and Eisenhardt, K.M. 1995. "Product Development: Past Research, Present Findings, and Future Directions," *Academy of Management Review* 20(2): 343–78
Burt, R.S. 1992. *Structural Holes: The Social Structure of Competition*. Cambridge, MA: Harvard University Press
Calori, R., Lubatkin, M., Very, P. and Veiga, J. 1997. "Modelling the Origins of Nationally-Bound Administrative Heritages: A Historical Institutional Analysis of French and British Firms," *Organization Science* 18(6): 681–96
Chwieroth, J. 2003. "Neoliberal Norms and Capital Account Liberalization in Emerging Markets: The Role of Domestic-Level Knowledge-Based Experts," paper presented at the American Political Science Association, Philadelphia, PA
Clark, K.B. and Fujimoto, T. 1991. *Product Development Performance*. Boston, MA: Harvard Business School Press
Cohen, W.M. and Levinthal, D.A. 1990. "Absorptive Capacity: A New Perspective on Learning and Innovation," *Administrative Science Quarterly* 35: 128–52
DiMaggio, P. and Powell, W. 1983. "The Iron Cage Revisited: Institutional Isomorphism and Collective Rationality in Organizational Fields," *American Sociological Review* 48(2): 147–60

Dougherty, D. 1992. "Interpretive Barriers to Successful Product Innovation in Large Firms," *Organization Science* 3: 179–202

Etzkowitz, H. 1983. "Entrepreneurial Scientists and Entrepreneurial Universities in American Academic Science," *Miverva*, 21: 1–21

Feldman, M.P., Feller, I., Bercovitz, J.E.L. and Burton, R.M. 2002. "Equity and The Technology Transfer Strategies of American Research Universities," *Management Science* 48: 105–21

Finnemore, M. and Sikkink, K. 1998. "International Norm Dynamics and Political Change", *International Organization*, 52(4): 887–918

Frank, R., Gilovich, T. and Regan, D. 1993. "Does Studying Economics Inhibit Cooperation?," *The Journal of Economic Perspectives* 7(2): 159–71

Haas, P. 1992. "Introduction: Epistemic Communities and International Policy Coordination," *International Organization* 46(1): 1–35

Koestler, A. 1990. *The Act of Creation.* New York: Viking

Locke, R. 1985. "The Relationship Between Higher Educational Management Cultures in Britain and West Germany: A Comparative Analysis of Higher Education from a Historical Perspective," in P. Joynt and M. Warner (eds.), *Managing in Different Cultures.* Norwich: Page Bros. Ltd., pp. 96–127

Mowery, D., Nelson, R., Sampat, B. and Ziedonis, A. 1999. "The Effects of the Bayh–Dole Act on U.S. University Research and Technology Transfer," in L.M. Branscomb, F. Kodama, and R. Florida, (eds.), *Industrializing Knowledge.* Cambridge, MA: MIT Press, pp. 269–306

Mowery, D. C. and Ziedonis, A. 1999. "The Effects of the Bayh–Dole Act on U.S. University Research and Technology Transfer: Analyzing Data from Entrants and Incumbents," paper presented at the Science and Technology Group, NBER Summer Institute. Cambridge, MA: National Bureau of Economic Research

Mowery, D., Sampat, B., and Ziedonis, A. 2002. Learning to Patent: Institutional Experience, Learning, and the Characteristics of U.S. University Patents After the Bayh–Dole Act, 1981–1992. *Management Science* 48(1): 73–89

Mowery, D.C. and Ziedonis, A.A. 2002. "Academic Patent Quality and Quantity before and after the Bayh–Dole Act in the United States," *Research Policy* 31(3): 399–418

National Science Foundation. 2000. "Graduate Education Reform in Europe Asia, and the Americas and the International Mobility of Scientists and Engineers," *Proceedings of an NSF Workshop,* NSF 00-318, Project officer, Jean M. Johnson (Avlington, VA: NSF)

Nelson, R.R. 2001. "Observations on the Post-Bayh–Dole Rise of Patenting at American Universities," *Journal of Technology Transfer* 26(1–2): 13–19

Saxenian, A.L. 2002. "Silicon Valley's New Immigrant High-Growth Entrepreneurs," *Economic Development Quarterly*, 16(1): 20–31

Schein, E.H. 1985. *Organizational Culture and Leadership: A Dynamic View.* San Francisco, CA: Jossey Bass

Schumpeter, J.A. 1934. *Theory of Economic Development*. (1911, in German; tr. 1934). Cambridge, MA: Harvard University Press

Shane, S. 2000. "Prior Knowledge and the Discovery of Entrepreneurial Opportunities," *Organization Science* 11(4): 448–69

Siegel, D., Waldman, D. and Link, A. 2003. "Assessing the Impact of Organizational Practices on the Productivity of University Technology Transfer Offices: An Exploratory Study," *Research policy* 32, No. 3–30

Slaughter, S. and Leslie, L. 1997. *Academic Capitalism: Politics, Policies and the Entrepreneurial University*. Baltimore: Johns Hopkins University Press

Thornton, P. 1999. "The Sociology of Entrepreneurship," *Annual Review of Sociology* 25: 19–46

Thursby, J.G. and Kemp, S. 2002. "Growth and Productive Efficiency of University Intellectual Property Licensing," *Research Policy* 31(1): 109–24

Thursby, J.G., Jensen, R. and Thursby, M.C. 2001. "Objectives, Characteristics and Outcomes of University Licensing: A Survey of Major U.S. Universities," *Journal of Technology Transfer* 26(1–2): 59–72

Venkatarman, S. 1997. "The Distinctive Domain of Entrepreneurship Research: An Editor's Perspective" in B. Katz (ed.), *Advances in Entrepreneurship, Firm Emergence, and Growth*. Greenwhich, CT: JAI Press

14 Modelling and measuring scientific production: a first estimation for a panel of OECD countries

Gustavo Crespi and Aldo Geuna

Introduction

There is increasing recognition in the OECD countries of the importance of public scientific research in providing the foundations for both innovation and competitiveness. At the same time, there is a lack of systematic evidence on how such investments can lead to increasing levels of scientific output, improved patenting and innovative output, better economic performance and, ultimately, to increased wealth for a country. Much of the available literature concentrates on examining either the effects of public basic research on the innovative activities of firms (see among others Jaffe, 1989; Mansfield, 1991; Klevorick et al., 1995; Narin, Hamilton and Olivastro, 1997; Arundel and Geuna, 2004) or the contribution of scientific research to productivity growth (Adams, 1990).[1] Very few studies have attempted to examine the relationship

The authors are grateful to Stefano Brusoni, Paul David, David Humphry, Ben Martin, Fabio Montobbio and Ed Steinmueller for their comments and suggestions. Earlier versions of this paper were presented at the Schumpeter Society Conference 2004 Milan; DRUID Conference 2004 Copenhagen, the Colloquium on Measuring the Impacts of Science 2004 Montreal. The comments and suggestions of participants at these meetings are much appreciated. The authors would also like to thank Evidence Ltd for supplying some of the data used in the econometric analysis. This chapter is derived from a report commissioned by the UK Office of Science and Technology, Department of Trade and Industry. All mistakes and omissions, along with the views expressed, remain the sole responsibility of the authors.

[1] In recent years there have been a large number of papers that have analysed university–industry relationships and university technology transfer and intellectual property rights developments (for an overview of the main trends see OECD, 2002a, 2003a; Geuna and Nesta, 2004; Mowery et al., 2004). Most of these works tend to be focused on the characteristics of the actors involved and not the relationships between inputs and outputs in the long recursive process between science investment and wealth creation. The contribution of public research spending is sometimes examined within broader analyses of productivity growth or economic growth of countries (Guellec and van

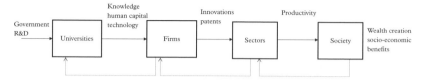

Figure 14.1   The linear model of investment in public scientific research

between investment in science and research outputs (of particular interest are the works of Johnes and Johnes (1995), Adams and Griliches (1998) and Bonaccorsi and Daraio (2003).
The context of limited resources for supporting basic research, and the need to justify the expenditure of these resources by the government have affected the way and focus of the analyses on the economic contribution of public scientific research. Though acknowledging the limitations of the linear model to the understanding of the relationships between science and technology, the push for quantification and measurement has constrained most of the studies to implicitly represent the process as a set of black boxes with the investment in public scientific research at the start and the creation of wealth or production of socio-economic benefits (see Figure 14.1) at the opposite end. A much richer and more complex set of interactions happen among the four major actors. The more towards the right end of the figure the outputs are considered in relation to the inputs, the more the model makes heroic assumptions on what happens in the boxes and between the boxes. Feedbacks, external factors, firm heterogeneity, industrial sectors knowledge bases variability, and so forth, are not included in the estimations of the socio-economic impact of the investment in public scientific research (resulting generally in an underestimation of the real contribution). Paradoxically, most of the literature has focus on the second and third boxes while very little is known about the first one although it is the one less effected by the modelling limitations.
The aim of this study is to contribute to the development of a better understanding of the relationships between governmental R&D funding and scientific production using a production-function approach. In doing that, we will define the limit of applicability of such approach, examining its limitations in the context of scientific production in which the inputs and outputs are very difficult to quantify and price. We frame the analysis in an international context considering the 'world' science

Pottelsberghe, 2001; OECD, 2003b). For a review of the literature on the economic return to public scientific research see Scott et al. 2001.

production function. We identify (not impose) the structure and length of the lag between the investment in R&D and the outputs. Finally, we devote particular attention to the analysis of international spillovers and to the impact of the U.S. science system on other countries' science systems.

Specifically, we focus on the determinants of scientific research production at the international level. We use a sample of fourteen OECD countries for which we have reliable information about Higher Education Research and Development (HERD). As proxies for outputs we have taken publications and citations from the Thomson ISI(R) National Science Indicators (2002) database. The inputs and outputs for this sample of countries have been recorded over a period of twenty-one years (1981–2002). On the basis of this panel dataset we aim to develop an empirical framework to address the following research questions:

(1) What is the profile of the time lag between the investment in HERD and the research output? Does it vary with the output being considered?
(2) What are the returns to national investment in science? Are there cross-country spillovers?

The chapter is organised as follows. Section 14.1 presents the model and data, and discusses their shortcomings. Section 14.2 makes a first attempt to develop a robust econometric estimation of the lag length and structure between science funding and scientific research outputs at the international level. International spillovers and the final version of our model are presented in Section 14.3. Finally, Section 14.4 offers concluding observations and suggests some avenues for further research.

14.1   Modelling and measuring scientific production

14.1.1   A knowledge production function model

Following Griliches (1979), a simple knowledge production can be specified as follows. Let $Y = F (X, K, u)$ be the production function connecting some measure of output, Y, at the micro (the researcher) or macro (the country) level, to the 'inputs' X, K and u; where X stands for an index of conventional inputs such as labour and other control variables, K is a measure of the current state of scientific knowledge, determined in part by current and past research and development expenditures and u stands for all other unmeasured determinants of output and productivity.

The standard approach is to assume that the production function to be a Cobb–Douglas and assume that the unmeasured factors u can be considered as random after the introduction of a time trend into the equation to represent the systematic component of the unmeasured factors. Then we can rewrite $F(.)$ as:

$$Y_{it} = A_i X_{it}^\alpha K_{it}^\beta e^{\lambda t + u_{it}}, \tag{14.1}$$

where A is a constant, i is a country index, t is a time index, e is the base of natural logarithms and α, β and λ are some parameters that we are interested in estimating. Most controversy arises around the specification about the knowledge capital (K_{it}). According to Griliches (1979), three major issues should be considered in the measurement of such capital: (1) the fact that the research process takes time and that current research and development may not have an effect on measured productivity until several years have elapsed, (2) past research and development investments depreciates and become obsolete; thus the growth in the 'net' stock of knowledge capital is not equal to the gross level of current or recent resources invested in expanding it and (3) that the level of knowledge for a given research unit (or country) is not only derived from 'own' research and development investments but is also affected by the knowledge of other units (or countries) through knowledge spillovers.

Regarding the time lag, we assume that the relationship between R&D investments and knowledge capital is far more complex than the traditional linear lag polynomial formulation. An alternative to the linear model, that implies perfect *substitution* between research and development expenditures carried out over different periods of time, is to assume that the old capital and the new investment are *complementary* inputs in the production of new knowledge capital (Klette and Johansen, 2000). The basic idea is that the greater initial knowledge will tend to increase the amount of knowledge obtained from a given amount of R&D. This is a particularly suited assumption in the case of science in which we 'stand on the shoulders of giants' to build new knowledge. Therefore, the more knowledge is produced, the more you can recombine it to produce new knowledge. Formally we will assume that:

$$K_{it} = \prod_{j=0}^{\infty} R_{it-j}^{w_j}. \tag{14.2}$$

Equation (14.2) has also the advantage that it makes the model estimation linear (in logs), allowing us to search rather than to impose the pattern of weights (w) in the lag structure.

The issue of the depreciation rate of knowledge capital is more complex. While it is clear that private knowledge capital depreciates sometimes quite fast (when new products and processes by competitors reach the market), much less is known about the impact of depreciation on the 'public' stock of knowledge. This sort of stock has the property of lasting quite a long period of time, and hence, of having much lower depreciation rates. In the estimations we will also assume that the lag length is finite but quite long.

The world of 'open' science rewards scientists for the quick disclosure of their discoveries and does not create incentives for keeping a discovery secret to appropriate the possible economic returns from its exploitation.[2] International collaborations are a much more common part of the scientific development compared with the situation in the industry. In this environment knowledge spillovers should be pervasive. The science production function (14.1) should be corrected to include cross-country knowledge spillover term; this paper devotes particular attention to its estimation.

A final issue is regarding the aggregation problems (Fisher, 1969; Griliches, 1979; Felipe and Fisher, 2003). Traditional estimations of industrial production function at the country level are affected by the way in which inputs and outputs are aggregated (the major problems are usually with the inputs). The basic assumption is an equilibrium situation at the micro level that allows one to sum up to the sectoral and then country level using prices. In the case of a science production function the research outputs (publications and citations) are an aggregation of a very diverse set of items: different publications, in different journals, in different scientific fields with different propensity to publish and different propensity to produce journal publications as their codified output. Clearly, we do not have 'prices' that could permit us to sum across the various categories in a homogeneous way. Does this mean that we cannot estimate a production function at the macro level for science?

We can express our output indicator as:

$$Y_{it} = \prod_{l=1}^{L} y_{ilt}^{\Omega_l} = A_i X_{it}^{\alpha} K_{it}^{\beta} e^{\lambda t + u_{it}}, \qquad (14.3)$$

where the 'aggregate' output Y_{it} is given by a number index form with weights Ω_l of some *unobserved* indicators of the quality of each type of

---

[2] This view of the 'open' science organisation system for the production of new knowledge (Dasgupta and David, 1994) is currently challenged by a more proprietary oriented model based on university property rights (see e.g. Mowery et al., 2004 for the discussion of the US situation and Geuna and Nesta, 2004 for the EU).

publication output. We suppose that there are L different types of research outputs. Because we do not observe Ω_l, we assumed a common weight for each type of publication (or scientific field). The deviations of the true weight regarding the average weight will appear as $\sum_{l=1}^{L} \varepsilon_{ilt} y_{ilt}$ in the residual of the knowledge production function as in (14.4)

$$\sum_{l=1}^{L} \bar{\Omega} y_{ijt} = a_i + \alpha x_{it} + \beta k_{it} + \lambda t - \sum_{l=1}^{L} \varepsilon_{ilt} y_{ilt} + u_{it}. \qquad (14.4)$$

We can assume that the within-country set of weights remains stable over time, that is to say we assume that there is no major change in the type or frequency of publications in a country (or that if there is it is similar across countries).[3] However, the size of the country's research output (Y_{it}) influences the estimation. If the relative differences in the scientific research size among countries remain constant over time (an acceptable assumption for this sample of countries given their highly developed science system), it would be possible to absorb this omitted factor in a country-specific fixed effect. Unfortunately, the lack of sufficiently detailed comparable data precluded us from directly testing this assumption. However, in order to further study if our results are affected by country's scale effect, we have tested for the stability of the knowledge production function across different sub-samples of countries; the results where stable.[4]

14.1.2 Data sources

We focus our analysis on the following fourteen countries: Australia, Belgium, Canada, Finland, Denmark, France, Germany, Netherlands, Spain, Italy, Switzerland, Sweden, the United Kingdom and the United States. We excluded other OECD countries, where although some information about HERD was available it was incomplete and/or inconsistent, and those countries that had a specific scientific research output (due to their size, history or other factors) and therefore were causing problems with the aggregation of the outputs.

---

[3] This assumption constrains the analysis to countries with similar levels of science development, so that we can assume that a change due, for example, to the arrival of genomic affects in a similar way the portfolio of research output of the various countries. This model would not be robust to the inclusion of developing countries in the sample.
[4] The results of this test are available on request from the authors.

In order to examine the relationship between science funding and scientific-research outputs we use the publicly available information on HERD expenditure and its components at country level. The OECD defines the HE sector as universities, colleges of technology and other institutions of post-secondary education, whatever their source of finance or legal status. This includes research institutes, experimental stations and clinics operating under direct control of, administered by or associated with the higher education institution (HEI) (OECD, 2002b). Because this sector does not usually directly match with an area in the System of National Accounts, it is difficult to provide clear guidelines that ensure internationally comparable data reporting. Universities and colleges of technology make up the core of the sector in all countries. Variations occur with respect to other post-secondary education institutions and even more so to institutes linked to universities, such as university hospitals and clinics and to public research centres – for example, CNRS laboratories are included in HERD in France while similar institutions in Italy, that is CNR laboratories, are included in the 'government' research and development expenditure category. Also, countries' reporting differs in the ways of classifying HE expenditure – for example, PhD students in Sweden and The Netherlands receive a state salary (they were or are public employees depending on the time period considered), but those in the United Kingdom and Italy only receive a grant (Geuna, 2001; Jacobsson and Rickne, 2003). These differences limit the validity of cross-country comparisons. A case-by-case analysis was carried out in order to first identify 'major' structural breaks in the series and second to select a set of comparable countries with comparable statistics.[5] However, because 'permanent' country-level differences in the way that information is collected persist, we need to control for these systematic differences in the estimations.[6]

The HERD figures that we used are expressed in millions of constant U.S.$ as reported by the OECD. R&D expenditure series were deflated using the implicit gross domestic product (GDP) deflator taken from the OECD National Accounts database. These national currency data at

---

[5] In particular all the models presented in this paper were first estimated on a country-by-country basis and then different dummy variables were included if there was some report about changes in the classification criteria. Only those countries for which results were robust to these breaks were considered.

[6] More specifically, a country level fixed effect was always introduced into the models. This fixed effect not only captures permanent differences relating to the functioning of the various national scientific systems, but also differences in how the information is collected. More particularly, for almost all the results shown below we worked with 'within' country information and 'averaged' to achieve a global estimation. We were able to deploy this approach due to the panel data nature of the dataset we built.

1995 prices were converted to US\$ using 1995 purchasing power parities (PPP).[7]

The scientific process produces several research outputs that can be classified into three broadly defined categories: (1) new knowledge; (2) highly qualified human resources; and (3) new technologies. This chapter focuses on the determinants of the first type of research output.[8] There are no direct measures of new knowledge, but several proxies have been used in previous studies. The most commonly used are (1) publications and (2) citations. The source of these two variables is the Thomson ISI(R) 'National Science Indicators' (2002) database on published papers and citations. These two measures have several shortcomings (Geuna, 1999), here we detail only those that directly affect the econometric estimations. First, they are incomplete and biased proxies for the production of new knowledge. Second, the Thomson ISI data are strongly affected by the disciplinary propensity to publish in journals, so they are a poor measure of the output of disciplines such as history or law. Third, the ISI includes an almost constant number of journal/pages in its archive (journals enter and leave, but the number is more or less constant at around 5,000). This clearly limits the possibility of output expansion and therefore biases our estimations in favour of decreasing returns.

In order to take account of the 'truncation problem' in the citations to most recent years, the citations variable has been adjusted.[9] One way of controlling for this is by using what Hall et al. (2001) called the fixed-effect approach. This method involves scaling citation counts by dividing them by the average citation count for a group of publications to which the publication of interest belongs. Using the same example as in Hall et al. (2001) this approach treats a publication that received say eleven citations and belongs to a group in which the average publication received ten citations as equivalent to a publication that received twenty-two citations, but happens to belong to a group in which the average was twenty. The groups were defined in terms of scientific fields and year and the scaling index was computed using the ISI dataset at world level.

---

[7] The fact that we are using the GDP deflator instead of the more correct HERD deflator induces a sort of omitted variable bias. If the HERD deflator shows a long run increasing trend compared to the GDP deflator, the time trend variable included in the econometric models will capture this.

[8] Preliminary estimates of the determinants of the second type of research output (highly qualified human resources) can be found in Crespi and Geuna (2004).

[9] The citation count is affected by the time span allowed for the papers to be cited: for example, papers published in 2000 can receive citations in our data just from papers published in the period 2000–2001, but in fact they will be cited by papers in subsequent years as well, but we do not observe them.

Both publications (and citations) and HERD expenditure are very persistent series over time. This greatly complicates the statistical inference because these series can be correlated owing to the presence of a third common variable (the time trend) leading to spurious or over-estimated results. In order to see what kind of stochastic process governs the series, we carried out several tests. Panel Data Unit root tests were used to determine whether it is better to work with first differenced observations (because they have a unit root) or to control by a time trend in the models (because they are trend stationary series). The results of these tests are omitted here; however, it is important to say that we can consider the series as stationary, but with a deterministic trend. In what follows we work with these series in levels and add a deterministic trend to the models.[10]

Acknowledging these inputs and output data limitations, as well as the modelling problems and strong assumptions discussed in the previous section, we recommend caution in interpreting the results of the econometric models. The aim of this chapter is not to provide accurate and robust estimates of the investment elasticities (a doubtful task given the poor quality of the data sources and the modelling problems), but to develop and critically assess the validity of an empirical approach to identify basic stylised facts about the production of science and its impact. The aggregate science production function is used in this context not as a tool for national account but as an instrument that allows to highlight interesting characteristics of the process of scientific production such as the role on knowledge spillovers.

In the following section we present the econometric model used to identify the structure and length of the lag between the investment in HERD and the HE outputs. Once we identified the lag structure we focused on the search for international spillovers. Section 14.3 presents the final estimation of our model (see Table 14.6), which includes both national and international HERD expenditures.

14.2   The polynomial distributed lag model

One (but not the only) way to search for both lag length and structure is to apply the technique known as Polynomial Distributed Lag (PDL) or the Almon Model (see Greene, 1993). The methodology can be applied

---

[10] For further details about the results of these tests see Appendix B in Crespi and Geuna (2004).

to our case. Let us define the following 'finite' distributed lag model:

$$y_{it} = \alpha_i + \sum_{j=0}^{q} \beta_j r_{it-j} + \gamma_i X_{it} + u_{it}, \quad i = 1, \ldots, N, \qquad (14.5)$$

where y_{it} is the log of given research output (publications and citations) and r_{it} is the log of HERD, respectively, for country i at time t. Although a model like (14.5) can in theory be estimated quite straightforwardly, there is the potential problem of very long lags in which case the multi-colinearity is likely to become quite severe. In such examples it is common to impose some structure on the lag distribution, reducing the number of parameters in the model. It is in this context that the PDL model can be useful. The approach is based on the assumption that the true distribution of the lag coefficients can be very well approximated for by a polynomial of a fairly low order.

$$\beta_j = \delta_0 + \delta_1 j + \delta_2 j^2 + \cdots\cdots + \delta_p j^p, \quad j = 0, \ldots, q > p. \qquad (14.6)$$

The order of the polynomial, p, is usually taken to be quite low, rarely exceeding 3 or 4. By inserting (14.6) into (14.5), one can estimate a transformed model where the estimated coefficients are deltas that can be put back into (14.6) in order to recover the original weights. In addition to the $p+1$ parameters of the polynomial, there are two unknowns to be determined: the length of the lag structure, q, and the degree of the polynomial, p. Here we follow the standard procedure for determining first the length of the lags and then the degree of the poly-nomial function.

The Table 14.1 shows the results of this exercise for publications and Table 14.2 gives the results for citations. Both tables show the lag structure for each alternative model and in the last three rows present the values for the information criteria and the long-run elasticity for domestic HERD. In the results for publications (Table 14.1) two information criteria have a minimum value at a five-year lag, while the remaining criterion reaches a minimum at six-year lag. Because of the potentially more serious consequences of omitting some relevant lag, we decided to keep six-year lags as the optimum lag length for publications.

To search for the lag length we start by taking a lag of ten years and reducing by one period down to 0. In each reduction we evaluate the information that is lost because we omit one additional lag with the

Table 14.1 *Unrestricted Polynomial Distributed Lag (PDL) model approach (fixed effects) publications*

	10	9	8	7	6	5	4	3	2	1	0
	Lpub	Lpub	Lpub	Lpub	Lpub	Lpub	Lpub	Lpub	Lpub	Lpub	Lpub
HERD	-0.104	-0.091	-0.085	-0.082	-0.076	-0.080	-0.104	-0.140	-0.157	-0.288	0.012
	0.121	0.119	0.116	0.115	0.114	0.114	0.116	0.124	0.132	0.137*	0.092
t-1	0.073	0.065	0.068	0.069	0.062	0.060	0.054	0.068	-0.009	0.366	
	0.145	0.146	0.145	0.145	0.144	0.146	0.151	0.157	0.158	0.140*	
t-2	0.030	0.040	0.041	0.039	0.043	0.040	0.052	0.000	0.326		
	0.138	0.139	0.139	0.138	0.138	0.139	0.145	0.140	0.115*		
t-3	0.103	0.103	0.098	0.098	0.094	0.097	0.043	0.306			
	0.131	0.131	0.131	0.131	0.130	0.131	0.132	0.102			
t-4	0.046	0.043	0.047	0.045	0.051	0.021	0.265				
	0.117	0.117	0.117	0.116	0.116	0.112	0.084*				
t-5	0.113	0.114	0.107	0.109	0.092	0.236					
	0.104	0.104	0.103	0.102	0.102	0.073*					
t-6	0.073	0.067	0.073	0.067	0.145						
	0.082	0.083	0.082	0.083	0.058*						
t-7	0.065	0.070	0.051	0.079							
	0.070	0.071	0.072	0.057							
t-8	-0.029	-0.047	0.028								
	0.078	0.079	0.053								
t-9	0.006	0.076									
	0.104	0.067									
t-10	0.073										
	0.085										

Table 14.1 (cont.)

	10	9	8	7	6	5	4	3	2	1	0
Non-HERD	0.000	0.000	0.000	0.000	0.000	0.000	0.000	-0.001	-0.001	-0.001	-0.001
	0.004	0.004	0.004	0.004	0.004	0.004	0.004	0.004	0.004	0.005	0.005
	0.015	0.016	0.017	0.017	0.018	0.020	0.023	0.027	0.030	0.034	0.036
	0.004***	0.004***	0.004*	0.004*	0.004*	0.004*	0.004*	0.004	0.004*	0.004*	0.003*
Constant	-23.419	-24.887	–	–	–	–	–	–	–	–	-61.193
	7.865***	7.680***	7.603*	7.442*	7.442*	7.349*	7.105*	7.203*	6.829*	6.468*	6.153*
Observation	168.000	168.000	168.00	168.00	168.00	168.00	168.00	168.0	168.00	168.00	168.00
AIC	-428.68	-429.89	–	–	–	–	-424.1	–	–	–	-386.47
SIC	-347.46	-351.79	-356	–	–	–	–	–	-347.1	–	-336.48
ICOMP	NA	-262.66	NA	–	–	–	–	–	–	–	-262.88
HERD L	0.45	0.44	0.43	0.42	0.41	0.37	0.31	0.23	0.16	0.08	0.01
	0.098***	0.096****	0.096*	0.096*	0.096*	0.098*	0.099*	1.104	0.100	0.096	0.092

Notes: Robust standard errors reported below each coefficient.
*significant at 10%; **significant at 5%; ***significant at 1%.

information that is gained because we have more degrees of freedom in the estimation.[11]

In our estimation we used two control variables: a time trend to capture the evolution of the general scientific opportunity and the proportion of non-HERD R&D in the total country research budget. This latter variable deserves a bit more attention. The rationale for including it derives from the fact that our 'observed' research output is the total (not only HE) country-level publications and citations. Even when more than eighty percent of the publications generated by a country are typically derived from the research being carried out in HE institutions, there is still a small proportion produced by firms and other non-university research centres. We expect that non-HERD institutions have a lower productivity in terms of publications and citations than universities. Publications and citations are a by-product of their innovation activities or research supporting government actions. An increased proportion of non-HERD R&D in total country's Gross Expenditures in Research and Development (GERD) would lead to a reduction in total publications and citations. In order to control for this, we built a new variable defined as the ratio between non-HERD R&D and GERD.

Table 14.2 summarises the results for citations. According to one criterion the optimum lag length is seven years, while for the other two it is six years. Again, taking a conservative approach, we chose the longer lag. It is interesting to compare both results in terms of the long-run elasticities implied by the sum of all the individual coefficients. In the case of publications, the long-run elasticity is 0.41 and is statistically significant, and for citations it is 0.51 and also significant. What is even more important is that we can see that in both cases the long-run elasticities reported above become quite stable to small variations in the lag length. In addition, in both models we find that the share variable non-HERD is negative although significant only for citations, while the time trend is positive and significant.

Assuming that we have been able to identify the right lag length we proceed by looking for the right polynomial function. We start by using a fifth degree function and proceed by testing sequential unit reductions in the degree. The results are shown in the last four rows of Table 14.3, where we can accept the reduction from fifth to fourth and from fourth to third but not lower. It is important to note that in order to keep the appropriate significance level in each step we used a very low individual

---

[11] We use three different statistics here: the Akaike Information Criteria (AIC), the Schwartz Bayesian Information Criteria (SBC) and the Bozdogan index of Information Complexity (ICOMP). These criteria are used to select the 'best' model by balancing an adequate goodness of fit against a small number of parameters. See Kolenikov (2000) for the source codes for STATA.

Table 14.2 Unrestricted Polynomial Distributed Lag (PDL) model approach (fixed effects) citations

Lag	10	9	8	7	6	5	4	3	2	1	0
	Lcit	Lcit	Lcit	Lcit	Lcit	Lcit	Lcit	Lcit	Lcit	Lcit	Lcit
HERD	-0.269	-0.257	-0.0251	-0.243	-0.229	-0.237	-0.271	-0.328	-0.354	-0.546	-0.129
	0.166	0.165	0.162	0.161	0.156	0.156	0.159*	0.168*	0.178*	0.179*	0.115
t-1	0.092	0.084	0.087	0.089	0.074	0.070	0.060	0.082	-0.041	0.508	
	0.206	0.205	0.205	0.204	0.200	0.202	0.213	0.226	0.232	0.198*	
t-2	0.013	0.022	0.023	0.017	0.027	0.021	0.038	-0.041	0.477		
	0.168	0.167	0.166	0.166	0.166	0.167	0.179	0.176	0.161*		
t-3	0.17	0.171	0.165	0.166	0.156	0.162	0.082	0.485			
	0.158	0.158	0.158	0.158	0.157	0.159	0.161	0.124*			
t-4	0.088	0.085	0.089	0.086	0.097	0.044	0.407				
	0.133	0.133	0.132	0.131	0.132	0.127	0.105*				
t-5	0.129	0.130	0.123	0.128	0.089	0.352					
	0.118	0.118	0.117	0.116	0.121	0.085*					
t-6	0.102	0.096	0.102	0.084	0.264						
	0.101	0.103	0.102	0.104	0.075*						
t-7	0.114	0.119	0.100	0.182							
	0.090	0.092	0.093	0.075**							
t-8	0.025	0.008	0.084								
	0.102	0.104	0.069								
t-9	0.011	0.077									
	0.111	0.079									
t-10	0.070										
	0.086										

	(1)	(2)	(3)	(4)	(5)	(6)	(7)	(8)	(9)	(10)	(11)
Non-HERD	-0.008	-0.008	-0.008	-0.008	-0.008	-0.009	-0.008	-0.010	-0.009	-0.010	-0.010
	0.005***	0.005**	0.005**	0.005***	0.005*	0.005*	0.005*	0.005*	0.005*	0.004*	0.004*
	0.013	0.014	0.015	0.017	0.021	0.026	0.032	0.037	0.042	0.045	
	0.005***	0.005**	0.005***	0.005*	0.005*	0.005*	0.005*	0.005*	0.004*	0.004*	
Constant	-18.836	-20.24	-21.787	-23.624	-27.831	9.005*	8.709*	8.822*	8.451*	7.802*	7.625
	9.650*	9.161**	9.107**	8.993***	8.882*						
Observation	168.000	168.000	168.000	168.000	168.000	168.00	168.00	168.00	168.00	168.00	168.00
AIC	-385.630	-387.070	-388.330	-389.450							
SIC	-304.410	-308.970	-313.360	-317.600							
ICOMP	-211.910	-219.300	-224.440	-230.040							
HERD	0.543	0.533	0.522	0.508	0.477	0.411	0.316	0.198	0.082	-0.037	-0.128
	0.125***	0.121***	0.121***	0.121***	0.122*	0.130*	0.131*	0.139	0.133	0.122	0.115

Notes: Robust standard errors reported below each coefficient.
*significant at 10%; **significant at 5%; ***significant at 1%

Table 14.3 *Unrestricted PDL and restricted Almon models (fixed effects)*

	Unrestricted	Restricted	Unrestricted	Restricted
	publications		citations	
HERD	−0.076	−0.017	−0.243	−0.032
	0.113	0.018	0.161	0.017*
t-1	0.062	−0.001	0.089	−0.021
	0.144	0.033	0.202	0.024
t-2	0.043	0.033	0.017	0.017
	0.139	0.020*	0.170	0.022
t-3	0.094	0.078	0.166	0.073
	0.131	0.013***	0.158	0.016***
t-4	0.051	0.109	0.086	0.124
	0.116	0.013***	0.130	0.012***
t-5	0.092	0.117	0.128	0.157
	0.102	0.017***	0.116	0.015***
t-6	0.145	0.085	0.084	0.158
	0.058**	0.015***	0.104	0.017***
t-7			0.182	0.111
			0.074**	0.014***
Non-HERD	0.000	0.000	−0.008	−0.006
	0.001	0.001	0.004*	0.003*
Year	0.018	0.018	0.015	0.012
	0.004***	0.003***	0.004***	0.003***
Constant	−28.829	−29.505	−23.624	−18.009
	7.354***	5.808***	8.982***	6.795***
Observations	168	168	168	168
HERD LR	0.410	0.405	0.508	0.587
	0.096***	0.071***	0.121***	0.081***
Constraints		1.54		1.06
Polynomial		Critical		Critical
5 to 4	1.26	6.63	3.710	6.63
4 to 3	1.45	7.83	4.580	7.83
3 to 2	12.53**	8.97	27.87***	8.97
2 to 1	59.27***	10.06	98.05***	10.06

Notes: Robust standard errors reported below each coefficient.
*significant at 10%; **sifnificant at 5%; ***significant at 1%.

significance level. The choice of a third-degree polynomial function is therefore accepted for both publications and citations.

The PDL model also implies a set of constraints on the unrestricted model (without a specified functional form for the lags) estimated above. For example, if the optimum lag length is six and we use a third-degree polynomial function, we are implicitly imposing three constraints. In addition to this there are endpoint constraints which allow the lag distribution to be 'tied down' at its extremes. These endpoint constraints capture the idea that there is no effect from R&D on the research outputs *before* the current period[12] and also that there is no effect from the research inputs after the maximum lag. That is, we need to impose:

$$\beta_{-1} = 0 \quad \text{and} \quad \beta_{q+1} = 0. \quad (14.7)$$

In total we have five constraints. One way of validating the PDL model is by testing whether these constraints are valid. As shown by the non-significant Chi test in Table 14.3, we could not reject any of them. In terms of long-run elasticities we found that their values are very similar to those for the unrestricted model for publications and slightly higher for citations. The time trend is positive and significant, while the non-HERD institutions have a negative effect only on citations.

It is important to compare the unrestricted weights with those obtained using the restricted model. The pattern is similar for both publications and citations. The impact of the first two years is always very low and not significantly different from zero. It is only at the end of the second year for publications (third year for citations) that we find the first positive impact. These impacts reach a peak at year 5 for publications (year 6 for citations).

In the Figures 14.2 and 14.3 are graphical representations of the lag structure implied by the restricted (dotted line) and the unrestricted models. What these figures clearly show is that only at the end of year 2 is it possible to see some positive impact from the investment in science. From the second year on, the returns from investment in science increase till the end of year 5 (or year 6 in the case of citations) and then decline till the end of year 6 (or year 7 for citations). Apparently, no significant returns can be expected after six years.

Figure 14.4 shows the evolution of the cumulative impact of HERD expenditure on scientific research outputs. In the case of publications, it is necessary to wait till year 4 (or year 5 in the case of citations) to gain fifty percent of the expected impact. We do not see any positive 'cumulative' impact until year 2 in the case of publications (year 3 in the case of citations).

The PDL model makes use of the lag structure of (the log of) past R&D. This implies a form of Cobb–Douglas knowledge-creation function where there is unit substitution elasticity between current and past R&D expenditure. It is important to say that this kind of function is

---

[12] This means that the research output does not react 'in advance' of an increase in the research inputs.

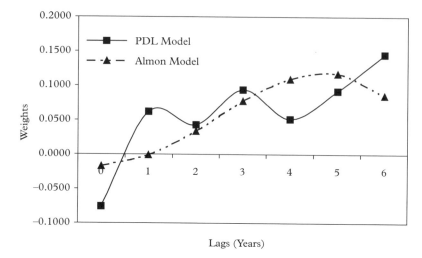

Figure 14.2   Unrestricted versus restricted pattern of weights (publications)

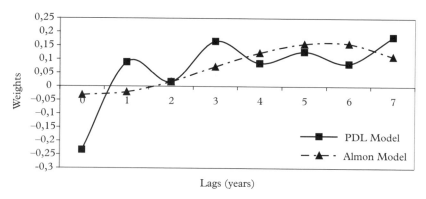

Figure 14.3   Unrestricted versus restricted pattern of weights (citations)

slightly different from the traditional one assumed in Adams and Griliches (1998) where they use the log of the weighted sum of past R&D. This implies a linear knowledge-creation function where there is a sort of perfect substitution between current and past R&D expenditure. Working with this sort of function in the context of panel data is very complex: it is not straightforward to cancel out the fixed effects by using either within or first-difference transformations when the underlying

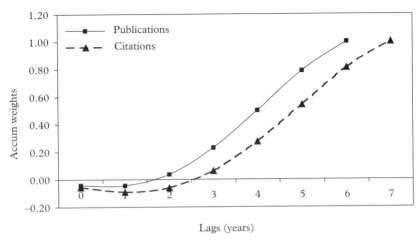

Figure 14.4   The cumulative patterns of weights (publications and citations)

function is not linear. Also, simply adding country dummies to the model complicates the non-linear estimation process. Having said this, we managed to calculate the non-linear estimate for the lag structure of publications. Both the optimum lag structure and the profile of the weights were similar to our previous linear estimations. Therefore, we decided to proceed with the simpler linear model.

14.3   The search for spillovers

In this section we investigate whether there are spillovers between countries. In this context, 'spillovers' means that part of the increase in the research output of a given country that is due to investment in HERD in other countries.

To the extent that an ever-growing external pool of knowledge available to each country generates these spillovers, the inclusion of a time-trend variable in the model partially captures them. As a consequence, we expect that the inclusion of a specific spillover variable will not affect to any significant extent the estimated domestic HERD elasticity. However, identification of a specific spillover effect allows us to calculate the total (domestic plus international) return to changes in the HERD investment for the system of countries considered.

In order to estimate the existence of spillovers, we need to assess the level of knowledge exchange or knowledge cooperation among countries. The higher the level of exchange/cooperation between countries i and l, the higher is the probability that some of the science investment in country i will affect the research output in country l and vice versa. To build up a matrix of knowledge proximity among countries, we used the information on international scientific co-authorship. The NSF *Science and Engineering Indicators* reports (various years) give the share of cross-country co-authorships in the 1980s and 1990s. We averaged the values and built a weight for each country as follows:

W_{il} = number of international co-authorships between countries i and l, divided by the total number of international co-authorships between country l and other countries in the sample.

This weight provides a proxy for the relative knowledge exchange or cooperation between two given countries in our dataset. Table 14.4 presents the resulting weights for the fourteen countries in our database. The United States is the most important country for collaboration for all the countries considered (the United States always has the highest weight). This indicates the special role played by the United States in the process of knowledge creation. The table also clearly indicates how geographical proximity and cultural and linguistic links, which apply to Belgium and France, or to the United Kingdom and Australia, affect co-authorship patterns.

After building these weights we defined the 'international' research and development relevant to each country as a weighted sum of the science budgets of all the other countries as follows:

$$S_{lt} = \sum_{i \neq l} w_{il} R_{it}, \tag{14.8}$$

where R_{it} is the HERD budget for country i. After constructing (14.8) we assume that the lag structure is the same as in the previous section.[13]

The model estimated in this section focuses on long-run spillover effects. In order to compare these effects with the long-run impact of domestic R&D, we redefined the stock of knowledge as a weighted sum of (the log of) R&D, where the weights are defined (as in Greene, 1993) as follows:

$$\omega_j = \frac{\beta_j}{\sum_{j=1}^{k} \beta_j} \tag{14.9}$$

and we use the weight ω_j in order to aggregate the lag for (the log of) R&D expenditure for each country in the dataset. In this way we

---

[13] We do not have enough observations to search for a different lag structure for the spillover variable.

Table 14.4 *Weighting matrix*

	AU	B	CA	Dk	Fin	F	D	I	NL	E	S	CH	UK	US
AU	0.000	0.012	0.087	0.017	0.007	0.051	0.088	0.025	0.028	0.008	0.031	0.024	0.214	0.409
B	0.011	0.000	0.035	0.015	0.014	0.192	0.118	0.065	0.120	0.035	0.036	0.050	0.107	0.201
CA	0.036	0.014	0.000	0.013	0.009	0.090	0.061	0.030	0.026	0.012	0.022	0.025	0.108	0.552
Dk	0.021	0.019	0.040	0.000	0.033	0.070	0.129	0.056	0.048	0.029	0.136	0.040	0.137	0.242
F	0.015	0.027	0.043	0.049	0.000	0.062	0.123	0.047	0.053	0.019	0.148	0.046	0.098	0.271
Fin	0.016	0.061	0.069	0.017	0.011	0.000	0.135	0.094	0.045	0.058	0.030	0.073	0.117	0.275
G	0.024	0.032	0.038	0.027	0.017	0.115	0.000	0.069	0.060	0.031	0.041	0.090	0.121	0.334
I	0.012	0.031	0.033	0.020	0.011	0.140	0.122	0.000	0.044	0.044	0.033	0.082	0.130	0.298
NL	0.019	0.080	0.041	0.025	0.018	0.094	0.148	0.060	0.000	0.029	0.036	0.049	0.147	0.255
E	0.007	0.036	0.028	0.021	0.009	0.186	0.114	0.093	0.043	0.000	0.024	0.040	0.156	0.241
S	0.024	0.029	0.040	0.083	0.061	0.075	0.118	0.053	0.042	0.020	0.000	0.042	0.114	0.300
CH	0.014	0.030	0.035	0.019	0.015	0.138	0.199	0.103	0.044	0.024	0.032	0.000	0.099	0.248
UK	0.056	0.029	0.068	0.028	0.014	0.098	0.118	0.072	0.059	0.041	0.039	0.044	0.000	0.335
US	0.053	0.026	0.171	0.024	0.018	0.113	0.162	0.082	0.050	0.032	0.050	0.055	0.165	0.000

Source: Weighting Matrix. Author's own elaboration based on NSF data.
Notes: Australia (AU), Belgium (B), Canada (CA), Denmark (Dk), Finland (Fin), France (F), Germany (G), Italy (I), Netherlands (NL), Spain (E), Sweden (S), Switzerland (CH), United Kingdom (UK) and United States (US).

Table 14.5 *Results using 6 (7) lags of RD for publications (citations) plus spillovers [dependent variable log publications (citations)]*

| | Publications | | Citations | |
	(1)	(2)	(3)	(4)
Non-HERD$_{it}$	0.001***	0.002**	0.002***	0.002***
Year	0.018	−0.000	0.014	−0.014
	0.002***	0.0020	0.003***	0.006**
HERD$_{it}$	0.475	0.447	0.536	0.499
	0.047***	0.045***	0.049***	0.047***
S$_{it}$		0.505		0.599
		0.116***		0.123***
Constant	−35.024	−2.859	−21.162	27.627
	4.444***	7.9410	6.187***	11.60**
Observations	224	224	210	210
R-squared	0.89	0.90	0.87	0.88
Test CRS (*P-Values*)		0.69		0.62

Notes: Fixed effects by country included.
Robust standard errors reported below each coefficient. Within R-squared reported.
*significant at 10%; **significant at 5%; ***significant at 1%.

dispense with the need to estimate short-run elasticities (which we assume to be known) and instead focus on long-run elasticities.

The results of estimating this model are presented in Table 14.5 (for both publications and citations). The first column shows the estimated parameters without spillover effects; these results are statistically equivalent to those in Table 14.3. When we include the variable S_{it}, which aims to capture the spillover effect, the magnitude of the long-run elasticity for HERD remains stable (dropping only marginally). Interestingly, the variable S_{it} is highly significant and has a large estimated parameter. In addition to this, the value of the time trend variable (year) drops and is no longer significant. This validates our conjecture that the time trend was in some way the capturing part of the spillover effect. In the case of citations (a quality adjusted measure of output) we obtain a positive and significant estimation for spillovers. It is interesting to note that in this case the time trend is negative and significant. This result can be interpreted as indicating an overall negative trend in the production of science output once it has been adjusted for impact. So, if we consider citations as a proxy for 'quality' of science and not just impact, the model indicates an overall decrease in the 'quality' of the scientific output at the world level.

Another interesting result relates to the magnitude of the coefficients for both domestic HERD and international spillovers (S_{it}). Their sum is very close to 1 for publications and just above that for the citation estimates. These results suggest the presence of decreasing returns to scale at the domestic level,[14] but constant or perhaps even increasing returns to scale at the global level. However, the null of constant returns to scale at world level was never rejected.

14.3.1   Testing for the contribution of proximity

The results in the previous sections rest on two assumptions: first that there are spillovers and, second, that the transmission mechanism is proximity. In our case the latter is measured on the basis of the international co-authorships between countries. In this section we question to what extent proximity is a real transmission mechanism. It is quite clear that because we are dealing here with public science, spillovers are expected 'almost by definition', to the extent that every author reads what authors around the world have published. This is the 'codified' component of the knowledge-dissemination process. However, we know that not all knowledge created during a research programme is 'codified'; an important proportion of it can be thought as being tacit. It is here that proximity or personal interactions among the authors can be relevant. Hence, this section can be interpreted as a test of the relative importance of codified versus tacit knowledge in science production.

In order to carry out this test we proceeded as follows. For each country in the sample we replaced each component of the weighting matrix by a random draw from an uninformative probability function. In this case, we use the uniform [0,1] density distribution, with the simulated coefficients standardised in order to add 1. After this we constructed the new stock of knowledge and ran the models as in the previous section. In order to guarantee independence of our results from each random draw, we reproduced the process 1,000 times.

If proximity matters for knowledge spillovers in science production, we should observe that the economic importance and statistical significance of the spillover coefficients obtained in our previous results are higher than those from the random matrix model. That is, we would expect that the majority of the spillover coefficients generated during our simulations be 'to the left' of the coefficient estimated using our proximity matrix.

---

[14] The result of decreasing to scale at domestic level could have interesting implications for the analysis of long-term economic growth. This result is consistent with the work of Jones (1995) showing that there is no correlation between the level of R&D investment and total factor productivity (TFP) growth in the long run (although there is in the short run).

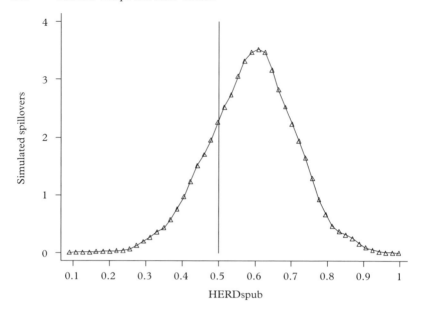

Figure 14.5   Simulated spillovers, uniform random matrix, publications

The results for publications are depicted in Figure 14.5. In this figure we present the empirical distribution of the simulated spillover coefficients and compare them with the previous result. This is represented by a perpendicular line on the graph. As can be seen from the figure, there is no evidence that proximity matters in the case of publications. Indeed, the opposite looks to be the case after about sixty percent of the runs generated coefficients higher than the previous ones. The results in case of citations are shown in Figure 14.6. The results are the same as for publications.

In summary, our results reject the idea that proximity matters as a transmission mechanism for spillovers in publicly funded research. According to our results, it seems that the 'codified' component of the knowledge is clearly dominant. Of course, it is possible that our measure of proximity is not the correct one, in which case our test becomes one for the 'correctness' of using international co-authorship as an index of proximity. Therefore, alternative indices of proximities should be developed. This is part of our future research agenda.

14.3.2 The importance of the United States

The results in Table 14.4 indicate that authors from several countries have a great propensity to publish with U.S. co-authors. In this section

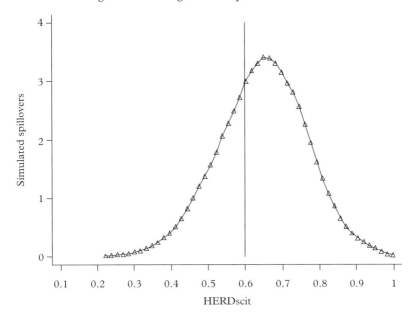

Figure 14.6   Simulated spillovers, uniform random matrix, citations

we explore further the importance of the United States as an independent source of spillovers. The idea here is to determine to what extent there is a strong dependence on R&D investments from the country that is the supposed world leader in several scientific disciplines. If this dependence is strong, the policy decisions of the rest of the world can no longer be taken in isolation. In other words, changes in U.S. scientific policy (e.g. discontinuation of stem cell research or investment in military science) will affect the scientific production of the rest of the world. If these other countries want to continue with their own research production despite some hypothetical reduction or change in the United States's research budget, they will have to make extra efforts to compensate for the United States's decline in certain fields.

In order to investigate these issues we re-estimated the spillover effect by splitting world-level investment into two groups: the United States and the Rest of the World. We estimated the following knowledge production function:

$$y_{it} = \alpha_i + \beta_0 W(r)_{it} + \beta_1 W(r)_{\text{USA},t} + \beta_2 W(r)_{-i-\text{USA},t} \qquad (14.9)$$
$$+ \gamma X_{it} + u_{it}, \quad i = 1, \dots, N,$$

Table 14.6 *Results using 6 (7) lags of RD for publications (citations) plus spillovers according to source of origin [dependent variable log publications (citations)]*

	Publications	Citations
Non-HERD$_{it}$	−0.003	−0.01
	1.51	4.56★★★
Year	−0.012	−0.021
	1.86★	2.38★★
HERD$_{it}$	0.418	0.468
	10.31★★★	11.27★★★
HERD$_{USA,t}$	0.494	0.585
	7.21★★★	8.08★★★
HERD$_{-I-USA, t}$	0.356	0.206
	2.20★★	1.22
Constant	24.536	39.166
	2.00★★	2.46★★
Observations	224	210
Number of Country Code	14	14
R-squared	0.92	0.91
Test CRS (P-Values)	0.07★	0.09★
Test HERD$_{USA,t}$ = HERD$_{-I-USA, t}$	0.49	0.07★

Notes: Fixed effects by country included.
Robust standard errors below each coefficient. With R-squared reported.
★significant at 10%; ★★significant at 5%; ★★★significant at 1%.

where y_{it} is the (log) output of the research 'intermediate' output (papers and citations) by country i (fourteen countries) and time t (twenty-one years). $W(r)_{it}$ is (the log of) a distributed lagged function of real past R&D expenditure and X_{it} is a vector of the control variables. The two new variables are $W(r)_{USA,t}$ and $W(r)_{-i-USA,t}$. The former captures the impact of the United States's stock of knowledge in the research outputs of each country, while the latter refers to the impact of the stock of knowledge from all the remaining countries, except the United States and country i. In building these two new stocks we have assumed the same lag structure and weights as in the previous sections. The results for the rest of the world were weighted assuming an equal contribution from each country to the non-U.S.A pool of knowledge (given that our proximity-personal interaction measure was non-significant).

As can be seen from Table 14.6, the source of the spillovers is important. For both publications and citations, spillovers from the United States are numerically more important than spillovers from the other countries. However, we also test if the differences between the spillover coefficients were statistically significant. The differences were not statistically significant for publications, while they where, although marginally, in the case of citations. This result suggests that only in terms of quality- or impact-adjusted research output the spillovers from the United States have a dominant role. Another interesting result is related to the impact on the returns to scale. Now the assumption of constant returns to scale is rejected both for publication and citations, pointing to the presence of increasing returns at the aggregate level.

14.4   Conclusions

Modelling and measuring scientific production is not an easy task given the fact that science inputs and outputs are difficult to quantify both in terms of quantity and quality. Is a citation a real impact- or quality-adjusted proxy for the output of scientific research? How do you model the cumulative process of knowledge creation? Can you assume some form of maximising behaviour for the researchers that will allow one to use a production function modelisation to examine scientific production?

This chapter does not provide answers to such questions, but it acknowledges their relevance and when possible it tries to take their implications into account. The chapter takes a pragmatic approach. It sets to develop a first attempt to use quantitative methods to examine the driving forces behind scientific production. Specifically, it develops econometric models based on the production function metaphor to relate a sub-set of inputs to two of the most common university research outputs: publications (as a proxy for the production of codified research) and citations (as an impact-adjusted proxy for codified research production). The aim of the chapter was not to provide accurate and robust estimates of the investment elasticities, a doubtful task given the poor quality of the data sources and the modelling problems, but to use more generalisable methods to develop some understanding on the process of scientific production.

We estimated a PDL model to calculate the pattern of the lag structure. After specifying the most appropriate model, we focused on the assessment of knowledge spillovers among the fourteen countries considered. We found some evidence of a strongly positive long-run relation between Higher Education R&D and the two research outputs examined. For both publications and citations we found evidence of decreasing returns to the domestic component of R&D. Finally, it should be noticed that the parameters of the knowledge production function were very stable and robust to different compositions of countries in the sample.

The above results refer to long-run impacts. However, there is also a long and quite complex lag structure with regard to the impact of domestic R&D on the different research outputs. These weights are quite different to those typically assumed in standard econometric models. There is no significant evidence of any positive impact before two years in the case of publications and three years in the case of citations. The total cumulated effects (the long-run elasticities) are spread over six and seven years respectively, and reach a maximum towards years 5 and 6, respectively. Evaluation of the impact of science policies on the different research outputs has to take account of this lag structure if erroneous conclusions are to be avoided.

The analysis of international spillovers indicates that for publications and citations there is evidence of a significant impact from the weighted investment in HERD in other countries. We studied the mechanism of transmission of international spillovers. We followed two different approaches. One based on an index of scientific proximity (personal interaction) between countries was rejected. It seems that the 'codified' component of the knowledge was clearly dominant. However, it is also possible that our measure of proximity (based on international co-authorships) is not the correct one. A second transmission mechanism was investigated by splitting the world pool of knowledge between the United States and the rest of countries. In this case results were consistent that spillovers from the United States (the scientific leader in many disciplines) are higher than those from the rest of the countries. The result, however, is only significant in the case of citations. For this model, the assumption of constant returns to scale is rejected both for publications and for citations, pointing to the presence of increasing returns at the aggregate level.

From a policy perspective our results highlight the importance of two phenomena. First, the lag between the investment and the creation of the research output is considerable; this is very problematic because it implies that there are a set of factors that we could not control for that may have a changing impact on the research outputs. Consequently, if it becomes difficult to link the inputs and outputs of the scientific production 'box', the task of connecting socio-economic benefits to certain scientific investments becomes a very doubtful enterprise. If quantitative approaches can be used at all in the assessment of public investment in science, the most promising area of development is the one that would focus on a better understanding of the relationships between the inputs and the most direct output of the scientific process.

Finally, this work provides same quantitative evidence in support of the view that science is an international enterprise characterised by major spillovers across countries. Approaching the funding, management and organisation of science only with a national focus would result in an at best incomplete or erroneous advance. In particular, the evidence of significant and extremely important spillover from the U.S. science system underscore how the other OECD countries cannot define their science priorities without taking into account the impact of changes in priorities in the United States. A reduction of U.S. public spending in stem cell research and an increase in military research as the one that has happened in the United States in recent years impact in a significant way not only the science output of the United States but also the one of all the other countries, requiring a significant increase in the public spending by the other countries to avoid a reorientation of their science output.

The results of our model estimations show that (if we exclude Canada), although there are knowledge spillovers across European countries, the impact of these is lower (at least in terms of impact- or quality-adjusted measurement) than the impact of U.S. spillovers. This may be due to the size of the science investment in the United States (142€ per capita against 89€ per capita in the EU-15 in 1999) but also to the fact that although the EU has similar or even higher publication output than the United States, the EU countries achieve excellence only in a small group of fields (EC, 2003). These results seem to support the policy view of the need of an European Research Council able to fund (at the needed level) excellent research at the European level.

References

Adams, J. 1990. 'Fundamental Stocks of Knowledge and Productivity Growth', *The Journal of Political Economy* 98: 673–702

Adams, J. and Griliches, Z. 1998. 'Research Productivity in a System of Universities', *Annales d'Economie et de Statistique* V0, 49–50, (Jan–Jun 1988): 127–62

Arundel, A. and Geuna, A. 2004. 'Proximity and the Use of Public Science by Innovative European Firms', *Economics of Innovation and New Technology* 13: 559–80

Bonaccorsi, A. and Daraio, C. 2003. 'A Robust Nonparametric Approach to the Analysis of Scientific Productivity', *Research Evaluation* 12: 47–69

Crespi, G. and Geuna, A. 2004. *The Productivity of Science*. Brighton: SRPU Report prepared for the Office of Science and Technology (OST), Department of Trade and Industry (DTI), UK. www.sussex.ac.uk/spru/publications/reports/CrespiOST.pdf

Dasgupta, P. and David, P.A. 1994. 'Towards a New Economics of Science', *Research Policy* 23: 487–521

DTI. 2003. *Forwardlook 2002*. London: OST

European Commission. 2003. *Third European Report on Science & Technology Indicators*. Belgium

Felipe, J. and Fisher, F. M. 2003. 'Aggregation in Production Functions: What Applied Economists Should Know', *Metroeconomica* 54: 208–62

Fisher, F. M. 1969. 'The Existence of Aggregate Production Functions', *Econometrica* 37: 553–76

Guena, A. 1999. The Economics of Knowledge Production Funding and the Structure of University Research. Cheltenham: Edward Elgar

Geuna, A. 2001. 'The Changing Rationale for European University Research Funding: Are there Negative Unintended Consequences', *Journal of Economic Issues* 35: 607–32

Geuna, A. and Nesta, L. 2006. 'University Patenting and its Effects on Academic Research: the Emerging European Evidence', *Research Policy* 35: 790–807

Greene, W. 1993. *Econometric Analysis*. Englewood Cliffs, NJ: Prentice Hall

Griliches, Z. 1979. 'Issues in Assessing the Contribution of Research and Development to Productivity Growth', *The Bell Journal of Economics* 10: 92–116

Guellec, D. and Van Pottelsberghe, B. 2001. 'R&D and Productivity Growth: Panel Analysis of 16 OECD Countries', *OECD Economic Studies* 33: 103–26

Hall, B., Jaffe, A. and Trajtemberg, M. 2001. 'The NBER Patent Citations Data File: Lessons, insights and methodological tools', NBER Working Paper Series No. 8498

Jacobsson, S. and Rickne, A. 2003. 'The Swedish "academic" paradox - myth or reality: How strong is really the Swedish "academic" sector?', paper presented at the Conference in Honour of Keith Pavitt, SPRU, The Freeman Centre, University of Sussex, November 13–15, 2003

Jaffe, A. 1989. 'Real Effects of Academic Research', *American Economic Review* 79: 957–70

Johnes, J. and Johnes, G. 1995. 'Research Funding and Performance in UK University Departments of Economics: A Frontier Analysis', *Economics of Education Review* 14: 301–14

Jones, C. 1995. 'R&D-Based Models of Endogenous Growth', *Journal of Political Economy* 103(4): 759–84

Kolenikov, S. 2000. 'Icomp: Information Complexity Measures', http://ideas.repec.org/c/boc/bocode/s410201.html, accessed on 02/06/2004

Klette, T. and Johansen, F. 2000. 'Accumulation of R&D Capital and Dynamic Firm Performance: a Not-so-Fixed Effect Model' in D. Encaoua, B.H. Hall, F. Lisney and J. Mairesse (eds.), *The Economics and Econometrics of Innovation*. Boston/Dordrecht/London: Kluwer Academic Publishers

Klevorick, A., Levin, A., Nelson, R. and Winter, S. 1995. 'On the Sources and the Significance of Interindustry Differences in Technological Opportunities', *Research Policy* 24: 185–205

Mansfield, E. 1991. 'Academic Research and Industrial Innovation', *Research Policy* 20: 1–12

Mowery, D., Nelson, R., Sampat, B. and Ziedonis, A. 2004. *Ivory Tower and Industrial Innovation*. Stanford: Stanford Business Books

Narin, F., Hamilton, K. and Olivastro, D. 1997. 'The Increasing Linkage between U.S. Technology and Public Science', *Research Policy* 26: 317–30

NSF. 2002. 'Science & Engineering Indicators'. National Science Board, Science and Engineering Indicators 2002. Arlington, VA: National Science Foundation (NSB-02-1)

OECD. 2003a. *Turning Science into Business. Patenting and Licensing at Public Research Organisations*. Paris: OECD

OECD. 2003b. *The Sources of Economic Growth in OECD Countries*. Paris: OECD

OECD. 2002a. *Benchmarking Industry-Science Relationships*. Paris: OECD

OECD. 2002b. *Frascati Manual: Proposed Standard Practice for Surveys on Research & Development Expenditures*. Paris: OECD, 254

Scott A., Steyn, G., Geuna, A., Brusoni, S. and Steinmueller, W.E. 2001. 'The Economic Returns to Basic Research and Benefits of University-Industry Relationships', A report commissioned by the OST-DTI, SPRU, University of Sussex, Brighton, 2001 www.sussex.ac.uk/spru/publications/econreturnsost.pdf.

Comments to Chapters 13 and 14

W. Edward Steinmueller

Schumpeter's great speculative work, *Capitalism, Socialism and Democracy*, raises many questions about the long-term evolution of the capitalist system, two of which have been the focus of recent interest for those who suspect that we have not yet reached the end of history. First, to what extent would the industrialist entrepreneur be supplanted by the growth of managerial capitalism? Schumpeter argued that the answer to this question would ultimately hinge on the extent that the process of innovation could be institutionalised, taming it and ultimately dispensing with the role of the entrepreneur. As Langlois (2003a) notes, this prospect was, for Schumpeter, an unwelcome one, as it would substitute the rationalist calculus of the bureaucrat for the boldness and imagination of the entrepreneur and thus, ultimately, would sap the vitality that Schumpeter viewed as responsible for the success of capitalism. The chapters in this section offer an intermediate solution to this paradox, one in which university researchers either become part of a larger network of innovation or enter, themselves, as entrepreneurs.

A second of the questions that are raised in *Capitalism, Socialism and Democracy* has also attracted renewed attention – what are the prospects for unlimited growth of the large business enterprise?[1] Efforts to answer this question, including those of Penrose, Coase, and even Berle and Means preserved the 'unit of analysis', the individual character of enterprises despite the long history of holding companies and trusts, conglomerate enterprises, and the rise of the multinational corporation. Schumpeter was, however, alert to the possibility that the prospects for managerial capitalism might well hinge on changes in industrial organisation (Langlois 2003a, Schumpeter 1947). Such changes were a

---

[1] See the debate 'Framing Business History' in the September 2004 issue of Enterprise and Society with the following articles; Langlois 2004, Lamoreaux et al. 2004, Sabel and Zeitlin 2004, as well as two of the earlier contributions that helped to provoke this exchange, Lamoreaux et al. 2003 and Langlois 2003b.

part of the historical process that Schumpeter regarded as an intrinsic element of economic analysis (Schumpeter and Schumpeter 1954).

If Schumpeter's vision of the future growth of these very large enterprises had been realised, our era would be marked by the convergence of socialism and capitalism; for Schumpeter, this would be a rather grey world managed by clerks and technocrats. Ironically, the large enterprises that Schumpeter suspected might eventually succeed in institutionalising technological innovation are, in our era, being transformed – a transmutation of managerial capitalism along a different axis than scale or scope. Instead of enlargement, and evoking Bourdieu's (Bourdieu 1983) vision of modern capitalism as a system of social networks, we observe the emergence of complex networks of organisations collaborating through market and negotiated exchange. The result is an 'outsourced' and vertically 'disintegrated' brand of capitalism. The essence of this new model of capitalism is that knowledge of productive and market opportunities may be flexibly re-configured in pursuit of the maximisation of shareholder wealth. Indeed, even within the vertically integrated companies that remain, substantial knowledge exchange is needed to operate and improve modern production systems that are characterised by a growing degree of complexity and that rely upon an expanding array of underlying technologies (Patel and Pavitt 1999). The chapters in this section suggest the role that universities might have in this new industrial structure.

Even more ironically, the forces of ideological socialism that Schumpeter detested have weakened and the various implementations of state socialism have either disappeared or become sponsors of capitalist enterprise. A peculiar consequence of this deconstruction has been a retreat from the resolve that marked the era commonly labelled the 'Cold War', during which the goals of competing in cultural production, joined with the prospect of finding new tools to employ against the adversary, amply justified funding of the republic of science as another client state. Science also provided some of the points of connection in which the competing societies could find common human purposes. In short, although the science of today is no less universalist or transcendent than it was in past eras, the features of the rationale for its public support connected to the competition between socialism and capitalism have evaporated. Science and the university continue in their roles of serving the states that have become their patrons, but 'service' has begun to assume a new meaning for the new era – servicing the needs of wealth creation, economic growth and job creation.

In this new economic world order, the role of the university is being re-examined and re-aligned (Etzkowitz and Leydesdorff 1998,

Geuna et al. 2003). A number of authors have suggested that universities are becoming sources of entrepreneurial activity in their own right (Etzkowitz et al. 2000; Etzkowitz 2001). In effect, two systems are in operation, one in which the older norms of open science offers 'priority' (status accorded to those who are first to discover), the second in which advances in knowledge become proprietary 'disclosures' sub-ject to patenting and the entrepreneurial efforts of researchers to com-mercialise their discoveries. Correspondingly these two systems may be viewed as competitive, complementary, or, perhaps, both. For example, Looy et al. (2004) argue for the possibility of a reciprocal 'Matthew effect' – success in one system might contribute to success in the other, that is, a complementary relationship between the systems, based on evidence from a single Belgian university. Mowery et al. (2001) finds little effect on academic publication patterns accompanying the expansion of patenting activity at three major universities in the United States and that much of the increase in patenting activity is related to biomedical invention. It has also been noted, however, that the patent-related system remains 'marginal' – deriving benefit from the larger structural funding of the publication-related system (making it easier for the newer system to generate benefits) and that the two systems inevi-tably compete for researcher time and effort, calling into question the long-term stability of their co-existence (David et al. 1999).

The focus on the 'entrepreneurial university' and the researcher as entrepreneur has directed attention away from and, to some extent, obscured views of the university as a repository of universal knowledge or as place of refuge from goal-directed action in which learning and contemplation is the first order of business. These roles, to which many still subscribe, are not high on the list of priorities in the re-examination of the university's role in society. Instead, the objectives appear to include harnessing those inhabiting the university to service in an extended network of knowledge production (Gibbons et al. 1994), crafting new configurations for this network involving universtiies, industry and government (Etzkowitz and Leydesdorff 1998, Etzkowitz 2001) and raising new expectations of these networks contributions to the new economic order (HM Treasury (UK) 2002). These trends have the effect of narrowing the scope of the university's role in society and confining public discussion of the university's 'mission' to its immediate contribution to economic growth and wealth creation.

The two chapters in this section address issues of university knowledge asset management, a symptom of the narrowing of con-temporary discussions concerning university missions. At one end of this axis, and preserving a more traditional 'public good' perspective,

Crespi and Geuna consider the management of public investments in the university with the aim of discerning what issues might need to be taken into account to assure an optimal flow of knowledge resources into the public knowledge-commons represented by the publications of scientific findings. In Crespi and Geuna's chapter, the university maintains a role in educating those who will eventually leave the university to make a more direct contribution to wealth creation while, at the same time, undertaking to deploy public resources to enlarge those resources that might be assimilated by the new networked enterprise. Viewed as a contributor to the science base of the societies in which they are embedded, universities are valuable assets for knowledge creation that should be appropriately managed.

Crespi and Geuna examine the relation between public investment in higher education research and patterns of publication, focussing on international co-authorship. Using an aggregate production function approach they establish the responsiveness (with substantial lag) of the level of publication to public investment. Their approach and data do not allow a 'tracing' of the effect of this investment on commercialisation activity. Instead, their principal concern is with the possibilities for international interdependence in the 'knowledge commons' as measured by international co-authorships. Historians of science (Collins 1974, 2001) have questioned a key assumption of the 'knowledge commons' – the assumption that published scientific outputs produce knowledge that can be used straight away by other researchers. Rejection of this premise would raise further questions about the 'public good' nature of public investments in science (Callon 1994).

Crespi and Geuna find that (1) a systematic pattern of interdependence exists between changes in national levels of funding and patterns of international co-authorship and (2) that 'proximity' (at least as measured by historical patterns of international co-authorship) is not a significant determinant of publications of citations when country effects and research and development expenditures are taken into account. These results suggest four interpretations, each of which may contribute to the explanation of their findings: (1) using the scientific outputs of distant researchers is less difficult than the results from individual case studies of breakthrough scientific advance would suggest, (2) scientists are equally well connected, across boundaries of language and culture, at a global level as they are with researchers in neighbouring countries, (3) that the larger scale of the United States makes it a research hub, drowning out other effects of proximity and/or (4) that a substantial share of scientific publication is devoted to 'normal science' – what Collins (2001) characterises as 'recognised knowledge'. The first two interpretations are consistent with the traditional view that the scientific community is 'universal' and that published scientific outputs are in fact taken up widely regardless of their origin. Crespi and Geuna's approach cannot distinguish between these two interpretations and their evidence does not contradict either of them. Crespi and Geuna are able to reject the third interpretation for publications, but not for citations – suggesting that US network centrality does have some degree of 'hub' effect on global scientific efforts. The fourth interpretation, which cannot be tested by the methods or data that Crespi and Geuna employ, raises further questions about the relation between scientific publication and scientific advance. The two may not be synonymous and, if they are not, a long tradition of employing publication and citation counts to examine not only the evolution of scientific output, but also the potential contribution of scientific activity to innovation may be questioned. An important conclusion drawn from their straightforward finding of interdependence is that the long-standing principle of attempting to raise and equalise scientific funding as a share of GDP appears to be in the collective interest of nations. In a world of universal positive spillovers, all participants benefit if others raise the level of their activity.

At the other end of new axis along which public investments and private incentives are discussed, Bercovitz and Feldman envisage universities as a deadweight loss to society unless their research efforts produce technology transfer, which can only be effectively achieved by the transformation of university researchers into entrepreneurs in the market for knowledge. Their interpretation goes beyond the 'triple helix' vision in which universities are portrayed as co-operating and interacting with industry and government to form a new entrepreneurial paradigm, with suitable qualifications as to the nascent qualities of this relationship. The role of universities are portrayed in Bercovitz and Feldman in a remarkable way:

After all, if individual faculty members do not disclose their inventions, then there can be no patenting and no subsequent downstream licensing and licensing revenue. Without disclosures, regardless of the amount of research dollars, quality of the faculty at the institution, or any other asset measure, the institution will not be productive in technology transfer. (Bercovitz and Feldman 2005).

This is a remarkable statement given the history of the research university in the United States where a collection of private and public institutions have long co-operated with industry in improving and commercialising technologies. In the area of agricultural technology alone, it suggests that the contributions of several generations of horticulturists, agricultural engineers and food science professionals were unproductive.

To cite just one example, consider the University of Wisconsin professor, Stephen M. Babcock, who took up the problem of measuring the butter fat content of milk. Professor Babcock used the solubility of the constituents of milk other than butter fat in sulphuric acid and a centrifuge to create an inexpensive and accurate test for butterfat content – a key factor in both the quality control and market value of milk. Professor Babcock refused to take out a patent on his invention claiming that, as the citizens of Wisconsin had been hired in their service, his invention should be used to their benefit (Henderson et al. 1949). For Bercovitz and Feldman, Professor Babcock's 'disclosure', as it did not result in patented knowledge, was unproductive, despite the fact that it has been universally employed for more than a century in the dairy industry.

Professor Babcock's adherence to the 'older norms of science' is contrary to the new, 'entrepreneurial', norm that Bercovitz and Feldman apparently believe requires that university researchers disclose their inventions in order to register patents. They appear oblivious to the consequences that such activity might have on the willingness of the citizens of Wisconsin or of any other jurisdiction that supports the universities to continue funding the activities of scientists. The scientists, in turn, are encouraged to abandon the 'old' and apparently mistaken view that their role was to serve humanity's needs by advancing scientific understanding. This may seem an ungenerous view of Bercovitz and Feldman, whose principal purpose is providing statistical estimations of characteristics shaping the entrepreneurial propensity of medical school faculty. Their exercise, however, is a logical step whose, direction has been set by the scholars who see university research as an unexploited resource and by the policy-makers willing to follow this path in the hopes of discovering another opportunity for public–private partnership in the hopes of reducing public expenditure.

Bercovitz and Feldman's contribution can also be viewed as an exercise in positive economics. The aim is to identify the features of researchers that enrol in the second system, where the focus is on 'disclosures' aimed at commercialisation. Bercovitz and Feldman confine their attention to medical school faculties at Duke and Johns Hopkins Universities. Their results indicate that individuals who hold both an MD and PhD degree, have appointments in more than one medical school department, who received training in institutions with a high propensity to patent and who received a degree from a non-US institution are significantly more likely than those who do not have these characteristics (singly or collectively) to make a disclosure. These are helpful results – at a minimum, they provide some useful human resource guidelines for medical schools. Applying these results to other possible joint appointments, such as in Computer Science and English, or to individuals holding multiple PhD degrees is more dubious. A higher patenting rate at the institutions where these individuals trained is interesting, although its (statistically significant) contribution to the propensity to disclose is very small.

The holding of a non-US degree may, however, suggest something more. To the extent that this is a measure of immigration (as Bercovitz and Feldman reasonably contend), it does capture the association between the initiative required to 'uproot' oneself from one's homeland. It may also, however, capture the diminution of ties to US social networks that would question entrepreneurial activity or provide higher rewards (in terms of status) to research conducted within the 'open science' system, an effect that is amplified by the negative effect of the researcher's age (measured by the years since last degree). In short, universities seeking a greater level of entrepreneurship might wish to enhance their recruiting efforts for young foreigners.

There seems little doubt that incentives will continue to favour the treatment of public investment in university research as an asset from which greater returns should be expected. Policies of this sort are part of the continuing effort to achieve fiscal stringency and to favour the growth of the market system. There is every reason, however, to monitor the effects of these activities, taking into account the possibility that they may produce untoward and unintended consequences as well as claims that they represent vital new opportunities for the university. A new system for managerial capitalism that provides a new set of answers to Schumpeter's classic questions concerning the institutionalisation of innovation and the limits to growth of the firm must surely have a role for the university. Exactly how that role is articulated will continue to occupy our attention in the coming years.

References

Bercovitz, J. and Feldman, M. 2005. 'Academic Entrepreneurs and Technology Transfer: Who Participates and Why?,' in F. Malerba and S. Brusoni (eds.), *Perspective on Innovation*. Cambridge: Cambridge University Press

Bourdieu, P. 1983. 'Forms of capital' in J. C. Richards, (eds.), *Handbook of Theory and Research for the Sociology of Education*. New York, NY: Greenwood Press, pp. 241–58

Callon, M. 1994. 'Is Science a Public Good? Fifth Mullins Lecture, Virginia Polytechnic Institute, 23 March 1993,' *Science, Technology and Human Values* 19: 395–424

Collins, H.M. 1974. 'The TEA Set: Tacit Knowledge and Scientific Network's, *Science Studies* 4: 165–86

Collins, H.M. 2001. 'Tacit Knowledge, Trust and the Q of Sapphire,' *Social Studies of Science* 31: 71–85

David, P. A., Foray, D., and Steinmuller, W.E. 1999. 'The Research Network and the New Economics of Science: From Metaphors to Organizational Behaviours,' in A. Gambardella, and F. Malerba, (eds.), *The organization of innovative activities in Europe.* Cambridge: Cambridge University Press

Etzkowitz, H. 2001. *The Second Academic Revolution: MIT and the Rise of Entrepreneurial Science.* London: Gordon Breach

Etzkowitz, H. and Leydesdorff, L. 1998. 'The Endless Transition: A 'Triple Helix' of University–Industry–Government Relations,' *Minerva* 36: 203–8

Etzkowitz, H., Webster, A., Gebhardt, C., and Terra B.R.C. 2000. 'The Future of the University and the University of the Future: Evolution of Ivory Tower to Entrepreneurial Paradigm,' *Research Policy* 29: 313–30

Geuna, A., Salter, A. J., and Steinmueller E. W. (eds.) 2003. 'Science and Innovation : Rethinking the Rationales for Funding and Governance,' in *New Horizons in the Economics of Innovation.* Northampton, MA: Edward Elgar

Gibbons, M., Limoges, C. Nowotny, H., Schwartzman, S., Scott, P., and Trow, M. 1994. *The New Production of Knowledge: The Dynamics of Science and Research in Contemporary Societies.* London: Sage

Henderson, M. G., Speerschneider, E. D., and Ferslev (eds). 1949. 'A Fortune Given Away,' in *It Happened Here: Stories of Wisconsin.* Madison: State Historical Society of Wisconsin, pp. 206–8

HM Treasury (UK). 2002. *Investing in Innovation: A Strategy for Science, Engineering and Technology.* London: www.hm-treasury.gov.uk/spending_review/spend_sr02/spend_sr02_science.cfm (Last Accessed: 15 March 05)

Lamoreaux, N. R., Raff, D.M.G., and Temin, P. 2003. 'Beyond Markets and Hierarchies: Toward a New Synthesis of American Business History,' *American Historical Review* 108(April): 404–33

Lamoreaux, N.R., Raff, D.M.G., and Temin, P. 2004. 'Against Whig History,' *Enterprise and Society* 5(3, September): 376–87

Langlois, R.N. 2003a. 'Schumpeter and the Obsolescence of the Entrepreneur,' *Advances in Austrian Economics* 6: 287–302

Langlois, R. N. 2003b. 'The Vanishing Hand: The Changing Dynamics of Industrial Capitalism,' *Industrial and Corporate Change* 12(April): 351–85

Langlois, R. 2004. 'Chandler in a Larger Frame: Markets, Transaction Costs, and Organizational Form in History,' *Enterprise and Society* 5(3, September): 355–75

Looy, B. V., Ranga, L. M., Callaert J., Debackere K., and Zimmermann, E. 2004. 'Combining Entrepreneurial and Scientific Performance in Academia: Towards a Compounded and Reciprocal Matthew-effect?,' *Research Policy* 33: 425–41

Mowery, D.C., Nelson, R.R., Sampat, B., and Ziedonis, A.A. 2001. 'The Growth of Patenting and Licensing by U.S. Universities: An Assessment of the Effects of the Bayh–Dole act of 1980,' *Research Policy* 30: 99–119

Patel, P. and Pavitt, K. 1999. 'The Wide (and increasing) Spread of Technological Competencies in the World's Largest Firms: A Challenge to Conventional Wisdom,' in A. D. Chandler, P. Hagstrom, and O. Solvell, (eds.), *The Dynamic Firm: The Role of Technology, Strategy, Organization, and Regions*. Oxford: Oxford University Press, pp. 192–213

Sabel, C. F. and Zeitlin, J. 2004. 'Neither Modularity nor Relational Contracting: Inter-Firm Collaboration in the New Economy,' *Enterprise and Society* 5(3): 388–403

Schumpeter, J. 1947. *Capitalism, Socialism and Democracy*, 2nd edn, New York, NY: Harper and Row

Schumpeter, J. A. and Schumpeter, E. B. 1954. *History of Economic Analysis*. Oxford: Oxford University Press

Part 8

Innovations and public policy

15 Innovation systems, innovation policy and restless capitalism

Stan Metcalfe

Introduction

In this chapter I outline the rationale for innovation policy from an evolutionary economic perspective, a perspective built around the dynamic properties of 'restless capitalism' and the concept of the adaptive policy-maker. The foundations of an evolutionary rationale stand in sharp contrast to the traditional 'market failure' and optimising policy-maker perspectives because, from an evolutionary viewpoint, markets are instituted devices 'designed' to promote the growth of knowledge and its application through innovation and the self-transformation of economic arrangements. The alleged failures are in fact the *sine qua non* of a market process. My starting point is Richard Nelson's view that market and non-market arrangements and processes are complementary elements in the innovative division of labour and that each sphere consists of an array of vastly different organisational forms and instituted rules that precludes any simple idea that markets can fail or that governments can fail (Nelson, 2002). Markets may be too extensive or too limited and the same is the case for their non-market alternatives; it is all a matter of the relative advantages of broad organisational form and thus where the boundaries should be drawn. In this context, I shall argue that the innovation systems concept is the natural frame in which to design adaptive policy initiatives but that these initiatives are necessarily general and facilitating and not specific and directing. Innovation is part of the complex dynamics of capitalism; it is the major source of business uncertainty and the basis of the open-ended unpredictable evolution of the system. Innovation policy of all policy arenas needs to be built from these fundamental facts and the highly ordered but non-equilibrium nature of the economic process. Capitalism is, as all economic systems are, a knowledge-based system, and the fundamental reason it cannot be in equilibrium is that no meaning can be attached to the notion of knowledge in equilibrium. It is, however, strongly ordered by an instituted frame of market and non-market relations that connect the market process with the process of innovation and the growth of knowledge more generally. Indeed for the system to progress at all it must be fundamentally unstable in the sense that the current constellation of activities can be invaded by new alternatives. If this were not possible, if the system were stable in the evolutionary sense, then enterprise, innovation and the diffusion of innovation would have no place in economic history and development. Certainly since Schumpeter (1934) published *The Theory of Economic Development*, economists should have known the essential veracity of this view and that growth and development cannot be squeezed out of an equilibrium framework. Yet change depends on order, and the key point is that the prevailing constellation of prices, quantities and activities in a market economy generate the opportunities, the incentives and tests that must be passed for innovations to invade and transform the system from within. It turns out that self-organisation and self-transformation are the two sides of the same market process. Finally, I will suggest that an innovation systems perspective provides the appropriate rationale for innovation policy and that these systems increasingly transcend national boundaries and increasingly call into question the idea of isolated national innovation policies and implicitly call for an assessment of international innovation policy conflicts and coordination problems[1].

Such a discussion is certainly timely in the context of enduring European concerns about the links between public science and commercial innovation, the competitive threat from the United States and emerging, large-scale, low-wage economies of India and China (Dosi et al. 2005). It is also of vital importance in relation to the challenge of economic development more generally, and the related claim that innovation provides the only sustainable route to material prosperity for all economies, irrespective of their level of per capita income. Whether a national target for R&D spending is the appropriate way to meet these challenges is an open question, since there is much more to innovation policy than the stimulation of scientific and technological research and development. Ulrich Witt (2003) has recently distinguished three approaches to evolutionary policy analysis in terms of what is done by policy-makers in practice, what could be done in particular circumstances,

---

[1] The innovation systems literature is vast and grows apace. I simply note here the classic essay by Chris Freeman (1987) that, at the outset, placed the concept in its international context.

and, what should be done to achieve particular policy objectives. My concern is with the last of these, the broad grounds under which innovation policy is justified, 'What is it that policy is meant to achieve from an evolutionary perspective?' A caveat to conclude: any attempt to address innovation policy in practical terms requires recognition of wide intersectoral variations in innovation conditions pertaining to the knowledge, technologies and markets in play, the institutions and the actors and their interrelations. This level of detail is beyond my brief but it cannot be forgotten in practice (Nelson, 1993; Malerba, 2004).

15.1   Evolving knowledge and the evolving economic order

If policy is intended to alter the operation of an economic system it is clear that an appropriate policy must rest on an accurate understanding of the system it seeks to influence. What is it about modern capitalism that an innovation policy must comprehend? The first fundamental point to grasp is that economies evolve because knowledge evolves and innovations are the vehicles that bring new knowledge into specific economic spheres. Innovations imply the application of new ideas and changes to the prevailing economic order, not only the entry of new activities, in established or new organisations but the changing relative importance of existing activities in general and the disappearance of some in particular. Firms disappear but more fundamentally so do entire activities, so that the system is transformed in its content over time and it is from these transformation events that economic progress emerges. Progress never occurs in a smooth pattern; innovation-led development is always, necessarily an uneven, unpredictable competitive process exactly because the growth of knowledge is an uneven, unpredictable process.

The second fundamental point is that changes in knowledge are not external to the economic system but are embedded within it. Here we need to say a little about the growth of knowledge in general since the relation between knowledge and belief is at the core of any model of economic action. Knowledge is necessarily a personal attribute; only individuals can be said to know but what they know depends on accumulated sensory experience and thus on their interaction with the environment and fellow human beings. Thus the growth of human knowledge has always depended upon the connectedness of individuals, for it is connection that makes possible the transmission of information and it is the transmission of information that challenges or reinforces existing beliefs. The development of institutional and organisational forms that permit information dissemination at multiple scales in multiple formats is precisely a central feature of modern capitalist economies. Science is a typical example of an information transmission system but so is the market system; they are each instituted devices for the flow of information and thus the stimulation of understanding in common and disagreement in particular. Indeed, the nature of restless capitalism is that it depends on processes to establish epistemic order, the correlation of understanding, and on processes to destroy that order from within through the emergence of discordant beliefs about the economic world. In part the underlying processes depend on calculation but in neither science nor enterprise is calculation sufficient; both depend in addition on the possibility of imagination and enterprise and institutions which keep the system open to action based on divergent conjecture. As Brian Loasby rightly insists the growth of knowledge is neither rational nor random but resides in a middle ground of guided variation (Loasby, 2002). Nowhere is this more transparent than in relation to innovation and enterprise. If economic beliefs were ever to fall into uniformity it would be the end of economic progress, hence the fundamental policy question becomes 'Is the system capable of generating and responding to new innovation conjectures to the appropriate degree?' All else is to a substantial degree secondary.

From the present perspective the answer depends directly on the institutions that promote business experimentation and their connection to experimentation in terms of science and technology. Two aspects of knowledge generation need to be distinguished. The first resides in the fact that substantial resources are devoted to the production of knowledge and dissemination of information in public research and education systems, and to some this is a defining characteristic of a modern knowledge-based economy. The second is more fundamental; it is that the market process is also a knowledge-generating process producing much of the practical knowledge that is essential to effective innovation. This is so not only in relation to the market incentives to invest in formal or informal R&D but more fundamentally in relation to the day-to-day conduct of business and trade. These changes in knowledge are generated within the economic process and they are the fundamental basis for its self-transformation. The conduct of business is a learning process, based on the combination of information flowing from within and without the market process, and new information will inevitably challenge prior beliefs in the system as to what defines an innovative opportunity. Thus a market order is not equilibrium, except in a trivial sense of the transitory consistency of plans and actions, but an evolving sequence of orders in which every ordered state conveys new information to stimulate the creation of new knowledge and beliefs. Indeed all innovations rest on a sufficiently strong belief, that the economic world can be ordered in a different way. Thus innovations are premised on differences in belief, and that some of these beliefs must turn out to be false precisely engenders further revision of beliefs over time.

This is a description of an entrepreneurial, experimental process, and it is a process that is ineluctably evolutionary in nature. That is to say it is based on processes of variation and selective retention of superior alternatives presented in the form of business conjectures. It is also a process in which the profit mechanism is vital and in which profits are the rents on superior business ability; they are as Schumpeter expressed it 'the child and victim of development'. Abnormal returns are not to be had by acting on the same, rational beliefs that others hold; profits are the result of disagreement on the best economic or business 'model' in specific circumstances. Thus innovative variation is the chief route to superior profitability in knowledge-based capitalism and markets are the context in which selection takes place to continually redefine the pattern of economic order via processes of differential growth, exit and entry. Innovation occurs in public enterprise too where markets are replaced by bureaucracy or quasi-markets, such as in many health care systems and where different constraints and incentives operate. Here we can see why central planning is orthogonal to the market process; it is not just a centralised version of perfect competition as the planning versus market debates of 1930s had it. Markets lead to a market process and the process is not only about the allocation of resources but the discovery of new allocations, new means and new ends (Nelson, 1981). To summarise what is significant about the capitalist market order is its transience, a temporal dimension that reflects open-ended self-transformation emerging out of economic order. It is on this particular knowledge dynamic that modern capitalism depends.

The third fundamental consequence of this view is that the evolution of the market order must be uncertain in the Knight or Shackle sense of the lack of basis for probability calculation. The open unpredictability of the system beyond immediate horizons generates many of the features of its ordering institutions. That one cannot know what others know or believe is simply a fact of existence, a fact which explains the surprise with which innovations are often received by the incumbents they challenge. This knowledge is only revealed in the course of the market process and only if agents experiment (Loasby, 2000, 2002). This is precisely the case why we need markets to provide a basis for ongoing adaptation to the opportunities created by innovations. In a stationary economic world, markets are redundant once the initial pattern of resources has been determined, for the corresponding pattern of activities will by definition repeat itself indefinitely. To put it more strongly, we have the market institutions we have because they have co-evolved to adapt to a system based on the internal growth of knowledge and innovation. Here lies a further important dimension of a market order, the stimuli to invasion by new types of activity, and thus the policy importance of maintaining open experimental conditions. This is one reason why competition policy and a proper concern with open markets is so important to innovation policy. It is not about keeping markets close to a perfectly competitive state so that resources are optimally allocated but in keeping them open to invasion and the structural changes, including the exit of marginal enterprises, that follow from innovation-led competition. All of these elements point to the futility of pretending that capitalism is a system that establishes and maintains market equilibrium. Quite the opposite its central dynamic is that it induces self-transformation out of the self-organising market order. Furthermore, it is not surprising to note that a system marked by ineluctable business uncertainty fails to develop markets for future activities, 'How exactly can a contract be written today for the supply of an unknown commodity produced by an unknown method to an unknown customer at an indeterminate date in the future?' Similarly, the possibilities of insurance markets to price and trade risks are greatly affected by information asymmetries and the scope for opportunistic behaviours that give rise to moral hazard. These characteristics of the market order are scarcely to be described as 'failures' when they arise out of the very phenomena that make a market process possible; those phenomena are innovation and the growth of knowledge. To eliminate the uncertainties that flow from innovation it would be necessary to eliminate innovation.

A similar question relates to the alleged public good nature of knowledge, and here insufficient attention has been paid to the distinction between personal knowledge and public understanding. That knowledge is indefinitely extensible that it may be used to produce any quantity of a good (the increasing returns aspect) or may be absorbed by indefinitely many minds (the correlation of understanding aspect) is of course correct. More telling is the point that the same knowledge may be used an indefinite number of times for the production of further knowledge (increasing returns in the production of knowledge), the fact essential for the combinatorial cumulativeness of knowledge production. However, this extended replicability property must not lead to the idea that all knowledge is accessible in the public domain without cost.

Information, the representations of knowledge accessible to the senses, is what is distributed publicly, but without absorptive capacity and channels of communication no link with a recipient's knowledge can be made. In fact, the generation of absorptive capacity may require, and typically does require, major investments in education, and prior R&D to acquire a capacity to understand and locate the information flow. Access costs are not to be equated with transmission costs and this is the potential flaw in treating knowledge along with information as inseparable public goods. Of course, secrecy is the extreme aspect of this point, the knowledge that is deliberately not placed in the public domain, often for good commercial reason[2].

From a policy perspective this leads directly to a series of problems. The first is in dealing with the consequences of uncertainty in relation to innovation and the market process. Thus it is perfectly rational for private firms not to invest in fundamental general knowledge if they consider the scope for exploitation too slender, and the general knowledge of science and technology, and education too, is of this kind. This is not market failure but the market process at work. If there is a failure it is in thinking that markets can deal efficiently with every kind of human activity independently of the breadth of the consequences. That government provides the bulk of the funding for these activities in modern capitalism is perfectly understandable in terms of the provision of 'general' goods and services with widely distributed consequences. It is equally understandable that some private firms do invest heavily in basic scientific and technological knowledge to build absorptive capacity when it is deemed profitable to do so (Rosenberg, 1990). The second follows from the conclusion that knowledge is a distributed system and that it changes in a manner dependent on the connections between the multiple actors; consequently, the generation and diffusion of innovation is a system problem (Edquist, 1999; Smith, 1999). This is the point that leads immediately to innovation systems and the possibility of system design and systems failure. A system may fail to operate in the desired way because knowledgeable actors are missing, because connections are absent or because system boundaries are drawn in the wrong place. Attention to these issues provides the basic rationales for innovation systems policy.[3]

---

[2] The overlap here with intellectual property rights is treated in the companion chapter by Geroski. See Nelson 2004 for a critical evaluation of recent developments in US patent policy that have shifted the market-non-market boundary in the domain of science.

[3] See Woolthuis et al. (2004) and Smith (1999) for more detailed elaboration.

15.2   A systems innovation policy perspective

I have argued that the growth of knowledge is a systemic problem. The component actors, ultimately individuals, are connected through instituted arrangements that permit the flow of information and facilitate the growth of knowledge necessary for innovation. This allows a distinction to be made between 'innovation ecologies', the set of individuals usually within organisations who are the repositories and generators of new knowledge and the 'system making' connections between the components that ensure the flow of information whether in general or directed at a specific purpose. Included in this ecology are those organisations that store and retrieve information as well as those that manage the flow of information in multiple formats. They define collectively a division of labour that is characteristic of the production of knowledge, and this is reflected, for example, within and between the universities and public and private research activities that are major components in any modern knowledge ecology. Ecologies are typically national in scope, with sub-national degrees of variation, which reflect rules of law and language, business practice and the social and political regulation of business. (Carlsson, 1997; Carlsson et al. 2002; Cooke et al., 2000). However, and quite crucially, an innovation ecology is the basis for a system but it is not a system of itself until subsets of the actors are connected with the intention of promoting innovation and the purpose of the connections is to combine multiple sources of knowledge through the flow of information. Hence, while there are national innovation ecologies, it is not at all obvious that there are national innovation systems in the sense usually meant. The logic of this view is that innovation systems are constructed to solve 'local' innovation problems (Antonelli, 2001) and that they are constructed around the economic problems that shape innovation and not only the problems that shape the growth of science and technology. Moreover, since the solution of one problem typically leads to different and new problems we would expect that as the problems evolve the actors in the system and their pattern of interconnection must also evolve and that while ecologies are more permanent the systems are more transient. Thus there is a close connection between the notion of trajectories of technological solutions within a particular technological paradigm and the dynamic notion of an innovation system (Dosi, 1982). Innovation systems will be a normal part of restless capitalism; they are a reflection of the multiple ways in which an innovation system can be instituted and are simultaneously embedded in a matrix of market and non-market relationships. The dynamism of an economy thus depends on the fluidity with which innovation systems are created, grow, stabilise and change as problem sequences evolve.

This line of reasoning immediately suggests that there is nothing inherently national about an innovation system as distinct from an innovation ecology. Depending upon the problems in hand there will be multiple innovation systems supported by the relevant ecology, reflecting the problem sequences in hand, the location of the actors at the leading edges of technological advance, particular links with the science base and the specific uses towards which the intended innovations are directed. Moreover, it follows naturally that the connections and actors can, and increasingly do, spread across national boundaries. It is common place to find firms collaborating with overseas suppliers or customers, to find them drawing on the skills of foreign universities or even setting up R&D facilities in overseas markets. Indeed it leads to the idea that competition for resources in the formation of innovation systems is a central feature of the modern capitalist dynamic in which information is an international commodity.

What are the ensuing policy issues? First, innovation systems are not naturally given; they have to be constructed and the actors define an extended division of labour in terms of what they know and how they know. Because they work in different organisations, it is not difficult to see that the respective absorptive capacities will not be the same and that the costs of the effective correlation of knowledge may be considerable. The differences in communication cultures between firms and universities are well known and the incentives to further or restrict the correlation of knowledge differ greatly across public and private organisations. Here the State emerges as the keeper of the potential for the formation of innovation systems; its role is to set in place the conditions for innovation systems to emerge and evolve.

The central policy problem is to, on the one hand, ensure that there is a rich knowledge ecology on which innovation processes can draw and, on the other hand, a set of rules of the game that openly facilitate the formation of innovation systems to solve problems. In practical terms this suggests a layer of policy themes that we may list as follows. First, general policies in relation to the education system and public research and development, primarily to provide the supply of trained minds whose imagination will be crucial to the experimental process and the growth of knowledge. These individuals are the basic building blocks of the innovation ecology and they require an appropriate supply of research organisations in which to work. The range of disciplinary skills available and their closeness to the world's best practice frontier will also determine the absorptive capacity to adapt to knowledge generated by foreign ecologies, for science and technology are global systems and the formation of innovation systems will reflect a search for the best partners wherever they are located (Harvey and McMeekin, 2004). Thus governments frequently create new elements of the innovation ecology, for example, creating new research organisations in new areas of science and technology or research organisations focused on a particular broad area of exploitation where it is necessary to combine together multiple disciplines to facilitate problem-solving (Kaiser and Prange, 2004). The Faraday Partnerships in the United Kingdom come to mind as do the Industrial Technology Centres in the United States. Second, there are general policies to further the absorptive capacities of private firms through the employment of qualified scientists and engineers and the conduct of R&D and thus facilitate communication with the public knowledge base as well as with other firms, whether suppliers or customers, wherever they are located. It is apparent that a principal reason why university investments in knowledge fail to stimulate innovation in the broad may be a lack of the requisite knowledge-absorbing capacity in the relevant firms. Third, we have policies not aimed at the ecology but at the connections between different actors in the innovation process and thus explicitly at system formation. These are bridging policies that do not take for granted a free flow of information and that recognise the fixed costs of forming network relationships. Of course, connections come in many forms ranging from informal exchanges of information, from participation in professional networks, from partnerships to develop particular projects, to deeper alliances for innovative collaboration. These connections form the external capital of the firm or other innovating unit and in some cases they may be internalised through the market for corporate control; indeed the ability to trade established bundles of business capability is one of the most important aspects of any innovation system. Furthermore, mobility of knowledgeable minds is surely one of the most effective contributors to the making of connections in innovation systems. Indeed, historically if not presently, the mobility of skilled individuals has been a principal form of international technology transfer and innovation diffusion. In distributed innovation processes network formation is crucial and the actors need to be open with respect to the exchange of information. As Nelson (2004) reminds us it is vitally important to keep the ecology open. It is not possible to predict which exact combinations of knowledge and individuals will solve a particular innovation problem; no one can know this in advance, the solution is emergent and emergence is a problem of unfathomable complexity. Universities as well as firms have to be receptive to collaboration in the innovation process and the barriers to collaboration need to be minimised. What kind of general policy instruments facilitate innovation system building?

The policy instruments may be, for example, to facilitate collaborative research, to incubate University ideas, to use public procurement to build networks or to stimulate the formation of clusters but in each case the point is to create connections that will not otherwise organise spontaneously.[4] Their principal purpose is to create opportunities and enhance innovative capabilities by stimulating innovation system formation (Metcalfe, 1995, 2003; EC, 2003). There is no general basis for government predicting which innovation systems will form or who the actors will be; that is a matter of spontaneous order and it implies an obvious corollary, that the connection between instruments and effects will be 'loose' with many unanticipated outcomes. Innovation systems are complex systems in which the growth of knowledge changes the actors involved so that learning effects continually shift the relation between policy cause and innovative effect (Ockruch, 2003). Thus the evolutionary policy-maker is not an optimising supplement to the market, correcting for imperfect price signals in such a way as to guide private agents to a better innovation mix. Rather the role is an adaptive one; the policy-maker is as boundedly rational as the agents that are the policy target and the policy aim is to raise the incentives to innovation by facilitating connections to a suitably rich knowledge ecology. This perspective may be contrasted with the traditional view of innovation subsidies or R&D incentives that took as given innovation possibilities and capabilities and thus encountered diminishing returns to R&D effort. This evolutionary perspective seeks to overcome diminishing returns by enhancing the innovation possibilities and capabilities and take advantage of and coordinate better the division of labour in the innovation process.

The distinction we have made between national innovation ecologies and spatially unconstrained innovation systems carries an implication as yet unexplored. It is that when innovation systems transcend national boundaries they are naturally influenced by the policy jurisdictions of multiple states. The possibility arises at least in principle for policy conflict to arise and inhibit innovation system formation. Put more sharply it may result in competition between nation states to have key elements of innovation systems within their national ecologies. The problem of innovation policy coordination, as for example within a

---

[4] See the detailed discussion in EC 2003 of European approaches to innovation systems policy. DTI 2003 covers related ground for the United Kingdom.

European innovation space, as well as more widely is worthy of much closer investigation.

It should now be clear that innovation policy directed in a narrow sense at innovation systems must be complemented by the wider range of policies that influence the innovation ecology and the propensity to make connections. Education policy and the supply of skills and the mobility of labour are important framing conditions and so is tax policy in relation to business experimentation, and so also is public procurement policy. Most important of all perhaps is a competition policy that fosters the competitive process, keeps the market order open to entrants and recognises that abnormal returns are more likely the result of transient innovative superiority rather than the exploitation of static market power. Indeed the relation is symmetric in that the best form of competition policy will be an effective innovation policy that maintains economic evolution.

15.3   Concluding remarks

The general thrust of this chapter has been to propose that a systems policy is the proper domain of attempts to enhance the rate of innovation. The broad rationale is system failure rather than the traditional market failure arguments. For the latter derive from an equilibrium theory of competitive resource allocation whereas the appropriate framework is one of a competitive process that is ordered but never in equilibrium. Indeed the purpose of innovation policy is to ensure that it never is in equilibrium but is continually challenged from within. Thus many of the alleged market failures, monopoly power needed to innovate, or information spillovers, for example, are necessary elements for the market process to work. In no sense are they market failures from a knowledge-based perspective. In relation to patents for example, their purpose is not simply to protect investments in knowledge but to encourage the disclosure of information to further guide the broad innovation process. At the root of these difficulties are three problems, the distorting mirror of the theory of competitive general equilibrium, the institutional neutrality of a theory of equilibrium states and the failure to place policy-making within a matrix of competing interests, including those of the policy-makers (Wegner, 2003; Witt, 2003). An evolutionary economic approach to innovative competition, embedded in co-evolving instituted frames of market and non-market arrangements provides the necessary understanding that innovation policy-makers require to deal with restless capitalism.

References

Antonelli, C., 2001. *The Microeconomics of Technological Systems*. Oxford: Oxford University Press

Carlsson, B. (ed.) 1997. *Technological Systems and Industrial Dynamics*. Dordrecht: Kluwer Academic Publishers

Carlsson, B., Jacobsson, S., Holmen, M. and Rickne, A. 2002. 'Innovation Systems: Analytical and Methodological Issues,' *Research Policy* 31: 233–45

Cooke, P., Boekholt, P. and Todtling, F. 2000. *The Governance of Innovation in Europe: Regional Perspectives on Global Competitiveness*. Cheltenham, U.K.: Edward Elgar

DTI 2003. *Competing in the Global Economy –The Innovation Challenge*. London: HMSO

Dosi, G. 1982. 'Technological Paradigms and Technological Trajectories - A Suggested Interpretation of the Determinants and Deviations of Technological Change', *Research Policy* 11: 147–62

Dosi, G., Llerena, P. and Sylos Labini, M. 2005. 'Science–Technology–Industry Links and the European Paradox: Some Notes on the Dynamics of Scientific and Technological Research in Europe,' Mimeo, S.Anna School of Advanced Studies, Pisa

Edquist, C. (ed.) 1999. *Systems of Innovation: Technologies, Institutions and Organisations*. London: Pinter

European Commission (EC) 2003. *Raising EU R&D Intensity: Direct Measures*. Brussels: Directorate-General for Research

Freeman, C. 1987. *Technology Policy and Economic Performance*. London: Pinter

Harvey, M. and McMeekin, A. 2004. 'Public – Private Collaborations and the Race to Sequence Agrobacterium tumefaciens', *Nature Biotechnology* 22(7): 807–10

Kaiser, R. and Prange, H. 2004. 'The Reconfiguration of National Innovation Systems – The Example of German Biotechnology', *Research Policy* 33: 395–408

Loasby, B. 2000. 'Market Institutions and Economic Evolution', *Journal of Evolutionary Economics* 10: 297–309

Loasby, B. 2002. 'The Evolution of Knowledge: Beyond the Biological Model', *Research Policy* 31: 1227–39

Malerba, F. 2004. *Sectoral Systems of Innovation: Concepts, Issues and Analyses of Six Major Sectors in Europe*. Cambridge: Cambridge University Press

Metcalfe, J.S. 1995. 'The Economic Foundations of Technology Policy' in Stoneman, P. (ed.), *Handbook of the Economics of Innovation and Technological Change*. Oxford: Oxford University Press

Metcalfe, J.S. 2003. 'Equilibrium and Evolutionary Foundations of Competition and Technology Policy,' in Pelikan, P. and Wegner, G. (eds.) *The Evolutionary Analysis of Economic Policy*. Cheltenham, U.K.: Edward Elgar

Nelson, R.R. 1981. 'Assessing Private Enterprise: An Exegesis of Tangled Doctrine', *The Bell Journal of Economics* 12: 93–111

Nelson, R.R. 1993. *National Innovation Systems*. New York: Oxford University Press

Nelson, R.R. 2002. 'On the Complexities and Limits of Market Organisation' in Metcalfe, J.S. and Warde, A. (eds.), *Market Relations and the Competitive Process*. Manchester, U.K.: Manchester University Press

Nelson, R.R. 2004. 'The Market Economy and the Scientific Commons,' *Research Policy* 33: 455–71

Ockruch, S. 2003. 'Knowledge and Economic Policy: A Plea for Political Experimentalism', in Pelikan and Wegner (eds.), *The Evolutionary Analysis of Economic Policy*. Cheltenham, U.K.: Edward Elgar

Rosenberg, N. 1990. 'Why Firms Do Basic Research (With Their Own Money)', *Research Policy* 19: 165–74

Schumpeter, J. 1934. *The Theory of Economic Development*. Oxford: Oxford University Press

Smith, K. 1999. 'Economic Infrastructure and Innovation Systems', in C. Edquist, (ed.), *Systems of Innovation: Technologies, Institutions and Organisations*. London: Pinter

Wegner, G. 2003. 'Evolutionary Markets and the Design of Institutional Policy,' in Pelikan and Wegner, *The Evolutionary Analysis of Economic Policy*. Cheltenham, U.K.: Edward Elgar

Witt, U. 2003. 'Economic Policy Making in Evolutionary Perspective', *Journal of Evolutionary Economics* 13: 77–94

Woolthuis, R.K. Lankhuizen, M. and Gilsing, V. 2004. 'A System Failure Framework for Innovation Policy Design', *Technovation* 25: 609–19

16   Intellectual property rights and competition policy

Paul A. Geroski

Introduction

Over the past decade or so, companies have begun to regard knowledge as a key source of competitive advantage, and they have learned to value – and protect – their knowledge-based assets. This, in turn, has led them to assert, and manage, their intellectual property rights more aggressively than hitherto.[1] Since knowledge is typically protected by granting inventors or innovators a monopoly over the use of their protected intellectual property for a finite period of time, the aggressive exploitation of intellectual property rights looks a lot like the aggressive exploitation of any other monopoly position on the face of it, and this has begun to attract the attention of those concerned with designing or implementing competition policy.[2]

At pretty much the same time, competition policy itself has risen up the policy agenda. In the United Kingdom, two major pieces of legislation have considerably strengthened the powers of competition authorities, increasing their independence, strengthening and deepening their powers, and increasing their resources. Much the same has occurred throughout Europe, particularly in those countries where competition policy has not, until now, taken deep root. And, on a global scale, competition policy has become more active and prominent as price-fixing cartels have been attacked in and between numerous countries.

The views expressed in this chapter are designed to stimulate discussion, and do not necessarily reflect the views of the Competition Commission, or any of it's staff or members. Needless to say, I alone am responsible for any omissions or errors of analysis.

[1] See, for example, Rivette and Kline, 2000, a popular business book which has called the attention of managers to underutilized intellectual property rights.

[2] For recent discussions of the law and practice on intellectual property, see FTC, 2003; Landes and Posner, 2003; Lerner and Jaffe, 2004; and others.

The question that both of these developments raise is whether intellectual property rights and competition policy conflict with each other. In fact, a moment's reflection suggests that the relationship between competition policy and intellectual property rights is not straightforward. On the one hand, patents and copyrights give monopoly like powers that can, in principle, be abused. This observation suggests that there may be serious conflicts between competition policy and intellectual property rights. On the other hand, keeping markets competitive is clearly one of the goals of competition policy, and not a few people think that competition stimulates innovation. In this second line of thinking, then, competition policy and intellectual property rights are complementary.

The truth is that both of these arguments are right, and that, as a consequence, there are some deep-seated tensions between both types of policy that need to be addressed if their complementarities are to be realized. In what follows, I would like to make three points.

- First, I will argue that competition policy and intellectual property rights do indeed have congruent goals. Accepting this requires one to believe that competition is more conducive to innovation – or to certain particularly valuable types of innovation – than monopoly, an argument that I will explore in a little bit of detail.
- Second, the apparent inconsistency between intellectual property rights and competition policy – one seems to promote monopoly while the other seeks to dismantle it – rests on a misunderstanding of the role that 'monopoly' plays in each. In particular, competition policy is often concerned most about what I will call monopoly *ex ante*, while intellectual property rights is concerned about monopoly *ex post*. It is quite possible – and, in my view, desirable – to promote innovation by removing *ex ante* monopoly through the application of competition policy whilst, at the same time, allowing some degree of *ex post* monopoly through judicious application of intellectual property rights.
- Third, it seems clear that there are circumstances where intellectual property rights can actually inhibit innovation, and where the application of competition policy can help solve this problem. I will explore these circumstances in some detail, and show that they can involve a complicated trade-off between current and future innovation.

My conclusion is that competition policy and intellectual property rights conflict only when intellectual property rights are abused, or where they inhibit future innovation, that is, when intellectual property rights do not work as they ought to.

16.1   Monopoly and innovation

The debate about the relationship between innovation and competition tends to turn on the rather different effects that monopoly has on the ability of particular firms to innovate on the one hand, and their incentive to do so on the other. Let me just remind you of just a few of the more familiar features of this hoary old debate.

It seems quite clear that monopolists are likely to be more able to innovate than firms who operate in the perfectly competitive markets much beloved of textbook writers. The existence of some degree of market power means that prices are likely to be set above variable costs and, if this occurs, the fixed costs of innovation will be covered (provided they are not too excessive). Further, a large monopolist may be able to more fully exploit economies of scale in R&D (if there are any) than smaller firms in more competitive markets. More generally, profitable monopolists are likely to be able to fund innovation out of retained profits, and this enables them to avoid borrowing from uninformed, risk-averse investors. Finally, a monopolist may be able to appropriate more of the returns to his or her innovation than would a firm in a very competitive market, and may find it easier to bring a new innovation to market than a firm lacking market power. All of this is to say that monopolists may be more able to innovate than firms operating in competitive markets.

The counter argument is that monopoly is likely to dull the incentive to innovate. In a long-run sense, the existence of particular monopoly positions, and the identity of the particular firms who enjoy them, cannot be taken to be exogenous. Firms who have a well-established market position have often acquired this position on the back of past innovations which they have successfully brought to market. Further, successful firms sustain their market position by developing both the initial innovation and various complementary assets to that innovation which enable them to utilize that innovation as fully as possible. They often make extensive investments in serving their market in a particular way, designing procurement systems, customer relationships, and pro-duction methods around their successful innovation. The incentives problem is sometimes an issue of arrogance or complacency, but it is often also basically a problem of rent displacement: new innovations are often disruptive, and when a new innovation disrupts a profitable activity (created by a previous innovation) that is undertaken by an established firm, it yields a lower net return to that firm than it would to a firm not involved in that existing activity.

For many people, the ability versus incentives discussion can only be resolved by empirical evidence, and for me the evidence is fairly clear: competition is often associated with more innovation than are more monopolistic market structures (holding all other things constant). Even more graphic – but possibly less persuasive – are a number of case studies that suggest that the introduction of competition in a market (e.g. by deregulation or through entry) stimulates a burst of innovative activity, and a further few that chart the lonely decline of well-established firms who innovate successfully once but never again. Needless to say, there are counter examples to every generalization, and well-established firms do sometimes innovate, even when they have considerable market power. It is also worth noting that virtually all of the evidence suggests that market structure is not obviously the most important driver of innovation.

16.2   Monopoly and types of innovation

All of this said, I am not sure that the "ability versus incentives" debate actually focuses in on the important question about the link between competition and innovation. Rather than asking whether monopolists are more innovative than competitive firms, it is probably more useful to ask whether monopolistic firms have incentives to introduce different types of innovations than competitive firms do. I find it useful to explore this particular question by distinguishing two types of innovation: what are sometimes called 'sustaining innovations' and 'disruptive innovations'.[3]

The key characteristic of sustaining innovations is that they offer consumers more of the same basic set of product characteristics, enabling them to do what they already do faster and more efficiently. New personal computers that offer more computing speed or more memory are examples, as are cars that drive faster, or more economically, than existing cars. Some definitions of sustaining innovations focus on the capabilities of firms – sustaining innovations often build on existing skills and they are, in a sense, 'competence-enhancing'. Disruptive innovations, on the other hand, offer consumers quite a different set of product characteristics, and enable consumers to do different things, or familiar things quite differently. The displacement of typewriters by personal computers with word-processing capabilities is a classic example of a disruptive innovation, as was the displacement of horses and carts by cars. Disruptive innovations often force firms to develop new skills and expertise, while at the same time reducing the

---

[3] This terminology derives from Christiansen, 1997.

value of existing skills and expertise. As a consequence, they are sometimes called 'competence destroying' innovations.

The basic argument is that a well-established firm with some degree of market power is much more likely to prefer to introduce sustaining innovations than disruptive innovations. Sustaining innovations build on, and develop, what it is already good at. No less important, they will help protect its existing business against the challenge of imitative rivals. Further, there is no rent displacement associated with sustaining innovations. Disruptive innovations, on the other hand, offer lower net returns for established firms than they offer for new players simply because new disruptive innovations displace existing activities, and the profits that they generate. For an established firm, the gain to introducing a disruptive innovation is the gross profits from doing so, less the loss of profits from activities that the innovation displaces. For new firms or fringe players, there is no existing activity to displace and so the net return to introducing disruptive innovation is rather higher: there are simply no rents to displace.

The bottom line, then, is that firms with market power may be more likely to be innovative than competitive, but only in those directions which suit their interests. This effectively means that they have much stronger incentives to introduce sustaining innovations than disruptive innovations. In contrast, entrants or fringe players anxious to expand have no major streams of rent arising from existing activities, and, as a consequence, they have a much stronger incentive to introduce disruptive innovations. If, as many people believe, sustaining innovations tend to be incremental innovations and disruptive innovations are more major, then this argument says that competition is likely to stimulate the production of major innovations.

16.3 *Ex ante* and *ex post* incentives to innovate

Needless to say, the incentive for a firm – whether it be competitive or monopolistic – to innovate is made all the stronger if the firm is likely to emerge with some degree of market power *ex post*. This observation suggests that there is another way to think about the relationship between competition policy and intellectual property rights. Suppose, for the sake of argument, that it were the case that monopoly stimulates innovation because it massively increases the ability of market leaders to innovate without, somehow, undermining their incentives to do so. This argument amounts to saying that it is monopoly *ex ante* which matters for innovation: it is firms who already have some market power that are (or are not) likely innovators. If this line of argument is correct, then any attempt to attack positions of market power is likely to retard innovation. Conversely, if competition stimulates innovation, this just says that the absence of monopoly *ex ante* stimulates innovation.

The point that I want to make is that this is a very different type of argument from that which is usually used to justify intellectual property rights. Intellectual property rights create monopoly *ex post* – that is, after the innovation has been made – in order to stimulate innovation. It is the promise of monopoly – not its actual fact – that stimulates innovation in this line of thinking. Any attack on existing monopoly positions will, in this view, not affect innovation since what matters is the expectation of having a monopoly (at least temporally) on the use of the innovation after it is made. This is perfectly compatible with having a good deal of competition in the market before the innovation is developed.

Thus, it seems that *ex ante* monopoly and innovation have only tenuous links, particularly in the context of disruptive or major innovations. There are grounds for thinking that some degree of monopoly in the market for, or created by, an innovation – that is, some degree of monopoly *ex post* – may stimulate innovation, but this is not the same thing as saying that monopoly market structures are more likely to produce new innovations than competitive ones. In this sense, then, there is clearly no obvious conflict between competition policy, which is designed to keep markets competitive *ex ante*, and intellectual property rights, which create temporary monopoly positions *ex post*.

In fact, this argument suggests that there is, or might be, an interesting trade-off in the application of competition policy to the problem of stimulating innovation. Let us suppose that innovation is more likely to occur the more competitive is the market *ex ante* and the less competitive it is *ex post*. Then, if competition policy limits intellectual property rights in some way, this will, in effect, increase competition *ex post* (that is, it will weaken the *ex post* monopoly granted by intellectual property rights like patents) and reduce innovation. To compensate, one would – in this line of argument – need to stimulate innovation by increasing competition *ex ante*. That is, it may be that the application of competition policy to *ex post* monopoly positions may have to be accompanied by an increased application of competition policy *ex ante* in order to preserve incentives to innovate.

16.4   Do intellectual property rights inhibit innovation?

The arguments that I have just outlined suggest that there is a good deal of congruence in the goals of competition policy and intellectual property rights. That is not, however, the whole story. There are at least two areas of tension between competition policy and intellectual property rights, and I want now to spend a little time exploring one of them in particular. The first arises when the temporary monopoly conveyed by a patent or copyright is abused by its holder in one way or another. There are a range of ways that patent licensors sometimes try to restrict the use of their patents by licensees – granting exclusive territorial rights, exclusive dealing, restricting price competition, and so on – and these practices are often just what they seem to be, namely restrictive practices designed to lessen competition in the market for the new good.[4]

The second source of tension arises when intellectual property rights impede innovation, and I want to focus on this in what follows.[5] The process of learning has one particularly important feature, namely that knowledge often builds on itself: new ideas are developed from old ideas, and new ideas often combine several old ideas in new and different ways. It follows that the process of innovation is likely to be more effective and more efficient if today's innovators are allowed free access to the results of yesterday's innovations. If, however, intellectual property rights impede the ability of innovators to use previous knowledge, then intellectual property rights can impede the rate of innovation. There are at least two ways in which this might happen.

The first is essentially the problem of the 'anti-commons'. Every innovator draws from a large pool of common knowledge, and when that knowledge is in the public domain no one is restricted in the assess to, or the use they can make of, that public information. Further, since information is essentially a public good – meaning that it is non-rival and non-exclusive in use – there is never likely to be a problem of congestion in the public domain, and hence there can be no real justification of restricting or regulating access to it. It follows that practices such as the issuing of overly broad patents, or allowing innovators to make claims for knowledge which they have drawn from the public domain are likely to pervert the purpose for which intellectual property rights were developed.

The second problem arises with complex innovations. When a new innovation draws on several different areas of technology, then the innovator will need to undertake a series of bilateral negotiations with existing intellectual property rights holders if his or her innovation is to see the light of day. This, in turn, means that any individual antecedent patent holder has the ability to hold up the new innovation, possibly

---

[4] See Gilbert and Shapiro, 1997, and FTC, 2003.
[5] See Shapiro, 2000; Scotchmer, 2004; and others.

using this bargaining power to extract most of the returns that it promises to produce for it's creator. These patent thickets essentially raise the transaction costs of innovating, and this is likely to inhibit the rate of innovation.

Clearly, when intellectual property rights are used to impose restrictive practices on licensees which have adverse effects on competition – grant backs and blocking patents are obvious examples – in the market for the new good or service covered by a patent, there is a potential conflict between intellectual property rights and competition policy. Equally, when intellectual property rights are used to inhibit subsequent innovation, there is also a potential conflict between intellectual property rights and competition policy. However, the two types of conflict are very different. In the former case, allowing competition policy to override intellectual property rights may undermine incentives to innovate, while in the latter case overriding intellectual property rights is more likely to stimulate innovation than not.

16.5   Current and future innovation

There is a slightly different way to think about the argument that intellectual property rights may, in certain circumstances, inhibit innovation. Whenever control over today's innovation can be used to retard the development of tomorrow's innovation – a problem most likely to arise when tomorrow's innovation is disruptive – then the effect of intellectual property rights is to stimulate current innovation at the cost of future innovation.

A system that allows firms to claim broad patents, or to make excessive claims that effectively remove knowledge from the public domain provides very strong incentives for firms to innovate. In essence, by protecting the initial innovator from the competitive effects of subsequent innovations, or by giving the initial innovator the option to monopolize subsequent innovations, the system essentially favours the present over the future. Conversely, anything that limits the ability of a patent holder to block or delay future innovation makes the initial patent less valuable and will, therefore, reduce the incentives for firms to innovate, at least initially.

The application of competition policy affects this trade-off in ways that we have already discussed. In particular, increasing competition *ex ante* is likely to stimulate innovation, and may help restore incentives to innovate which are reduced if broad patenting or excessive claims are prevented. Similarly, limiting intellectual property rights *ex post* is likely to simulate future innovation at the expense of current innovation.

This is likely to seem to be very sensible for everyone who thinks that innovation is a stream of activity over time (rather than a one-off event).

16.6   Conclusion

We live in a decade where both competition policy and intellectual property rights have, for quite different reasons, become more prominent. For many, this seems like a good example of public policy which is not joined up: the left hand of policy creates monopolies while the right hand of policy dismantles them. I have tried to argue today that this view of policy conflict is too simple.

In particular, there is a good deal of congruence between both types of policy: both can reasonably be characterized as being concerned with stimulating innovation and long-run growth. At base, this is because competition *ex ante* is likely to stimulate innovation, particularly the major disruptive innovations which transform markets. That said, there still remain tensions with any system of protection that awards innovators *ex post* monopoly rights. Some of these tensions arise because intellectual property rights can actually reduce innovation, and solving such problems can have at least the appearance of reducing competition. To put the matter in a nutshell, intellectual property rights conflict mainly when intellectual property rights do not work as they ought. And, when this occurs, the application of competition policy will almost certainly help restore incentives to innovate.

References

Christensen, C. 1997. *The Innovators Dilemma*. Cambridge, MA: Harvard Business School Press

Federal Trade Commission (FTC) 2003. *To Promote Innovation: A Proper Balance of Competition and Patent Law and Policy*. Washington DC: FTC

Gilbert, R. and Shapiro, C. 1997. 'Anti-Trust Issues in the Licensing of Intellectual Property', *Brookings Papers on Economic Activity*, Washington DC: The Brookings Institution, pp. 283–349

Landes, W. and Posner, R. 2003. *The Economic Structure of Intellectual Property Law*. Cambridge, MA: Harvard University Press

Lerner, J. and Jaffee, A. 2004. *Innovation and Its Discontents*. Princeton, NJ: Princeton University Press

Rivette, K. and Kline, D. 2000. *Rembrandts in the Attic*. Cambridge, MA: Harvard Business School Press

Scotchmer, S. 2004. *Innovation and Incentives*. Cambridge, MA: MIT Press

Shapiro, C. 2000. 'Navigating the Patent Thicket', in A. Jaffee, J. Lerner and S. Stern(eds.), *Innovation Policy and the Economy*. Cambridge, MA: MIT Press

17 The policy-shaper's anxiety at the innovation kick: how far do innovation theories really help in the world of policy?

Paraskevas Caracostas

Introduction

If he was alive today Schumpeter would need to include a new actor in his analysis of the dynamics of knowledge-based capitalism: the 'policy-shaper', that is, a civil servant working in the Ministry of Research or Industry in charge of supporting the policy process by preparing decisions.[1] These decisions might concern, for example, budgets to allocate to research and how to distribute them between research domains (or between different types of organizations), the design of incentive structures such as fiscal measures (to convince private enterprises to invest in knowledge creation) or the regulations that define the professional status and career structure of researchers.

Unlike the macro-economist working in the Treasury or the micro-economist working in the Competition Commission, the 'policy-shaper' involved in research and innovation policy is a strange animal. His or her education background is diverse: he or she is either a scientist from natural or social sciences, a management specialist, a lawyer or an engineer. His or her professional experience varies as well: he or she may come from the world of research that he or she has abandoned some time ago, have a business experience or a career in the public service as a generalist.

The knowledge base he or she is drawing upon to understand the world in which he or she operates in is thus quite heterogeneous and eclectic because of the above-mentioned lack of professional codification but also because the theories and related evidence he or she needs to mobilize are fragmented, sometimes contradictory and often in flux.

---

[1] This paper updates the approach developed in a previous analysis of the evolution of EU research policy (see Caracostas and Muldur 1997, Caracostas and Soete 1997). Its title is modelled on Peter Handke (1970). Views expressed in this paper are those of the author and do not reflect official positions of the European Commission.

Very often general theoretical frameworks that are dominating for a specific period the intellectual scene co-exist with fragments of older explanations which stick to practices of the various actors he or she is interacting with.[2]

Nevertheless, there are in most industrialized and emerging countries similar trends that shape the environment in which research and innovation policies are evolving:

- Restrictions on public budgets in virtually all the industrialized countries mean that choices have to be made and this fosters debate on which are the 'best' public policies; the globalization of information also means that experiments conducted elsewhere can be compared and assessed; reference models influence the way people see the world; the United States, for example, is since the beginning of the 1990s mentioned in Europe as the model for modern science policy.
- Since the Second World War, public opinion no longer regards the scientific world as awesome and unassailable; the alliance between research and warfare and, more recently, a number of industrial and technological disasters, have led to a powerful political tide of mistrust and criticism of progress in science and technology; the research community has reacted by promoting new activities to improve the 'public understanding of science' or to involve social actors in the research enterprise.
- Debate on the various forms of capitalism, which widened after the fall of the Berlin wall, has highlighted the complex nature of the factors which in some cases lead to growth and jobs and in other cases to stagnation and joblessness; research and innovation then become part of a more complex set of factors (which explain the success of the Asian countries, for example).

The present paper is written by an economist who has spent more than fifteen years in the Research Directorate General of the European Commission (EC), mostly in policy-related jobs. It is a very limited attempt to confront innovation theories with the mindset of the 'policy-shaper' working in the research and innovation policy field. A few examples are used to provide a partial answer to the following question: How far do innovation theories really help policy development? The policy-shaper will try to exploit available units of knowledge to answer

[2] Susan E. Cozzens has brilliantly analysed the profession of science and technology policy shapers (that she calls 'science and technology policy professionals') in her keynote address to the 2001 AAAS Workshop on Science and Technology Policy Careers (see www.aaas.org/spp/nextgen/2001/cozzens.html).

the questions he or she needs to address in the day-to-day process of policy development, implementation and assessment.

The structure of the chapter, inspired by a sequential perspective on activities that are, in the real world, evolving simultaneously, is the following:

- First, the policy-shaper is faced with the question of the object of the policy, that is, of the boundaries of this reality he is trying to influence.
- Second, he or she must develop the rationale behind public intervention in order to legitimize it: 'Why does the State need to intervene?' is the typical question here.
- He or she will then be called to identify the problems that need to be addressed and to set objectives for public interventions.
- Finally, innovation theories should help policy-shapers in defining policy instruments.

For each of these phases or types of activity I will try to give examples of the contribution that economics and sociology of research and innovation or political sciences research can make and to identify questions the policy-shaper would like to see addressed in a concrete manner. Issues for further research will be mentioned too. In doing so I might have been biased by two elements of my experience: the fact that I have worked in the area of research policy and that the context of my work has been the one of European Union research activities. I will nevertheless attempt as much as I can to quote a few examples from national research policy contexts (in European countries) as well.

This paper is therefore not aiming at drawing the implications of evolutionary or systems of innovation approaches for designing and implementing research and innovation policies but rather at describing the ways policy-shapers combine them in an eclectic manner to other, different and sometimes contradictory, approaches in order to underpin their professional practice.

17.1   Defining the policy field and its boundaries

In most member states of the European Union, the structure of government departments includes a Research (or Higher education and Research) Ministry and an Industry (or Economic Affairs) Ministry. Very often other ministries (e.g. Agriculture or Defense Ministries) manage important research programmes and regulate science and technology. Sometimes a specific body (or a person like the 'Chief Scientist' supported by a specific body) is established at the level of the Prime Minister or attached to a senior Minister to coordinate the research activities across government departments. This structure of government departments reflects the way policy-shapers in the past have segmented the reality that government policies attempt to influence. What about now and in the future?

If one assumes that the notion of 'national system of innovation' (NSI) (Lundvall, 1992; Nelson, 1993; Edquist, 1997) is the most popular reference framework for the current generation of policy-shapers in Europe, one needs to question each of its three components, that is, to explain the field of policy ('innovation' versus other social activities), to delineate its spatial boundaries ('national' versus regional, global or European) and to understand its consistency ('innovation system' versus other looser notions such as 'innovation environment').

The first question concerns the boundaries of the field of innovation policy. While not claiming that they follow a simple historical sequence, Lundvall and Borras (2005) attempt to clarify the issue by making a distinction between science, technology and innovation policies, each of them focusing respectively on scientific knowledge, technical knowledge and the innovative performance of the economy as a whole.

Science policy focuses on scientific knowledge. It concerns the funding of public research institutions (such as universities or public laboratories), the creation of tax incentives to firms and the definition of intellectual property rights. Technology policy aims at the advancement and commercialization of sectorial technical knowledge through public procurement, aid to specific sectors or technologies, the creation of public–private partnerships or the support to standardization. Finally, innovation policy intervenes on the conditions that influence enterprise behaviour and focuses on the overall innovative performance of the economy through regulation, corporate law, the support to organizational innovation, competition policy or cluster policies at regional level. According to the authors there has been a truly 'innovation policy turn' since the late 1990s in a number of industrialized countries, especially in Europe. When one looks at the details of Lundvall and Borras's analysis, the distinction between these three policy fields become somewhat problematic: are tax incentives an instrument of science or of innovation policy? Does public procurement belong to technology or to innovation policy? Is it obvious to consider that education and training policies are instruments of innovation policy and not of science policy?

Instead of distinguishing between those inter-related fields, other authors integrate them into a single policy framework in which they interact. The configuration of this interaction varies from one country to the other or from one spatial level to the other (i.e. national versus regional innovation policies). Stefan Kuhlman (2003), for example,

'The "research and innovation system" of a society comprises, according to the internationally accepted understanding, the "cultural landscape" of all those institutions, which are engaged in scientific research, which accumulate and disseminate knowledge, educate employees, develop technology, which create and distribute innovative products and processes; in this category also belong appropriate regulatory regimes (standards, norms, laws) as well as government investments in appropriate infrastructures. The innovation system covers schools, universities, research institutes (education and science system), industrial enterprises (economic system), the policy-administration and intermediary instances (political system) as well as the formal and informal networks of the actors of these institutions. As a "hybrid system" it represents a sector of society, which radiates far into other areas, via the education system, or via entrepreneurial innovation activities as well as their socio-economic effects. No innovation system is identical to another, just as no society is identical to another.'

This very broad, 'interactive' and 'problem-oriented' vision of the role of research and innovation in our societies first implies that no area of action should be barred *a priori* to public policies in this field. From the most basic research to industrial design or social experiment and market research, nothing, in principle, may be excluded from the scope of government intervention. Combining scientific, technical, organizational, commercial and social dimensions and involving a multitude of public and private players, research in the broad sense encompasses basic research, the testing of various technological options, exploration of the relevant organizational and socio-economic dimensions and assessment of the regulatory and institutional frameworks which will enable it to deploy its potential to best effect. These various processes of invention, investigation or creativity can no longer be seen as juxtaposed or successive; they usually take place in interaction with one another and simultaneously.

How to set up organizational structures at the heart of government that are designed to manage such a level of complexity? Two examples illustrate recent attempts at joined-up government in European countries.

In Finland the Science and Technology Policy Council set up in 1987 is chaired by the Prime Minister. The membership consists of the Minister of Education and Science, the Minister of Trade and Industry, the Minister of Finance and a few other Ministers. In addition to them the membership includes ten other members well versed in science and technology. The members must include representatives of the Academy of Finland, the Technology Development Centre, universities and industry as well as employers' and employees' organizations.

The Council has been assigned the following tasks:

- *'To direct S&T policy and make it nationally compatible and to prepare relevant plans and proposals for the Council of State.*
- *To deal with the overall development of scientific research and education, to prepare relevant plans and reviews for the Council of State, and to follow up the development and the need of research in the various fields.*
- *To deal with, follow up and assess measures taken to develop and apply technology, and to prevent or solve eventual problems involved in this.*
- *To deal with important issues relating to Finland's participation in international scientific and technological co-operation.*
- *To issue statements on the allocation of public science and technology funds to the various ministries, and on the allocation of these funds to the various fields.*
- *To handle the most important legislative matters pertaining to the organisation and prerequisites of research and the promotion and implementation of technology.*
- *To take initiative and make proposals in matters under its competence for the Council of State and its ministries.'[3]*

In the Netherlands, in August 2003 the Prime Minister initiated the Innovation Platform along the lines of the Finnish Science and Technology Policy Council. The Innovation Platform is installed by Royal Decree for a three-and-a-half year period, from 1 January 2004 to 1 July 2007. Chaired by the Prime Minister, the Platform includes the Minister for Education, Culture and Science, the Minister for Economic Affairs and many personalities from the world of research, business and education. It aims *'to strengthen the innovation potential of the Netherlands in order to secure a leading role for this country in the European knowledge economy of 2010. To achieve this, the Netherlands should recover/regain values such as excellence, ambition and entrepreneurship.'*

The Innovation Platform started its activities in September 2003. Since then, five working groups have been set up, each headed by a member of the Platform. The following working groups or projects can be quoted:

- Dynamics of the Dutch Innovation System
- Long-term Choices
- Moving Up in Higher Education
- Consultation Groups
- Innovation in Public Governance

---

[3] See the Council's website at www.minedu.fi/tiede_ja_teknologianeuvosto/eng/structure.html

These groups may propose policy changes that are discussed in government and implemented.

Organizational innovation in government has therefore focused in the recent years on the creation of trans-departmental organizations in order to promote coordination of public interventions. The aim is to cope with an evolving and complex sphere of human activity. Applied research is needed to understand under which conditions such organizational innovations can be made sustainable and conversely the economic, social and historic reasons for organizational inertia in government structure.

We now turn to the second issue, the notion of 'national' versus other territorial scales.

Nelson (1993, p. 517) answered the question: 'What remains national about innovation systems?' as follows:

- There will be increasing internationalisation of technology and management styles in industry;
- but, in most countries, resident firms will be largely national and the same applies to university research, publics labs, laws, policies, relations between government and business.

Niosi and Bellon (1994) have developed the notion of 'open national systems of innovation'. For them *'all NSIs are open to a different degree, and . . . . the links between national systems and the dynamics of their interdependence are keys to understanding their national characteristics'*. They argue that three types of innovation systems (regional, national, international) coexist and compete with each other. *'Internationalisation grows but does not suppress local and national networks; it modifies their functioning, however, since some previously regional or national activities are transferred to international networks.'*

For Gregersen, Johnson and Kristensen (1994), national boundaries still matter for defining innovation systems because they characterize entities where specific learning processes and institutional building take place and evolve.

Gaffard, Bruno, Longhi and Quéré (1993), in their study of the diversity of innovation systems in Europe, have highlighted the relations between firms and industrial relations as determining the territorial innovation dynamics at sub-national level. Four types of local innovation systems (industrial districts, metropolitan areas, territorialized agglomerations and territories in transition) have been identified. For these authors territoriality is the result of strategic behaviour of firms in relation to institutions.

For those who focus on NSIs, the European integration process is difficult to analyse since it does not correspond to a coherent institutional set-up but to a set of different and unequally developed institutions. For example, in Boyer and Saillard (1995, p. 290) post-national 'regulation'[4] is seen, as in the previous cases, as an aborted national one because: *'Community interventions confront old regulations with severe constraints and lead to short term negative adjustments instead of establishing at European level sufficiently robust institutional forms to compensate the loss of autonomy of nation-states.'*

Di Ruzza (1995) tries to address the issue when he talks about a 'superposition' of 'spaces on the move':

- the global space which is the field for competition between large (multinational) firms;
- national spaces which are still the level for political and legal governance;
- between the first two types, 'intermediary spaces' like the EU.

European integration – mainly a political process – aims at constraining large enterprises to re-centre their activities in Europe, by integrating national productive systems and making the impossibilities of national solutions irreversible. But enterprises are not more 'European' today than in the 1950s. For Di Ruzza the articulation of the three levels mentioned above cannot be assumed to be coherent *ex ante* since actors (firms, trade unions ...) do not share the same spatial reference models. The concept of 'economic space' used in his analysis is based on a distinction between *'the international economic relations which are internal to a productive system from those which are external to it and which articulate it to the other productive systems'*. It therefore goes beyond the boundaries of a national economy as it includes the non-national elements necessary for reproducing a coherent system of production. The French school of the 'Théorie de la régulation', in its diversity, has thus tried to confront this problem of articulating different levels of territorial reference frameworks with sectoral approaches (see also Boyer, 1995b).

More recent analyses of sectoral systems of innovation (Malerba, 2004; 2005) are more innovative in this respect as they point at transformation dynamics that affect all spatial dimensions, therefore de-structuring the reference scale of the earlier NSI approaches: *'While national innovation systems take innovation systems as delimited more or less clearly by national boundaries, a sectoral system approach would claim that sectoral systems may have local, national and/or global dimensions. Often these three different*

---

[4] For a complete explanation of the concepts developed by the 'Théorie de la régulation' see Boyer and Saillard (1995, in particular pp. 539–51).

dimensions coexist in a sector'(Malerba, 2004). The policy implications are that innovation-enhancing policies should be based on a thorough analysis of the specifics of each sector and that *'policy has to consider the coexistence of different geographical dimensions of sectoral systems.'*

Political science, in a somewhat similar way to economic theory, is confronted with the need to reconcile in a coherent vision the simultaneous and contradictory processes of globalization, localization and ethnic fragmentation. One might thus need to borrow from recent developments in political science in order to go beyond characterizations of territoriality centred on a main spatial reference, the nation-state. For example, Badie (1995) attempts an original analysis of European post-national institutional building as *'an ensemble of interdependent units which are agglomerated to degrees variating according to different domains, and which find themselves more and more deprived from their authority but without a symmetrical displacement into a central authority'*. Formal national institutions are destructured by deregulation and this leads to a progressive and continuous decoupling between territory and sovereignty .

This illustrates how a new approach to territoriality seems increasingly necessary to understand its role in innovation. In this context the consideration of territoriality in the framework of the Systems of Innovation approach might benefit from the analysis of post-national institutions such as the EU. There is thus a growing interest in the study of the emergence of a 'European system of innovation' (see for example, Caracostas and Soete, 1997; Borras, 2004). Another research field which would lead to the re-appraisal of territoriality is the analysis of scientific networks across boundaries and the related institutions, which cannot be thought of in terms of national, local and sectoral systems. Exceptional or complex cases are often the source of theoretical innovations. Following on Di Ruzza's, Badie's and Malerba's approaches referred to above, policy-shapers are in need of applied research on the interplay of spatially determined and sectoral systems of research and innovation and of tools that allow them to reduce uncertainties and better delineate the realities they are supposed to influence.

Concerning the third issue, implicit in the concept of 'systems' and in the Systems of Innovation (SI) approach is the assumption that there is some coherence between the various elements of national or sectoral SI. In particular, institutional set-ups are normally stable over time. The analysis of the European post-national institution building process made by Muller (1994) has shown that it creates a structural uncertainty and instability for firms and governments. Because the process of selecting institutions is very open, national institutions may be strongly affected by its results (institutional 'hybrids').

The example of the impact of EU research policy on Greece is briefly analysed hereafter to illustrate this point.

The 'national impact study' concerning Greece (Tsipouri and Xanthakis, 1993) has concluded that the EC Framework Programmes have influenced substantially national R&D policy and the R&D system in the country. The influence of EC research priorities on national ones is not the result of a deliberate attempt of the EC to harmonize national science and technology policies but rather of '*the inherent inadequacies of the national structures*' and of the fact that '*the public sector does not have its own policy to fight for*'.

The national R&D system has been affected through the '*creation of a new subsystem in the local scientific community*', the so-called '*excellence system*' regularly benefiting and being based on EC support. This subsystem works only marginally for the Greek economy (except in the case of software or laser technologies where some local impact has been pinpointed). Its 'clients' are mostly European firms. The authors conclude by saying that '*whether the excellence part integrated into the European system will diffuse to the local system depends on the absorptive capacity of the average and low performing actors, as well as on the (industrial and RTD) policies that the Greek state will adopt in the future.*'

The Greek example raises the question of the coherence of national systems of innovation in late industrialization countries. One traditional way of looking at this is to see if the cooperation patterns induced by transnational institutions reinforce virtual circles of cooperation within the country between the R&D system and the manufacturing base.[5] Another perspective would be to consider the 'excellence subsystem' as a new sector centred on producing and exporting knowledge products and services on European or global markets. Such a sector would require specific policies in terms of knowledge accumulation and diversification and sophisticated intellectual property protection capabilities.

The emergence of an 'excellence subsystem' linked to the EU in the Greek NSI does not mean that something similar did not exist before. Probably the Greek system comprised already a space of international scientific relations (mainly looking towards the United States). The EU research framework programmes together with Structural Funds support for the establishment of high-level research centres (e.g. in Crete)

---

[5] Caracostas (1987) suggests that, in the absence of national institutions built to complete the links between national and foreign organizations by national networks (between research and private organizations for example), there is an important trend towards a differentiation between two different subsystems in the national system of innovation, something very similar to what the Greek impact study highlights.

have produced a transformation of this 'internal space of external relations', recentring it towards Europe.

The concept of system, even coupled with evolutionary approaches, therefore entails the risk of transporting an inherent notion of coherence. The System of Innovation approach, when incorporating evolutionary features (selection, path dependency, diversity generation . . .) and taking on board socio-historical research results and methodologies, may need to reconsider its systemic epistemology and to focus on 'non-systemic' reconfiguration, stability and change, that is, on complex and heterogenous sets of different, conflicting or disconnected processes.

Boyer (1995b) has stressed this non-coherent, unintentional/character of institutions:

'Insofar institutions are constituting the social link and coordination procedures, the promotion of economic efficiency is for them only a secondary objective . . . It is only in a second phase that each institutional form needs to prove its compatibility with an economic reproduction sustainable in the longer term . . . In other words, one has to be attentive not to fall into a naive functionalism: at best it is possible to observe an ex post and approximative coherence after a long process of trial and error . . . '

Beyond its potential for an *ex post* description of a variety of inter-related institutional developments, the innovation systems approach is for policy-shapers not yet in a position:

- to define precisely what is the scope of innovation policy;
- to explain why it varies from a country or region to another and how the various territorial and sectoral dynamics interact;
- and to be operationalized for identifying the key interactions in an innovation system that deserve policy intervention.

In addition, it lacks a forward-looking dimension. For the policy-shaper who is asked to prepare decisions related to a constantly evolving science, technology and productive fabric, there is an expectation to decipher the transforming dynamics of research (emerging transdisciplinary configurations) and to identify the factors and actors behind the 'not-yet-born industries' of tomorrow. It would be extremely useful in this sense to explore new ways of using the different developments in SI approaches for devising methods for assessing alternative policy options for the future. 'SI-based foresight methodologies' would help policy-shapers especially when designing 'structural' or 'horizontal' policy measures.[6]

[6] The FutuRIS initiative in France (2003–2004) is one of the first attempts at 'structural foresight' as it focused on systemic features of the French research and innovation system (see www.operation-futuris.org/dyn_menu.asp).

Despite these important limitations, the innovation systems approach is nevertheless used by the policy-shaper as a proxy of a theory for underpinning joined-up government in a world where the actors and activities concerned by a policy addressing research, education and innovation challenges are numerous and interacting. In this sense it indicates where to search when policy develops at the interface between those various actors and activities.

17.2   Legitimizing an increase in resources for public intervention: the charm of the 'linear model'?

Whether he or she likes it or not a policy-shaper trying to defend the need for more funds for R&D relies implicitely on the famous 'linear model of innovation'. This view of innovation sees the relations between research and markets as forming a 'chain', a straight line extending either from research to the market ('technology push') or from the market to research ('market pull'). Despite the fierce criticism they have attracted from the more popular systemic approaches, these linear models paradoxically continue to influence thinking amongst decision-makers and public opinion because they have the virtue of being simple (or of appearing to be so).

A decision-maker, whether he or she is a Minister, Member of Parliament, someone managing a company, an organization or research programme, has to come up with very good arguments to justify why he or she is calling for increased funding or is choosing to allocate scarce resources in a given way. In situations like that it has to be said that experience shows that it is often a self-regulating system of research (on the German or U.S. model) which has in the past engendered inventions which produced huge economic and social effects (e.g. the laser or World-Wide Web). This has produced a good case for investing in basic research. Similarly, one may suppose, looking at the length of cycles of innovation and growth, that the new technologies will in the long term create more jobs than they destroy when first disseminated in society. The simplicity of these observations makes the linear model attractive at a time when the systemic uncertainty and the pressure of rapid change in the socio-economic and technological scene require players in the public and private sectors to take urgent decisions.

When it comes to justifying public action in the field of research, it should therefore be noted that the neo-classical theory of 'market failures' is the objective ally of the linear approach. Here too, its simplicity may account for the weight it carries in government thinking. The argument most frequently used to justify public action is that of the imperfection of markets which are recognized as being unable to accept the risk of intangible investment and to take a collective and long-term view. The validity of this argument is challenged less and less, either as economic theory or as practical public action. Its corollary, the right and need of governments to remedy the problem by public funding is tacitly acknowledged, provided the risk of *'government failure'* is not greater than that of *'market failure'*.[7]

The neo-classical economists who developed 'endogenous growth' theories (P. Romer, R. Lucas, R. J. Barro) have shown that sustained growth depends in the longer term on the accumulation of four main factors: physical capital, technology, human capital and public capital (communication and transport structures but also education and research).[8]

Theories of endogenous growth have major implications for economic policy. Guellec and Ralle (1995) identify two ways in which this requires the state to act:

- Firstly, government must manage the externalities created by the three factors of accumulation, i.e. physical, technological and human capital; it must fund basic research and create conditions in which intellectual property rights will be observed in respect of applied research; it must extensively co-finance education and training and guarantee that even the poorest have access to financial markets to fund training; where none exist, it may create markets or other institutions enabling private players to co-ordinate their decisions; but the way in which governments act will depend on the specific exogenous measures to be managed (there are no ready-made formulae for the areas of research or training policy);
- secondly, the state may provide public goods to help improve private productivity; numerous studies point to the positive effects of infrastructure investment on growth, but questions of the marginal efficacy of public investment and relationships between rates of taxation and private behaviour are not solved.

Certain scholars have questioned whether or not scientific know-how belongs in the category of public goods. Callon (1994) suggests that government aid to basic research should be justified instead as an investment, allowing technical and economic networks to be reconfigured and revamped. The main objective of public funding is thus now to promote diversity, in setting up networks and in the range of the scientific options open to companies.

---

[7] OECD, 1995, "National systems for financing innovation". See also: Muldur, U., 1991.
[8] For an overall presentation of the new theories of growth cf. Guellec and Ralle (1995). Kuttner (1996) examines these theories in the context of the United States political debate on 'scientific and technological interventionism'.

A second motive which prompts governments to fund science and technology development is that it is perfectly normal for governments to act in fulfilment of their socio-political remits. In the face of social problems such as the increasing number of infectious or viral diseases or pollution of the environment, governments cannot simply remain passive and wait for markets to decide whether or not additional investment in these areas is profitable and necessary. The work of the Organisation for Economic Cooperation and Development (OECD) suggests, indeed, that an extremely large proportion of the industrialized nations' public R&D budgets is set aside to fund social and/or political objectives (see OECD, 2004).

These three different approaches to the question of why it is appropriate to use public funds to finance scientific and technological progress have been used in turn throughout the post-war period to justify intervention in this area by governments in the industrialized countries.

The only major differences are in the choice of mechanisms and financing instruments, and points of friction between the industrialized nations only relate to a number of well-known industries.

Despite the fact that neo-classical theories of endogenous growth are orthogonal to evolutionary theories (see the chapter by S. Metcalfe, in this book), they continue being used by policy-shapers because they are convenient for legitimation purposes. As a recent document from the U.K. government illustrates it. (H. M. Treasury, 2004), they are often combined with references to evolutionary and systemic approaches.

The spread of impact assessments of policies from the environment policy field to all policy fields is another trend to quote here. For example, within the framework of the Better Regulation Package and the European Sustainable Development Strategy, the European Commission (EC) has introduced a new impact assessment method in 2002, integrating and replacing previous single-sector-type of assessments.[9]

Impact Assessment (IA) is a process aimed at structuring and supporting the development of policies. It identifies and assesses the problem at stake and the objectives pursued. It identifies the main options for achieving the objective and analyses their likely impacts in the economic, environmental and social fields. It outlines advantages and disadvantages of each option as well as synergies and trade-offs. As the recent impact assessment of the Commission's proposals on the Seventh Framework Programme for Research (2007–2013) shows,[10] these new assessments can be quite

---

[9] For more details see http://europa.eu.int/comm/secretariat_general/impact/index_en.htm.
[10] See http://europa.eu.int/comm/research/future/pdf/comm_sec_2005_0430_1_en.pdf
See: European Commission, 2005.

extended in scope and sophisticated in their argumentation. Many options are presented and discussed and their long-term (mainly economic) impacts analysed with the help of an adapted version of the NEMESIS econometric model.[11] For policy-shapers such analytical approaches substantially strengthen their capacity to influence or rationalize decision-making.[12]

17.3   Delineating the problems to be solved and setting objectives

Once governments have defined the scope of research and innovation policies and developed a rationale with the help of social sciences, one approach is to identify in a rational manner the concrete problems public action needs to tackle and its specific objectives. Research on NSIs should in theory at least be of great help.

Edquist (1997:14) defines systems of innovation as *'all important economic, social, political, organizational, institutional and other factors that influence the development, diffusion and use of innovations'*. In a more recent paper (Edquist, 2004), he lists the activities that can be expected to be important in most systems of innovation:

- *'Provision of research and development (R&D), creating new knowledge, primarily in engineering, medicine and the natural sciences.*
- *Competence building (provision of education and training, creation of human capital, production and reproduction of skills, individual learning) in the labour force to be used in innovation and R&D activities.*
- *Formation of new product markets.*
- *Articulation of quality requirements emanating from the demand side with regard to new products.*
- *Creating and changing the organizations required for the development of new fields of innovation, for instance, enhancing entrepreneurship to create new firms and intrapreneurship to diversify existing firms, creating new research organizations, policy agencies and so on.*
- *Networking through markets and other mechanisms, including interactive learning among different organizations (potentially) involved in the innovation*

---

[11] More information on NEMESIS can be found on: www.nemesis-model.net. See also Fougeyrollas, Le Mouël and Zagamé (2002).

[12] In an editorial in the May 20 issue of Science magazine in 2005, the director of the White House Office of Science and Technology Policy in the United States, John Marburger, has similarly called for a *'new "science of science policy"'* and added, *'We need econometric models that encompass enough variables in a sufficient number of countries to produce reasonable simulations of the effect of specific policy choices'*. In an era of complexity, uncertainty and limited public resources, quantitative models are perceived to be powerful guiding tools.

processes. This implies integrating new knowledge elements developed in different spheres of the SI and coming from outside with elements already available in the innovating firms.

- *Creating and changing institutions – for instance, intellectual property rights laws, tax laws, environment and safety regulations and R&D investment routines – that influence innovating organizations and innovation processes by providing incentives or obstacles to innovation.*
- *Incubating activities, for instance, providing access to facilities, administrative support and so on for new innovating efforts.*
- *Financing of innovation processes and other activities that can facilitate the commercialization of knowledge and its adoption.*
- *Provision of consultancy services of relevance for innovation processes, for instance, technology transfer, commercial information and legal advice.'*

The fact that he very openly recognizes that *'these activities should ideally be related to the propensity to innovate', that 'the ambition should be to reveal which activities are important (…) and, if possible, how important they are'* shows that the NSI approach is very difficult to use by policy-shapers when they try to identify key systemic failures and to define specific objectives of public actions.

While, as we have seen above, market-failure arguments and endogenous growth theories and models are useful to them for legitimizing public investments in R&D, the enumeration of the subsystems and their relations, of actors and of activities in NSIs is not sufficient for defining a hierarchy of problems, of instruments and of resources to allocate to these instruments.

Moreover the difficulties embedded in the national systems of innovation approach referred above (related to the determination of scope, spatial boundaries and systems coherence) make it difficult for them to claim that their policy work is founded on solid theoretical ground and robust evidence.

Nevertheless, according to Technopolis (2001: 14–15) more recent developments in theory do not invalidate but extend this idea of market failure. We can think of these failures as belonging to four types:

- *'Capability failures. These amount to inadequacies in companies' ability to act in their own best interests, for example through managerial weakness, lack of technological understanding, learning ability or 'absorptive capacity' to make use of externally generated technology.*
- *Failures in institutions. Not only companies but also other social institutions such as universities and research institutes, patent offices and so on need to work well if the NRIS (National Research and Innovation System) is to*

facilitate innovation and growth. Rigid disciplinary orientation in universities and consequent inability to change with changes in knowledge would be an example of such an institutional failure. Failure to provide adequate investment in knowledge institutions would be another.

- *Network failures. These relate to problems in the interaction among actors in the innovation system, and can themselves be of several types.*
  - *Inadequate amounts and quality of interlinkages, as where there is low trust among companies or where universities are isolated from their social context.*
  - *'Transition failures' and 'lock-in' failures, where clusters or innovation systems fail or take on board new technological opportunities or remain locked into old ones.*
  - *Various problems in industry structure, such as too intense competition or monopoly, which stifle innovation, or the absence of key complementarities (such as when a cluster's development is stifled by the lack of a crucial type of producer).*
- *Framework failures. Effective innovation depends partly upon regulatory frameworks, health and safety rules etc. as well as other background conditions, such as the sophistication of consumer demand, culture and social values.'*

Deficiencies in these frameworks can have negative effects on innovation and economic performance. These failures justify state intervention not only through the funding of research, but more widely in ensuring that the NSI performs as a whole – always provided that in an individual case the state is actually capable of reducing failure. Because systems failures and performance are highly dependent upon the interplay of characteristics in individual systems, there can be no simple rule-based policy as is possible in relation to the static idea of market failure.

An alternative for policy-shapers would be to accept a looser conceptual framework drawing on evolutionary economics and systems of innovation approaches and to engage in a learning experiment with key actors such as, for example, firms. In such a context policy is embedded in the system it tries to influence and co-evolves with it.

Teubal (1995, 1997) speaks of a 'technology policy framework' or 'policy subsystem':

'It is simplistic to assume that policy consists exclusively of a set of exogenously determined tools associated with monetary incentives. Rather these tools are the result of a complex policy process involving the above mentioned priorities, the coordinated design and implementation of policies in the various priority areas; and policy evaluations. Therefore, it is appropriate to refer to a policy subsystem involving government bureaucrats, stakeholders, and academic and other

experts. (. . .) It is best in our opinion to consider this subsystem as consisting of a set of institutions, capabilities and incentives' (Teubal, 1997, pp. 351–52).

Design and implementation of a varied range of public policy instruments should, at least in theory, be determined by a precise strategic vision of the issues in a given domain, the relevant players and the way in which they behave, the cost–benefit relationships on which the various methods of public intervention depend, and so on. In other words, the objectives, instruments and context of the action are intricately linked. This is reflected in the idea of the 'public policy cycle'. The 'evolutionist' view of technology policy developed by Teubal must of necessity include: *'explicit consideration of learning by government as far as the policy is concerned, including the whole issue of policy capabilities'* (Teubal, 1996). And he adds:

'A necessary condition for successful take-off and consolidation of firm-based R&D and the possibility of effecting the above-mentioned policy restructuring is the adoption of a 'learning' rather than a 'planning' approach to government policy . . . A major component of such approach is the transformation of individual experience with R&D/innovation – including that associated with identification, selection and management of projects within firms – into collective experience, i.e., into a body of more or less codified knowledge about the process available to all firms' (Teubal, 1996, p. 456).

It is not possible in this paper to explore all the issues raised by an 'entrepreneurial' and 'learning' view of public action. At most one might mention, for research policy, the possibility of devising – together with players and users – procedures, which are based on a policy 'life cycle': for example, during an initial phase low-intensity consultation and co-ordination would aim at stimulating collective thinking on the part of participants; the identification together with them of specific problems and objectives would lead, in a second phase, to a more intensive mobilization of public funds for research partnerships; lastly, during a third phase, the government would support the players themselves (e.g. through feasibility studies) in order that they gradually co-fund on their own resources the common knowledge infrastructures that need to be exploited and further developed (e.g. in the form of a 'sectoral' levy collected and used to set up an association or industrial research programme collectively managed by the companies in the 'sector').[13]

---

[13] See Romer (1993) and the research experience of the European Coal and Steel Community which used a similar 'levy' method for fifty years.

This is just one example of the need to develop the objectives, substance and methods of public action through a process of shared learning and against a given time perspective. This learning implicitly entails a measure of 'as-you-go' codification of the knowledge acquired so that it can be disseminated throughout the system of public action.

Another example of a process aiming at embedding the system of public action in the wider context of economic evolution is foresight. Foresight (see High Level Expert Group for the European Commission, 2002) emerged in the last decade in the policy environment of European countries as a way of constructing the object of policy and to define policy objectives and instruments in a collective system of agenda-setting. Inspired by recent work in political sciences in the field of governance theory (see Le Galès, 2004) foresight initiatives explicitly recognize that public interventions cannot be shaped by policy-shapers and civil servants alone. In sectors characterized by high uncertainty they assume that bounded rationality of public and private actors alike imply the joint endeavour of developing the common knowledge for collective and coordinated action.

At the level of the European Union research policy, new organizational entities such as 'Technology Platforms' (see EC, 2004) mimic the participative features of national or regional foresight processes. Technology Platforms mobilize heterogeneous sets of public and private organizations, of national and EU funding bodies, around a broad research problem or opportunity (for example hydrogen-based energy generation or nano-electronics). They aim at devising a broadly accepted '*strategic research agenda*' which will be subsequently implemented by research activities carried out by diverse organizations at national (and if relevant regional or global) levels and will be co-funded by the EU. Such collective learning processes pave the way to coordinated action between hybrid groups of actors in the emerging European innovation system. They would gain from a stronger interaction with academic research on specific systemic failures in specific fields of social activity such as the one on sectoral systems of innovation quoted above.

17.4   Inventing policy instruments: the example of research programme design options

When designing complex policy instruments such as research programmes – with the view of articulating demands from various social actors within the research process itself – the policy-shaper may borrow its supporting rationale from the work on 'social shaping of technology'.[14] The notion of 'the social shaping of technology' may be defined as follows:

- *'It explores the social processes related to technological change;*
- *negotiation between different social groups and actors is a focal point, emphasizing concepts like flexible interpretation of technology and technological controversy;*
- *it highlights the choices between different technical options potentially available at every stage in the generation and implementaion of new technologies'* (Cronberg and Sørensen, 1995).

This work started from a two-fold critique:

- At theoretical level, it questioned 'technological determinism' for which social change is the result of new technologies which have developed, so to speak, outside society.
- At policy level, it challenged the traditional rationalist and techno-cratic science and technology policies, inspired by the linear model of innovation.

The way in which research programmes are organized may indeed to a greater or lesser extent favour the shaping of social demand in the process of research itself.

For the sake of simplicity, the various options for structuring research activity could be classified into four types:

- 'discipline-oriented' programmes designed to develop a particular field of scientific research (e.g. molecular biology, particle physics), whether with specific end uses in mind or not;
- 'technology-oriented' programmes which, driven by engineering science, channel human and financial resources into technical objectives of varying precision (e.g. microelectronics, nanoelectronics, clean cars);
- 'problem-oriented' programmes, more systematic and interdisciplin-ary in nature, which seek to bring research efforts together to solve problems of major socio-economic importance (e.g. research con-cerning the environment, mobility systems, problems of urban living, and so on.);

---

[14] The studies on this subject are too numerous and too varied to be quoted here. For Europe, Cronberg and Sørensen (1995) give a full list of work of this kind done in Belgium, Denmark, Finland, France, Germany, the United Kingdom, the Netherlands, Norway and Switzerland.

- 'structural' programmes which focus financial and human effort on objectives relating to the functioning of the research system (e.g. interdisciplinarity or international mobility of researchers, development of university–industry partnerships).

'Sectoral' structuring (where measures are tailored to the needs of a specific sector of industry or the economy) is less and less fashionable in government circles in the industrialized countries because governments are afraid of being captured by short-term policy agendas in a field that, by its nature, ought to be geared to the medium and long term. But to be accurate, sectoral considerations are not always lacking in the first three types of programme described above.

The shaping of social demand in the research process requires indeed not only participation in the research effort by the end-users or intermediate users of the results, but also the development of specialist socio-economic research to accompany the development of scientific knowledge and technical artefacts/simulations. A topic such as 'clean, safe and low-energy transport' cannot be managed without specialist work on the costs, ergonomic aspects or rhythms of work involved in transport flows. Even a 'hard' subject such as the development of parallel computer architectures should incorporate the investigation of psycho-sociological constraints on use or the long-term skills required.

It would appear obvious at first sight that the third option ('problem-oriented' programmes) is the one best able to combine the different approaches provided by the supply of research (disciplines, combination of social and natural sciences, technologies) with the demands of society (various users involved). A hybrid between the hard sciences and social sciences in particular may indeed be achieved more easily by tailoring and combining research efforts for addressing the challenge of a specific socio-economic problem.

However, the necessary scientific and technological creativity, underpinned by the first two options, can also marry itself with the expectations or needs of various groups in society if they wish to benefit. In other words, the way in which research activity is structured does not in itself ensure 'decompartmentalization' or, indeed, the opposite.

Despite their analytical power – based on an increasing number of case studies – the 'social shaping of technology' approaches and the 'science studies' more generally need from the policy-shaper's point of view further refinement and codification in order to be able to provide concrete advice on the problems to be tackled and the way policy instruments such as research programmes should be designed. By capitalizing on the rich accumulated experience of various participative processes (such as foresight or technology assessment) that are implemented in European countries, they could develop both a theoretical understanding of the way different sets of heterogeneous actors shape the evolution of policy priorities and instruments and specific experimental tools for and in close relation with policy-shapers and programme managers.

17.5   Conclusions

The chapter has shown that different theoretical frameworks may be exploited by policy-shapers in different situations. Research on the social shaping of technology has been instrumental in designing and experimenting new ways of organizing research programmes. Endogenous growth economics have strengthened the arguments in favour of increasing investments in intellectual and human capital. Systems of innovation approaches have inspired and accompanied the diversification of research and innovation policies and the development of instruments aiming at linking different actors and organizations. Far from questioning the importance of efforts of those like Edquist (2004) who plead in favour of improving the conceptual clarity of the systems of innovation approach, I limited myself to highlight some of the problems which policy-shapers have to come to grips with. The implications of such empirical observations are two-fold:

- The professional status of policy-shapers in fields such as research and innovation policies is in need of greater codification; in a society which the political elites view as a 'knowledge-based society', the profession of policy-shaper must be continuously enriched through novel education and lifelong learning programmes connected to the latest developments in innovation theories; policy-shapers in this field may be socially recognized as intermediaries or 'translators' of a specific kind.
- Conversely, social scientists doing research on knowledge, research and innovation may learn from a closer interaction with the prime users of their results; while many business schools and some economics or STS (Science and Technology Studies) departments in universities have trained many young professionals in technology-management courses and initiated stable linkages with managers in business, the exchanges with practitioners in the system of public action is more recent and deserves more attention. More research and more experimentation is required at the interface between innovation research and related sciences on one hand and policy development on the other. New institutions and new organizational innovation are

needed as well in order to renovate both policy research and research policy in the knowledge society but without falling in the illusion of policy-related research perfectly adapted to policy.

As Bartzokas (2002) has underlined, the idea of relevance and/or usefulness of policy research is, among others, related to the issue of what is considered to be 'knowledge', but also to the issue of where that knowledge is coming from, its validity and reliability. Scientists and policy makers often diverge on what is useful policy knowledge, as well as on how that knowledge is to be developed or obtained.

Within the knowledge-utilization literature 'the enlightenment model' (Weiss, 1991) has gained considerable agreement. Weiss's enlightenment model illustrates the idea that knowledge gained through research can enlighten or broaden the existing knowledge base of policy-makers. This, over time, can create a gradual shift of conceptual thinking and, therefore of the policies this conceptual thinking supports.

It is also perhaps one of the most realistic uses of research since it rests on the idea of the accumulation of knowledge through the aggregation of findings that promotes a gradual shift in concepts and paradigms. In relation to such shifts, Weiss sees the role of research as clarifying, accelerating and legitimizing changes in opinion and that this may be the most important contribution social research can make to the policy process (Weiss, 1977, p. 535). Data from recent studies suggest that the major use of social research is not the application of specific data to specific decisions. Rather, policy-shapers and government decision-makers tend to use research indirectly, as a source of ideas, information, and orientations to the world. This is why multi-diciplinary research on research and innovation policies in their socio-economic context is so important. This is why as in other fields such research may benefit from a closer interaction with practitioners.

References

Badie, B. 1995. *La fin des territoires, Essai sur le désordre international et sur l'utilité sociale du respec*. Paris: Fayard

Bartzokas, A. 2002. 'Innovation Policy Instruments: A review of EU Trends and Relevant literature', Final Report of the STRATA consolidating workshop, Session 4, Brussels, 22–23 April (available at www.cordis.lu/improving/strata/workshop.htm)

Borras, S. 2004. 'Systems of Innovation Theory and The European Union', *Science and Public Policy* 31 (6): 425–433

Boyer, R. 1995a. 'Vers une théorie originale des institutions économiques?' in Boyer and Saillard (eds.)1995. *Théorie de la régulation, l'état des savoirs*. Paris: La Découverte

Boyer, R. 1995b. 'La théorie de la régulation dans les années 1990' in N° Spécial Revue Actuel Marx, Théorie de la régulation, théorie des conventions, N°17, 1e semestre 1995

Boyer, R. and Saillard, Y. (eds.) 1995. *Théorie de la régulation, l'état des savoirs.* Paris: La Découverte

Callon, M. 1994. 'Is Science a Public good? Fifth Mullins Lecture, Virginia Polytechnic Institute', *Science Technology and Human Values*, 19: 395–424

Caracostas, P. 1987. Two tier Europe as European Technology Community? Paper presented at the 'European Conference on industrial integration strategies: Pandora's box?', 9–11 February 1987, Erasmus University Rotterdam, the Netherlands.

Caracostas, P. and Muldur, U. 1997. *Society, the Endless Frontier.* Brussels and Luxembourg: European Communities Publication Office

Caracostas, P. and Soete, L. 1997. 'The Building of Cross-Border Institutions in Europe: Towards a European System of Innovation?', in Edquist, C. (ed.) 1997. *Systems of Innovation; Technologies, Institutions and Organizations.* London: Pinter

Cronberg, T. and Sørensen, K.H. 1995. 'Similar Concerns, Different Styles? A note on European Approaches to The Social Shaping of Technology', in COST Social sciences A4, Proceedings of the COST A4 in Ruvaslahti, Finland, January 13 and 14, 1994

Di Ruzza, 1995. 'Théorie des systèmes productifs et recomposition de l'économie mondiale' in N° Spécial Revue Actuel Marx, Théorie de la régulation/théorie des conventions, 1e tr. 1995

Edquist, C. (ed.) 1997. *Systems of Innovation; Technologies, Institutions and organizations.* London: Pinter

Edquist, C. 2004. 'Reflections on The Systems of Innovation Approach', *Science and Public Policy* 31 (6): 485–489

European Commission 2004. *Technology Platforms, from Definition to Implementation of a Common Research Agenda.* Fagerberg, J., Mowery, D.C. and Nelson, R.R. (eds.) 2005, *The Oxford Handbook of Innovation*, Oxford: Oxford University Press.

European Commission 2005. 'Commission Staff Paper, Annex to The Proposal for the Council and European Parliament Decisions on the 7th Framework Programme (EC and Euratom), Main Report: Overall Summary, Impact Assessment and Ex Ante Evaluation', COM(2005) 119 final

Fougeyrollas, A., Le Mouël, P. and Zagamé, P. 2002. *The NEMESIS model: New Econometric Model for Environment and Sustainable Development Implementation Strategies.* Brussels: ECOMOD

Gaffard, J.L., Bruno, S., Longhi, C. and Quéré, M. 1993. *Cohérence et diversité des systèmes d'innovation en Europe.* Rapport de synthèse, FAST Dossier, Continental Europe: Science, Technology and Community cohesion, Vol. 19, Brussels

Guellec, D. and Ralle, P. 1995. *Les nouvelles théories de la croissance*, Paris: Repères, La Découverte

Gregersen, B., Johnson, B., and Kristensen, A. 1994. 'National Systems of Innovation and European Integration', paper presented to the EUNETIC conference on 'Evolutionary Economics of Technical Change: Assessment of Results and New Frontiers', European Parliament, Strasbourg, October 6–8

Handke, P. 1970. *Die Angst des Tormanns beim Elfmeter* (The Goalkeeper's Anxiety at the Penalty Kick)

High Level Expert Group for the European Commission 2002. *Thinking, Debating and Shaping the Future: Foresight for Europe*; available at ftp://ftp. cordis.europa.eu/pub/foresight/docs/for_hleg_final_report_en.pdf

H. M. Treasury, DTI and Department for Education and Skills 2004. *Science and Innovation Investment Framework 2004–2014, Annex A, The Economic Case for Investment in Science and Research*

Kuhlman, S. 2003. 'Evaluation of research and innovation policies: a discussion of trends with examples from Germany', *International Journal of Technology Management* 26(2/3/4)

Kuttner, R. 1996. *Everything for Sale: The Virtues and Limits of Markets*. New York: Alfred A. Knopf

Le Galès, P. 2004. Article 'Gouvernance' in Boussaguet, L., Jacquot, S. and Ravinet, P. (eds.) *Dictionnaire des politiques publiques*. Paris: Presses de Sciences Po

Lundvall, B.A. 1992. *National Systems of Innovation: Towards a Theory of Innovation and Interactive Learning*. London: Pinter

Lundvall, B.A. and Borras, S. 2005. 'Science, Technology and Innovation Policy', in Fagerberg, J., Mowery, D.C. and Nelson, R. (eds.) *The Oxford Handbook of Innovation*, Oxford University Press

Malerba, F. (ed.) 2004. *Sectoral Systems of Innovation: Concept, Issues and Analyses of Six Major Sectors in Europe*. Cambridge: Cambridge University Press

Malerba, F. 2005. 'Sectoral Systems, How and Why Innovation Differs across Sectors', in Fagerberg, J., Mowery, D.C. and Nelson, R. (eds.) *The Oxford Handbook of Innovation*, Oxford University Press

Muldur, U. 1991, 'R&D Funding at the crossroads of industrial, financial and political logic', Fast Working Paper, European Commission

Muller, P. 1994. 'La mutation des politiques publiques', N° Spécial de la revue Pouvoirs sur 'Europe, de la Communauté à l'Union', Avril

Nelson, R. (ed.) 1993. *National Systems of Innovation: A Comparative Study*. Oxford: Oxford University Press

Niosi, J. and Bellon, B. 1994. 'The Global Interdependence of National Innovation Systems: Evidence, Limits and Implications', *Technology in Society* 16 (2): 173–97

OECD 1995. *National Systems for Financing Innovation*. OECD, Paris

OECD 2004. *Science and Innovation Policy: Key Challenges and Opportunities*. OECD, Paris

Romer, P. 1986. 'Increasing Returns and Long-Term Growth', *Journal of Political Economy* 98: 71–102

Romer, P. 1993. 'Implementing a National Technology Strategy with Self-Organising Industry Investment Boards', *Brookings Papers on Microeconomics* 2: 345–99

Technopolis 2001. *A Singular Council, Evaluation of the Research Council of Norway*; available at http://www.technopolis.co.uk/downloads/243_RCN_synthesis.pdf

Teubal, M. 1995. 'A Catalytic and Evolutionary Approach to Horizontal Technology Policies', discussion paper, the Jerusalem Institute for Israel Studies, Jerusalem

Teubal, M. 1996. 'R & D and Technology Policy in NICs as Learning Processes', *World Development* 24 (3): 449–60

Teubal, M. 1997. 'A Catalytic and Evolutionary Approach to Horizontal Technology Policies (HTPs)', *Research Policy* 25 (8): 1161–1188

Tsipouri, L. and Xanthakis, M. 1993. *Impact of the EC Science and Technology Policy on the Greek S/T Policy*. University of Athens

Weiss, Carol 1977. 'Research for Policy's Sake: The Enlightenment Function of Social Science Research', *Policy Analysis* 3 (4): 531–45

Weiss, Carol. 1991. 'Policy research as advocacy: Pro and con', *Knowledge and Policy* 4 (1/2): 37–56

Index

in innovation and production function, 172
heterogeneous firms, 181
Hewlett Packard, 123
higher education (HE) sector, OECD definition of, 405
high-tech industries, spinoff entry in, 187
hiring firm, 365
history-friendly models, 52
Hitachi, 98
Hong Kong, 241
horizontal differentiation, 53
  type models, 53
HP and ST coinnovation
  alliance design, 313
    actual cost data sharing, 314
    internal performance incentives, 313
    shadow of the future, 313
    virtual joint venture, 313
  collaboration and its interpretation, 311
  evolution, 316
human knowledge, 33
human practice, 33, 34, 35, 37

IBM, 115–16, 146, 243
  clone competition, 122
  creative destruction waves
    consequences, 124
    precedents, 123
  divided technical leadership, 120
impact assessment, 477
immigrants, 387–94
impresarii, 318
incentives, 457
incremental innovations, 46, 48, 49, 56
India, 442
individual incentives
  effect on innovation, 84
  extrinsic incentives, 77
  impact
    on innovative activity and firms performance, 87
  implications
    management, 94
    policy, 96
  intrinsic incentives, 78
  social incentives, 78
  and technological change, 76
  technologists, 88, 89
industrial dynamics, 219
  growing literature on, 37
industrial evolution
  statistical regularities, 153
industrial structures, 158
information, 446

information asymmetry, 77
information transmission system, 444
innovation, 1, 24, 47, 49, 59, 273, 381, 441, 442, 443
innovation ecologies, 448–9
innovation-enhancing policies, 472
Innovation Platform, 469
  working group, 469
innovation policy, 441, 443, 452
  evolving knowledge and evolving economic order, 443
innovation policy field
  impact assessment, 477
  innovation vs other social actions, 467
  national vs other territorial scale, 470
  Systems of Innovation (SI) approach, 472
innovation process
  actors, connection between, 450
innovation size, 56
innovation system, 447–8, 448, 451, 470
  central policy, 449
  policy instruments, 451
  policy issues, 449
  policy themes, 449
innovative variation, 445
institutional change, 228
  mechanisms, 238
    dynamic institutional complementarities, 241
    overlapping social embeddedness, 240
    Schumpeterian dis-bundling and new-bundling, 239
institutional complementarities, 237
  dynamic version, 241
institutional failure, 479
institutions, 231, 279
  game-theoretic frame, 229
  primitive domains, 234
    2-person symmetric domain, 234
    3-person asymmetric domain, 235
    linkages, generic modes, 236
    N-person symmetric (asymmetric) domain, 235
  Schumpeterian innovation, 227
Intel, 85, 94, 111, 131
Intellectual Property Agreements (IPA), 268
intellectual property
  ownership, 262
intellectual property rights, 455–60
  and competition policy, 456
  current and future innovation, 462
  innovation impede rate
    anti-commons, 461
    complex innovation, 461